Rudolf Janoschek (Ed.)

Chirality

From Weak Bosons to the α-Helix

With 80 Figures, 18 Tables
and 95 Schemes

Springer-Verlag
Berlin Heidelberg New York
London Paris Tokyo
Hong Kong Barcelona
Budapest

Editor

Professor Rudolf Janoschek
Institut für Theoretische Chemie
Karl-Franzens-Universität Graz
Mozartgasse 14
A-8010 Graz

ISBN 3-540-53920-4 Springer-Verlag Berlin Heidelberg New York
ISBN 0-387-53920-4 Springer-Verlag New York Berlin Heidelberg

Library of Congress Cataloging-in-Publication Data
Chirality: from weak bosons to the [alpha]-helix / R. Janoschek, ed. p. cm.
Includes bibliographical references.
ISBN 3-540-53920-4 (Springer-Verlag Berlin Heidelberg New York).
ISBN 0-387-53920-4 (Springer-Verlag New York Berlin Heidelberg)
1. Chirality. 2. Stereochemistry. I. Janoschek, R. (Rodolf), 1939- . QD481.C55
1991 541.2'23-dc20 91-14364 CIP

Printed in Germany

Typesetting: Springer TEX in-house system

02/3140 543210 - Printed on acid-free paper

Preface

Les hypothèses, n'en déplaise à mon contradicteur, sont l'âme des progrès de la science.

Louis Pasteur

The concept of chirality, established 100 years ago, plays an important role in almost all domains and dimensions of our recent scientific view of life. Chiral properties can be found in fundamental nuclear particles, in molecules, and in the macroscopic world of living nature (plants and animals) and inanimate nature (crystals). In particular, chirality, or more precisely chiral excess, is evident in human beings. For example, the expected symmetry of the hands turns out to be functionally non-existent. Consequently chirality occurs in the technical sphere, where screws are the best-known examples, since most of them are made for right-handed people. Chirality is not confined to static objects but influences processes such as chemical reactions.

The occurrence of chiral objects on different dimensional scales has been treated in the past in mutually independent frameworks. There were, however, two remarkable events from which the conclusion can be drawn that the appearance of chirality in various fields has a common cause. On the one hand, physicists found evidence that the well-known biomolecular homochirality can be traced back to the chirality of weak bosons. At the same time, on the other hand, the so-called thalidomide tragedy occurred when thalidomide molecules of a certain chirality, taken by pregnant women, caused deformed children.

Spectacular events like these are reason enough for a group of authors to compile a survey on important aspects of chirality in a book which comprises topics from nuclear particle physics and various fields of chemistry to pharmacy. The authors agreed that they would not write for specialists in their respective fields of research but for anybody with a sound scientific education.

Although it is chemistry which dominates in this book, chirality is introduced in the chapter of fundamental-particle physics. There are two reasons for this. On the one hand, physicists carefully define the notions that will be applied later. On the other, the chirality of certain fundamental particles seems to be the origin of biomolecular homochirality as mentioned before. Corresponding theories on the basis of molecular kinetics are introduced in the second chapter. The third chapter deals with the mathematical treatment of molecular chirality. Two crucial experimental methods for the determination of absolute stereochemistry, circular dichroism and anomalous X-ray diffraction, are presented in the fourth and fifth chapters.

The second half of the book is dedicated to the synthesis and separation of enantiomers of chiral chemical compounds. After a general introduction to chiral phenomena in organic chemistry, the main strategies for the production of chiral compounds are reviewed. These are enzymatic catalysis, synthesis using prochiral auxiliary compounds, and catalysis by means of chiral transition-metal complexes. Finally the separation of enantiomers by the technique of liquid chromatography is described. These contributions cover to a large extent the requirements in organic chemistry, biochemistry, and pharmaceutical chemistry. The closing chapter presents a study of biopolymeric structures, in particular the α-helix, which is the final point on our scale of dimensions for chiral objects.

The authors owe their gratitude to Dr. Marion Hertel (Springer-Verlag). The appearance of the present book would have been impossible without her unending patience and support.

Rudolf Janoschek

Contents

List of Authors

Henri Brunner
Institut für Anorganische Chemie
Universität Regensburg
Universitätsstr. 11
W-8400 Regensburg

Gerhard Derflinger
Institut für Statistik
Wirtschaftsuniversität Wien
Augasse 2–6
A-1090 Wien

Kurt Faber
Herfried Griengl
Institut für Organische Chemie
Technische Universität Graz
Stremayrgasse 16
A-8010 Graz

Siegfried Hoffmann
Biotechnikum
Bioorganische Chemie
Martin-Luther-Universität
Halle-Wittenberg
Weinbergweg 16a
O-4050 Halle/S.

Rudolf Janoschek
Institut für Theoretische Chemie
Universität Graz
Mozartgasse 14
A-8010 Graz

Christoph Kratky
Institut für Physikalische Chemie
Karl-Franzens-Universität Graz
Heinrichstr. 28
A-8010 Graz

Heimo Latal
Institut für Theoretische Physik
Karl-Franzens-Universität Graz
Universitätsplatz 5
A-8010 Graz

Wolfgang Lindner
Institut für Pharmazeutische
Chemie
Karl-Franzens-Universität Graz
Universitätsplatz 1
A-8010 Graz

Günther Snatzke
Lehrstuhl für Strukturchemie
Fakultät für Chemie
Ruhr-Universität Bochum
Postfach 102148
W-4630 Bochum 1

Ekkehard Winterfeldt
Institut für Organische Chemie
Universität Hannover
Schneiderberg 1B
W-3000 Hannover 1

1 Parity Violation in Atomic Physics

H. Latal

1.1 Introduction

Physicists have been convinced for a long time that the laws of nature do not distinguish between left and right, and therefore parity (mirror symmetry) is conserved in all interactions. The discovery in 1956 that parity is not conserved for the weak interaction governing β-decay had an immediate and profound influence on nuclear and elementary particle physics. For atomic physics, however, it was of no importance, since the weak interaction is not involved directly in atomic processes. Only through the progress in our understanding of the connections between the various interactions, which finally led to a unified theory of electromagnetic and weak forces in the "Standard Model" of Glashow, Salam and Weinberg at the end of the 1960s (for reviews, see e.g [1–3]), have parity-violating effects in atoms become predictable. The effects are caused by an interference between the photon as carrier of the electromagnetic force and the heavy intermediate boson Z^0, which mediates the so-called "neutral currents" of the weak interaction. In the meantime these predictions have already been demonstrated experimentally (for reviews, see [4–8]).

After a short introduction into the concept of parity the relevant facts about elementary particles and their interactions are collected. A central concept there is that forces are mediated by the exchange of (virtual) particles. Then the notions of spin and helicity are introduced and it is pointed out that the origin of parity violation in weak interactions is due to the experimentally established fact that neutrinos occur only with left-handed helicity. A brief review of the essentials of the "Standard Model" of the unified theory of weak and electromagnetic interactions and its implications on atomic transitions are given. Finally the results of various experiments on parity violation in atoms are presented.

1.2 Parity

Only in this century was it recognized that the familiar conservation laws of physics are connected with certain inherent symmetries of nature. This connection has also some esthetic attraction since symmetry is fundamentally associated with the concept of order. Some of these symmetries in physics are of a geometrical character whereas others are quite abstract.

One speaks of a symmetry operation if this operation brings a system into a state identical to the one in which it was before – the system is then said to be invariant with regard to the symmetry operation. Thus, invariance is the principal feature of symmetry. If the symmetry is not perfect, one speaks of a broken symmetry; especially in particle physics this symmetry breaking is of considerable importance. The connection between invariance and conservation laws has been formulated mathematically in Noether's Theorem.

Familiar examples for the relation between symmetry (invariance) and conservation laws are: conservation of momentum is connected with the homogeneity of space (invariance under spatial translations), conservation of angular momentum with the isotropy of space (invariance under rotations), conservation of energy with the uniformity of time (invariance under time translation). These conservation laws of classical physics are all associated with so-called continuous symmetries. In quantum mechanics there exist additional conservation laws which are connected with discontinuous symmetries: they concern parity, charge conjugation, and time reversal. The parity operation $\mathcal{P}$ is the symmetry under spatial reflections which changes $\mathbf{r}$ into $-\mathbf{r}$. The second symmetry concept is the charge conjugation $\mathcal{C}$, which is a "mirror" operation working on the charge; for example, it converts an electron into a positron. The last concept is that of time reversal $\mathcal{T}$, which reverses the sense of time in physical processes. All of them are involved in the fundamental $\mathcal{PCT}$-Theorem which states that nature is invariant under the simultaneous application of these three operations. In the following we shall be concerned with the parity operation only.

One can easily convince oneself that the parity operation (reflection of all three coordinates at the origin) is equivalent to a mirror operation (reflection at a plane) plus a rotation of the coordinates by 180° (see Fig. 1.1). As a result, a right-handed coordinate system becomes left-handed, and vice versa. The invariance of a system under this transformation therefore expresses the symmetry of space under mirror reflections – one cannot distinguish between "right-handedness" and "left-handedness". In classical mechanics, this invariance does not lead to a conservation law, in contrast to quantum mechanics. The parity operator $\mathcal{P}$ has the property

$$\mathcal{P}\psi(\mathbf{r}_1, \mathbf{r}_2, \ldots) = \psi(-\mathbf{r}_1, -\mathbf{r}_2, \ldots) , \tag{1.1}$$

namely, that in operating on a function, it changes each of the position variables into its negative. It is easy to find the eigenvalues P of this operator, which are determined by the equation

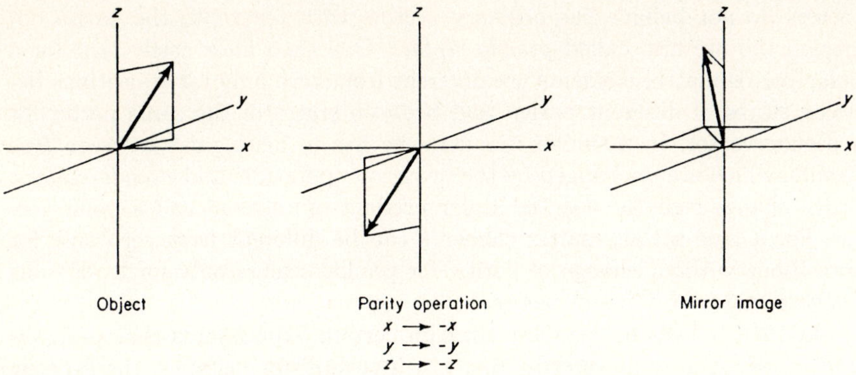

Fig. 1.1. The parity operation

$$\mathcal{P}\psi = P\psi \,. \tag{1.2}$$

If we apply the parity operator once again, we get

$$\mathcal{P}^2\psi = P^2\psi \,. \tag{1.3}$$

However, it is apparent that a double application of the parity operator converts a function back into itself; in other words, we have

$$\mathcal{P}^2\psi = \psi \,, \quad \text{i.e.} \quad P^2 = 1 \,, \quad \text{thus} \quad P = \pm 1 \,. \tag{1.4}$$

Thus the eigenfunctions of the parity operator are either unchanged or change sign when acted upon by this operator. In the first case, the wave funtion (or state) is said to be *even*, and in the second it is said to be *odd*. The invariance of the Hamiltonian $\mathcal{H}$ (the energy operator) under inversion (i.e. the act that the two operators $\mathcal{H}$ and $\mathcal{P}$ commute) thus expresses the law of conservation of parity: if the state of a closed system has a given parity (even or odd), then this parity must not change in the course of time. Choosing simultaneous eigenstates of the Hamiltonian and the parity, one can characterize the various states of different energy by their parity – for example, the wave functions of the simple harmonic oscillator can be catalogued as being either even or odd. The angular momentum operator is also invariant under inversion, which changes the sign of the coordinates and the differentiation operators at the same time. This means, that a system can have a definite parity simultaneously with definite values of the total angular momentum and its z-component.

There are specific parity selection rules for the matrix elements of various physical quantities. Let us first consider vectors: here one must distinguish between *polar* (ordinary) vectors, which change direction under a coordinate inversion, such as position, momentum, electric field. *Axial* vectors, on the other hand, remain unchanged, for instance the angular momentum vector, being the (cross) product of the two polar vectors $\mathbf{r}$ and $\mathbf{p}$. Because axial vectors do not behave like ordinary vectors with regard to the parity operation the are also called *pseudo-vectors*. One then finds easily that for a polar vector matrix elements are different from zero only for transitions between states of different parity, and between states of the same parity for a pseudo-vector. In a similar manner, one has to distinguish between *true scalars*, which are unchanged by the inversion operation, and *pseudo-scalars*, which change sign, as, e.g. the scalar product of an axial and a polar vector. For a true scalar, matrix elements can be different from zero only for transitions without change of parity, for pseudo-scalars only for transitions between states of different parity.

In 1956 it became manifest through certain experiments that parity is *not* conserved in weak interactions, which govern, for instance, the β-decay of unstable nuclei. Here, then, we have a conservation law that holds in some interactions (in the strong and electromagnetic) but not in all. The nonconservation of parity in the weak interactions also implies that nature actually *does* distinguish between left and right. For a better understanding of the concepts involved in this problem, we shall review shortly the main facts about elementary particles and their interactions, as one sees them today, in the following Section.

1.3 Elementary Particles and Forces

1.3.1 Leptons and Quarks

Elementary particle physics deals basically with the study of the ultimate constituents of matter and the nature of the interactions between them. In the past 100 years this search has revealed four layers of structure: all matter has been shown to consist of atoms; the atom itself has been found to have a dense nucleus surrounded by a cloud of electrons; the nucleus in turn has been broken down into its component protons and neutrons; and more recently it has become apparent that the proton and neutron are also composite particles – they are made up of smaller entities called quarks. At this level, matter is built out of just two classes of elementary particles: the leptons, such as the electron and the neutrino, carrying integral electric charges, and the quarks with fractional electric charges, which are constituents of the proton, the neutron and many related particles. These basic constituents of matter have half-integral spin and are called fermions, because they obey Fermi-Dirac statistics (a consequence of which is the famous Pauli principle). Table 1.1 represents our current understanding of the fundamental particles. It is not yet understood why there exists this triplication of "generations", since the universe, as we see it today, seems to be constructed predominantly from just two types of quark, "up" and "down" – the proton being the combination *uud*, the neutron *udd* – and one charged and one neutral lepton (electron and its associated neutrino). The masses given in this Table for the quarks

Table 1.1. Basic constituents of matter

LEPTONS			
Particle Name	Symbol	Restmass (MeV/c^2)	Electric Charge
Electron Neutrino	ν_e	$\simeq 0$	0
Electron	e^-	0.511	-1
Muon Neutrino	ν_μ	$\simeq 0$	0
Muon	μ^-	106.6	-1
Tau Neutrino	ν_τ	< 164	0
Tau	τ^-	1,784	-1

QUARKS			
Particle Name	Symbol	Restmass (MeV/c^2)	Electric Charge
up	u	310	+2/3
down	d	310	$-1/3$
charm	c	1,500	+2/3
strange	s	505	$-1/3$
top	t	$> 20,000$	+2/3
bottom	b	$\simeq 5,000$	$-1/3$

should be taken as indicative only, since quarks have not been observed as free particles and are permanently confined in the strongly interacting particles.

1.3.2 Forces and Interactions

Four fundamental forces act between the elementary particles: gravitation and electromagnetism have long been familiar in the macroscopic world; the weak force and the strong force are observed only in subnuclear events. Classically, interaction at a distance is described in terms of a potential or field due to one particle acting on another. In quantum theory, it is viewed in terms of the exchange of quanta specific for the particular type of interaction. Since the quantum transmits energy and momentum, the conservation laws can be satisfied only if the exchange process takes place over a time limited by the uncertainty principle, i.e. $\Delta t \leq \hbar/\Delta E$ ($\hbar$ = Planck's constant). Such transient quanta are said to be *virtual* particles, they possess integral spin, obey Bose-Einstein statistics and thus belong to the group of bosons. There is no restriction on the number of bosons (photons, for example) which may exist in the same quantum state – an example of this is the laser.

The equivalence of the two descriptions (field/virtual particle) on a macroscopic scale can be illustrated by considering the electrostatic field between two point charges q_1 and q_2, placed a distance r apart. In the classical case the force $\mathbf{F}$ on q_2 is ascribed to the electric field $\mathbf{E}(r)$ due to

q_1: $\mathbf{F} = q_2\mathbf{E}(r) = q_1 q_2 \mathbf{r}/r^3$. Quantum-mechanically, the force between the charges is ascribed to the exchange of virtual photons of momentum Δp, the change of momentum of the charge as it emits or absorbs such a photon. The uncertainty principle links this momentum with the linear dimension of the system: $\Delta p \simeq \hbar/r$. Since photons travel with the speed of light c, the momentum transfer takes place in a period $\Delta t = r/c$. Thus each photon gives rise to a force $|\mathbf{F}| = \Delta p/\Delta t = \hbar c/r^2$. The number of photons emitted and absorbed by the charges is assumed to be proportional to the product of the charges, so that we obtain Coulomb's law $|\mathbf{F}| = q_1 q_2/r^2$ as in the classical case.

The exchange of massive bosons, in contrast to massless photons, gives rise to the so-called *short-ranged* forces. Suppose that the quantum to be exchanged has a mass M, then the uncertainty principle restricts the time of its existence to $\Delta t \leq \hbar/Mc^2$, during which it could travel at most a distance $R \simeq c\Delta t \leq \hbar/Mc \equiv \lambda\!\!\!^{-}_c$. Thus the range of the force is given by the Compton wavelength $\lambda\!\!\!^{-}_c$ of the exchanged quantum. The static potential associated with such an exchange is called the Yukawa potential and has the form

$$V_{\mathrm{Y}}(r) = \frac{g}{4\pi r}\, e^{-r/R} , \tag{1.5}$$

where the quantity g is identified with the strength of the source. The analogous expression in electromagnetism is the familiar Coulomb potential

$$V_{\mathrm{C}}(r) = \frac{q}{4\pi r} . \tag{1.6}$$

If a particle is scattered by such a potential, the effect will be observed through the angular deflection of the particle or, equivalently, the momentum transfer $\mathbf{p}$. The potential $V(\mathbf{r})$ is associated with a scattering amplitude which is the Fourier transform of the potential. The scattering amplitude due to single boson exchange then turns out to be proportional to the product of the coupling of the boson to the scattered and the scattering particle, and a *propagator* of the general form

$$\frac{1}{p^2 + (Mc^2)^2} \quad \text{for } V_{\mathrm{Y}} , \tag{1.7a}$$

$$\frac{1}{p^2} \quad \text{for } V_{\mathrm{C}} . \tag{1.7b}$$

A summary of the characteristics of the basic forces of nature together with their associated exchanged quanta is given in Table 1.2. We see that gravity, although the force most familiar to everyone, is by far the least important of the four interactions on the scales involved in particle physics. Electromagnetic interactions account for most extranuclear phenomena in physics (because of their long range) and lead to the bound states of atoms and molecules. Weak interactions are exemplified by the extremely slow process of radioactive β-decay of nuclei. Strong interactions are supposed to hold together the quarks in a proton, and their residual effects apparently account

Table 1.2. Fundamental interactions

Interaction	Gravity	Electro-magnetic	Weak	Strong
Source	Mass	Electric Charge	"Weak Charge"	"Color Charge"
Range (m)	∞	∞	$\leq 10^{-18}$	$\leq 10^{-15}$
Strength at 10^{-15} m	10^{-38}	10^{-2}	10^{-13}	~ 1
Field quantum	Graviton	Photon	Intermediate Bosons $W^{\pm}$, Z^0	Gluons
Restmass	0	0	80–90 GeV/c^2	0
Spin	2	1	1	1

for the complex nuclear binding force. Both weak and strong interactions are of short range.

An understanding of nature at this level of detail has been the remarkable achievement of particle physics in the last decades; nevertheless, it is possible to imagine that there exists a still simpler theory explaining the multiplicity of particle generations and fundamental interactions. For the latter, some real progress has been made: there are good grounds for supposing that some, perhaps all, of the interactions are *unified*, i.e. different aspects of one single interaction. The weak and electromagnetic interactions appear to have the same intrinsic coupling of fermion constituents to the respective mediating bosons – they are different aspects of a single *electroweak* interaction, as formulated in the so-called "Standard Model" of Glashow, Salam and Weinberg (see Sect. 1.3.4). Compared with electromagnetism, the weakness of the weak interaction is ascribed to its short range due to the exchange of the extremely massive intermediate bosons $W^{\pm}$ and Z^0. At high enough energies and momentum transfers, well above such a mass scale, electromagnetic and weak interactions should have the same actual strength. This can be seen from the propagators given above: for $p^2 \gg (Mc^2)^2$ they become indistinguishable. The important point in this context is that the strengths of the different interactions are not fixed once and for all; they depend on energy scales (or, equivalently, on distance). At high energies (short distances) even strong interactions appear to grow weaker, and they may merge with the electroweak interaction at the unimaginable energy of 10^{15} GeV (10^{-31} m).

1.3.3 Spin and Helicity (Chirality)

As mentioned in Sect. 1.3.1, the fundamental constituents of matter are fermions, which carry spin 1/2. Such particles are described quantum mechanically in terms of the *Dirac equation*, the relativistic analogue of the Schrödinger equation. As is well known, the transition from classical mechanics to quantum mechanics is achieved by replacing the energy E and the components of the momentum $\mathbf{p}$ by the differential operators

$$\begin{aligned} E &\rightarrow i\hbar\partial/\partial t \,, \\ p_k &\rightarrow -i\hbar\partial/\partial x_k \,, \quad k = 1,2,3 \,. \end{aligned} \tag{1.8}$$

In the nonrelativistic case, the classical expression for the total energy (kinetic plus potential energy) of a particle with mass m

$$E = \frac{\mathbf{p}^2}{2m} + V \,, \tag{1.9}$$

then yields the Schrödinger equation

$$i\hbar \frac{\partial \Psi}{\partial t} = -\frac{\hbar^2}{2m} \Delta\Psi + V\Psi \,. \tag{1.10}$$

Its solution, the wave function Ψ, contains the complete information about the quantum mechanical system in question.

If we make the replacement (1.8) in the relativistic energy-momentum relation for a free particle of mass m,

$$E^2 = c^2\mathbf{p}^2 + m^2c^4 \,, \tag{1.11}$$

then we would get a differential equation of second order in time, the *Klein-Gordon equation*

$$\left[\frac{\partial^2}{\partial(ct)^2} - \Delta - \left(\frac{mc}{\hbar}\right)^2 \right] \Psi = 0 \,, \tag{1.12}$$

which doesn't lead, however, to a positive-definite probability density in the usual form, and admits negative energy solutions. Therefore Dirac tried a linearized form of (1.11)

$$E = c\boldsymbol{\alpha} \cdot \mathbf{p} + \beta mc^2 \,, \tag{1.13}$$

whose square should give back (1.11). This, however, can only be achieved if the four coefficients $\alpha_i (i = 1,2,3)$ and β in (1.13) are not simple numbers but 4×4 matrices. Again inserting the replacement (1.8) in (1.13) we eventually arrive at the *Dirac equation*

$$\left[i\gamma_\mu \frac{\partial}{\partial x_\mu} - \frac{mc}{\hbar} \right] \Psi = 0 \,. \tag{1.14}$$

Here we are using the covariant notation of special relativity, i.e., summation over the index μ from 0 to 3 is implied. The new coefficients γ_μ are related to the α_i and β of (1.13) by

$$\gamma_0 = \beta \,, \quad \gamma_i = \beta\alpha_i \quad (i = 1,2,3) \,, \tag{1.15a}$$

and form a Lorentz-four-vector. In a certain representation the 4×4 matrices γ_μ involve the 2×2 Pauli spin matrices σ_i. In addition to the identity matrix, which behaves like a scalar under Lorentz-transformations, the pseudoscalar quantity

$$\gamma_5 = i\gamma_0\gamma_1\gamma_2\gamma_3 \tag{1.15b}$$

is of importance ("*chirality operator*", see below). Since (1.14) is a matrix equation, the wave function Ψ has to be a column matrix with four components, a so-called *spinor*.

The assertion that the Dirac equation describes particles with spin 1/2 can best be visualized by considering its nonrelativistic limit in the presence of an external electromagnetic field, characterized by a potential $A^\mu = (\Phi, \mathbf{A})$. The relevant equation is obtained from the free Dirac equation through the "minimal coupling" prescription

$$\frac{\partial}{\partial x_\mu} \to \frac{\partial}{\partial x_\mu} + i\frac{e}{c}A^\mu . \tag{1.16}$$

We first write the spinor Ψ in terms of two-component spinors ϕ and χ,

$$\Psi = \begin{pmatrix} \phi \\ \chi \end{pmatrix} . \tag{1.17}$$

In a nonrelativistic limit, ϕ and χ are related through

$$\chi = \frac{\boldsymbol{\sigma} \cdot \mathbf{p}}{2mc}\phi \ll \phi , \tag{1.18}$$

thus justifying the use of the terms "large" and "small" components for ϕ and χ, respectively. The large component ϕ, in this limit, satisfies the *Pauli equation* (with $\mathbf{B} = \operatorname{rot} \mathbf{A}$)

$$i\hbar\frac{\partial\phi}{\partial t} = \left[\frac{\mathbf{p}^2}{2m} - \frac{e}{2mc}(\mathbf{L} + 2\mathbf{S}) \cdot \mathbf{B} + e\Phi\right]\phi . \tag{1.19}$$

Here $\mathbf{L} = \mathbf{r} \times \mathbf{p}$ is the orbital angular momentum, and $\mathbf{S} = \hbar\boldsymbol{\sigma}/2$ is the spin operator with eigenvalues $\pm\hbar/2$. The coefficient of the interaction with the magnetic field $\mathbf{B}$ gives the correct magnetic moment corresponding to a (gyromagnetic) g factor of 2. The two components of ϕ suffice to accomodate the two spin degrees of freedom.

The expression (1.11) holds for both positive and negative values of E. This additional degree of freedom is represented by the other two components of the spinor Ψ: the negative-energy solutions are then reinterpreted as describing antiparticles, again with two possible orientations of the spin. In conclusion therefore, the Dirac equation, as a relativistically covariant wave equation, automatically leads to half-integer spin particles and their antiparticles. Since the Dirac equation is a differential equation of first order in the space coordinates, the spinor wave function $\Psi(-\mathbf{r}, t)$, obtained simply by replacing $\mathbf{r}$ by $-\mathbf{r}$ (space inversion), is not a solution of (1.14), but the product $\gamma_0\Psi(-\mathbf{r}, t)$ is. Therefore the positive-energy states (particles) and the negative-energy states (antiparticles) possess opposite intrinsic parity.

Mathematically the extra twofold degeneracy implies that there must be another observable whose eigenvalue can be taken to distinguish states with the same energy. A possible choice is the "*helicity operator*"

$$h = \boldsymbol{\Sigma} \cdot \hat{\mathbf{p}} = \begin{pmatrix} \boldsymbol{\sigma} \cdot \hat{\mathbf{p}} & 0 \\ 0 & \boldsymbol{\sigma} \cdot \hat{\mathbf{p}} \end{pmatrix} , \quad \hat{\mathbf{p}} = \mathbf{p}/|\mathbf{p}| , \tag{1.20}$$

which measures the spin component in the direction of motion, with eigenvalues $\lambda_h = \pm 1$. The state with $\lambda_h = +1$, i.e., where the spin vector points into the direction of motion, is said to be "*right-handed*", the state with $\lambda_h = -1$ is called "*left-handed*". In general the handedness ("*chirality*") of a massive particle can be reversed simply by bringing it to rest and accelerating it in the opposite direction without changing its spin. Thus massive particles have both left-handed and right-handed components. The handedness of a *massless* fermion, however, can never change, since it always travels at the speed of light and cannot be stopped. Such particles are the neutrinos; they are described by the Dirac equation (1.14) without the mass term. In this case the Dirac equation decouples into two separate equations for two-component spinors (*Weyl equation*), one representing a left-handed neutrino (ν_{L}) and its antiparticle, a right-handed antineutrino ($\bar{\nu}_{\mathrm{R}}$), the second one describing the other helicity states. In this representation γ_5 is called the *chirality operator*, since the operator $(1 - \gamma_5)$ projects out just ν_{L} (and $\bar{\nu}_{\mathrm{R}}$) from the four-component Dirac spinor. Thus for massless fermions chirality equals helicity.

In fact, experimentally only left-handed (negative chirality) neutrinos and right-handed (positive chirality) antineutrinos have been observed in nature, their oppositely spinning counterparts are presumed not to exist. As far as we know, neutrinos experience only the weak interaction, and since their coupling to leptons must involve the projection operator $(1-\gamma_5)$, we see that parity is violated (maximally) in weak interactions, γ_5 being a pseudoscalar quantity.

1.3.4 Unified Theory of Weak and Electromagnetic Interactions ("Standard Model")

The weak interactions take place between all the quark and lepton constituents; each of them has, so to speak a "weak charge". This weak charge is unusual, however, in that it is assigned on the basis of handedness. Only left-handed particles and right-handed antiparticles bear a weak charge; the right-handed particles and the left-handed antiparticles are neutral with respect to the weak force and do not participate in weak interactions. This is another way of expressing the experimental fact that weak interactions maximally violate parity. An example for such a process is the β-decay of the neutron, $n \rightarrow p + \mathrm{e}^- + \bar{\nu}_{\mathrm{e}}$, where it was observed in 1956 that the electron comes out spinning always in the left-handed sense and the antineutrino in the right-handed sense. Fermi's original theoretical explanation of the β-decay of 1932 in terms of a current-current interaction (see Fig. 1.2a) was then modified accordingly, leading to the so-called $V-A$ (vector – axial vector) structure of the weak currents [containing the combination $\gamma_\mu(1-\gamma_5)$]. They couple together particles of different electric charge (e.g., neutron and proton or electron and neutrino, respectively); one therefore speaks of these as the "charged weak currents". The existence of "neutral weak currents" (in analogy to the electromagnetic current) was not revealed until 1973, when neutrino events of the type $\bar{\nu}_\mu e^- \to \bar{\nu}_\mu e^-$, among others, were observed.

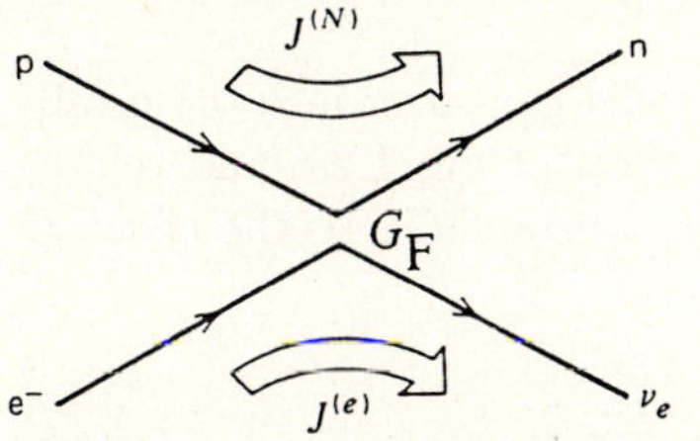

Fig. 1.2a. Current-current theory of weak interactions (Fermi)

In 1967–1968 Weinberg and Salam, extending a proposal first made by Glashow in 1961, formulated a theory unifying weak and electromagnetic interactions, based on the (gauge) symmetry group SU(2) × U(1) – the so-called "*Standard Model*" of electroweak interactions. The fundamental carriers of the interaction are four massless vector bosons: a triplet representing SU(2) and a singlet for U(1). Then a process called "spontaneous symmetry breaking" is invoked, as a result of which three bosons (denoted W^+, W^-, Z^0) acquire mass, and one (the photon γ) remains massless. These physically observable intermediate vector bosons are certain linear combinations of the original fundamental massless bosons. The symmetry breaking is related to the fact that the weak charge is not invariably conserved: since the weak charge is tied to handedness, for massive particles it depends on their motion (compare Sect. 1.3.3). It could be conserved only if the leptons and quarks were all massless (*chiral fermions*). The interaction energy (usually represented by the so-called Lagrangian energy density) of fermions with the vector boson fields is the product of the fermion currents with the fields and consists of three parts: one representing the weak charged currents coupled to the charged vector bosons $W^\pm$, one for the weak neutral current coupled to the neutral vector boson Z^0, and one for the electromagnetic (neutral) current coupled to the photon. As a result the model connects the electric charge e to the effective weak coupling g_w by

$$g_w = \frac{e}{2\sqrt{2}\sin\theta_W} \,. \tag{1.21}$$

The angle θ_W, one of the free parameters of the model, is called the weak mixing angle (or Weinberg angle). From this Lagrangian density matrix elements for the various weak processes can be derived, which then involve propagators of the form (1.7a) representing the exchanged intermediate vector bosons (see Fig. 1.2b). In the low energy limit of $p^2 \ll (Mc^2)^2$ they obviously reduce to

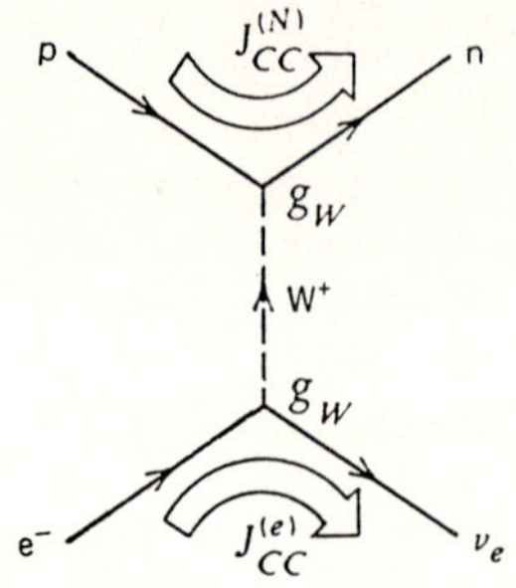

Fig. 1.2b. Charged currents in the Standard Model

$1/(M^2c^4)$. Thus, in this limit, the Standard Model goes over into the established current-current picture of weak interactions of Fig. 1.2a, and relates the experimentally determined Fermi coupling constant G_F to the effective weak coupling g_w through

$$\frac{G_\mathrm{F}}{\sqrt{2}} = \frac{g_w^2}{M_\mathrm{W}^2 c^4} . \tag{1.22}$$

The model also predicts the masses of the intermediate vector bosons as

$$M_{\mathrm{W}^\pm}c^2 = \frac{37.4}{\sin\theta_\mathrm{W}}\,\mathrm{GeV} , \quad M_{Z^0}c^2 = \frac{M_{\mathrm{W}^\pm}c^2}{\cos\theta_\mathrm{W}} = \frac{75}{\sin 2\theta_\mathrm{W}}\,\mathrm{GeV} . \tag{1.23}$$

A compilation of the most recent experimental data from various processes yields the following values

$$M_{\mathrm{W}^\pm}c^2 = 81.0\,\mathrm{Gev} , \quad M_{\mathrm{Z}^0}c^2 = 92.4\,\mathrm{GeV} , \quad \sin^2\theta_\mathrm{W} = 0.230 . \tag{1.24}$$

1.4 Parity-Violating Effects in Atoms

Parity-violating effects in atomic systems basically result from the interference between electromagnetic and weak amplitudes, where the first ones are due to the exchange of (virtual) photons while the second ones are caused by the exchange of the intermediate vector boson Z^0. Experimentally one observes these effects in photon-induced transitions between atomic states. Very schematically such transitions proceed through the three diagrams shown in Fig. 1.3. There the exchange of photons and Z^0's between the electron and the nucleus symbolizes the fact that the electron is bound to the nucleus by the Coulomb force (γ) as well as by the weak force (Z^0). The interference of diagram 1.3a with 1.3b and 1.3c generates pseudoscalar terms in the transition rate that violate parity.

Since we are dealing with a bound electron-nucleus system, the calculation of parity-violating effects in atomic systems proceeds in four steps:

i) An effective potential for the atomic electron moving in the Z^0-field of the nucleus is derived from the general weak Lagrangian of the Standard Model.

ii) By means of standard stationary-state perturbation theory the effect of this potential on the atomic wave function is determined. As a result, states are generated which are mixtures of (unperturbed) states of different parity.
iii) Matrix elements of photon-induced transitions between these perturbed atomic states are then calculated using the usual electron-photon interaction. The existence of opposite-parity admixtures thus results in pseudoscalar terms in the transition rate.
iv) Finally, these matrix elements are used to calculate observable parity-violating effects, as e.g., optical rotation.

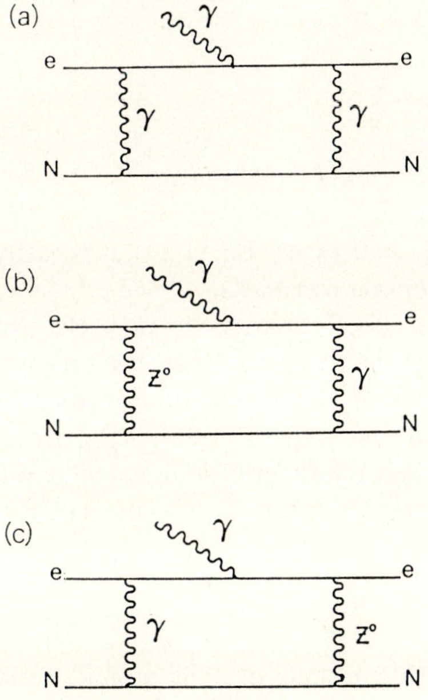

Fig. 1.3. "Diagrams" for the absorption of a photon by an atom. The interference of diagram (**a**) with diagrams (**b**) and (**c**) gives pseudoscalar terms in the absorption cross section. From [5]

1.4.1 Phenomenology

At the level of atomic physics the Standard Model is adequately approximated by the current-current picture of weak interactions (see Fig. 1.2a; here, however, only weak neutral currents have to be considered). The respective weak Lagrangian of the electron-nucleus system is therefore the product of an electron current with a nuclear current,

$$\mathcal{L}_{\mathrm{w}} = \frac{G_{\mathrm{F}}}{\sqrt{2}} J_{\mu}^{\mathrm{e}} \sum_{\mathrm{p,n}} (J_{\mu}^{\mathrm{p}} + J_{\mu}^{\mathrm{n}}) \,, \tag{1.25}$$

where the sum extends over all protons (p) and neutrons (n) in the nucleus. Each of these currents is the sum of a vector (V) and an axial vector (A) component, and only those terms which couple a vector to an axial vector current violate parity. The parity-violating effective Lagrangian thus is given by

$$\mathcal{L}_{\mathrm{PV}} = \mathcal{L}_{\mathrm{PV}}^{(1)} + \mathcal{L}_{\mathrm{PV}}^{(2)} = \frac{G_{\mathrm{F}}}{\sqrt{2}} \bar{\psi}_{\mathrm{e}} \gamma_\mu \gamma_5 \psi_{\mathrm{e}} \sum_{\mathrm{p,n}} (C_{\mathrm{Vp}} \bar{\psi}_{\mathrm{p}} \gamma_\mu \psi_{\mathrm{p}} + C_{\mathrm{Vn}} \bar{\psi}_{\mathrm{n}} \gamma_\mu \psi_{\mathrm{n}})$$
$$+ \frac{G_{\mathrm{F}}}{\sqrt{2}} \bar{\psi}_{\mathrm{e}} \gamma_\mu \psi_{\mathrm{e}} \sum_{\mathrm{p,n}} (C_{\mathrm{Ap}} \bar{\psi}_{\mathrm{p}} \gamma_\mu \gamma_5 \psi_{\mathrm{p}} + C_{\mathrm{An}} \bar{\psi}_{\mathrm{n}} \gamma_\mu \gamma_5 \psi_{\mathrm{n}}) \,. \tag{1.26}$$

The first term couples the electron axial current to the nucleon vector current, and the second one the electron vector current to the nucleon axial current. In the Standard Model the coupling parameters C_{Vp}, C_{Vn}, C_{Ap}, C_{An} are related to the weak mixing angle θ_{W} and the axial coupling constant of neutron β-decay $g_{\mathrm{A}}(\approx 1.25)$ by

$$\begin{aligned} C_{\mathrm{Vp}} &= \frac{1}{2}(1 - 4\sin^2\theta_{\mathrm{W}}) & C_{\mathrm{Vn}} &= -\frac{1}{2} \\ C_{\mathrm{Ap}} &= \frac{1}{2} g_{\mathrm{A}} (1 - 4\sin^2\theta_{\mathrm{W}}) & C_{\mathrm{An}} &= -\frac{1}{2} g_{\mathrm{A}} (1 - 4\sin^2\theta_{\mathrm{W}}) \end{aligned} \tag{1.27}$$

Because of the experimental value of $\sin^2\theta_{\mathrm{W}} \approx 0.23$, see Eq. (1.24), it is easily seen that C_{Vn} is the largest of the four coupling constants.

From this Lagrangian one may obtain an effective parity-violating electron-nucleus potential as the Fourier transform of the respective scattering matrix element – this represents the inverse procedure of the reasoning leading to (1.7) Let us consider, for the moment, the first part of (1.26), $\mathcal{L}_{\mathrm{PV}}^{(1)}$. In an atomic nucleus, the nucleons are certainly non-relativistic, so only the $\mu = 0$ component of the nuclear current contributes. In addition, the nucleus can be considered as pointlike, so that

$$\bar{\psi}_{\mathrm{p}} \gamma_0 \psi_{\mathrm{p}} = \bar{\psi}_{\mathrm{n}} \gamma_0 \psi_{\mathrm{n}} \approx \delta^{(3)}(\mathbf{r}) \,,$$

$\mathbf{r}$ being the position of the electron. For calculations in light atoms – and as a first approximation in heavy atoms as well – we may treat the electron also non-relativistically. With the help of the non-relativistic representation of the γ-matrices (cf. Sect. 1.3.3) we then obtain for the relevant component of the axial electron current ($m_{\mathrm{e}} \ldots$ electron mass)

$$J_0^{\mathrm{Ae}} \approx \phi^+ \left[\frac{\boldsymbol{\sigma} \cdot \overleftarrow{\mathbf{p}}}{2 m_{\mathrm{e}} c} + \frac{\boldsymbol{\sigma} \cdot \overrightarrow{\mathbf{p}}}{2 m_{\mathrm{e}} c} \right] \phi \,. \tag{1.28}$$

Here the first momentum operator $\mathbf{p} = -i\hbar\nabla$ in the brackets acts on the wave function ϕ^+ on the left-hand side, and the second one on the wave function ϕ on the right-hand side, as indicated by the arrows.

Combining all these expressions we arrive at the final result for the potential

$$V_{\mathrm{PV}}^{(1)} = \frac{G_{\mathrm{F}}}{4\sqrt{2}} \frac{Q_{\mathrm{W}}}{m_{\mathrm{e}} c} \left[\boldsymbol{\sigma} \cdot \overleftarrow{\mathbf{p}} \, \delta^{(3)}(\mathbf{r}) + \delta^{(3)}(\mathbf{r}) \boldsymbol{\sigma} \cdot \overrightarrow{\mathbf{p}} \right] \,, \tag{1.29}$$

where the arrows above the momentum operators again indicate on which wave function they act. The factor Q_W is proportional to the sum of the vector couplings C_{Vp} and C_{Vn} of the nucleons, and for a nucleus consisting of Z protons and N neutrons it is equal to (in the Standard Model)

$$Q_\mathrm{W} = 2\sum_{\mathrm{p,n}}(C_{\mathrm{Vp}} + C_{\mathrm{Vn}}) = Z(1 - 4\sin^2\theta_\mathrm{W}) - N\,. \tag{1.30}$$

Since $\sin^2\theta_\mathrm{W} \approx 0.23$, it is easily seen that the electron couples primarily to the neutrons, and the factor N in Q_W enhances parity-violating effects in heavy nuclei (see also below).

The second part of the effective potential, coming from the coupling of the electron vector current with the nuclear axial current, i.e., $\mathcal{L}^{(2)}_{\mathrm{PV}}$ in (1.26), is suppressed by an overall factor of $(1 - \sin^2\theta_\mathrm{W}) \ll 1$ [cf. Eq. (1.27)], and also contains no enhancement factor Q_W. Therefore its effect can safely be neglected.

The modifications of the atomic wave functions due to the parity-violating potential (1.29) are then calculated using standard stationary-state perturbation theory. In the absence of parity violation atomic states $|P,n\rangle$ are labeled by their parity $P = \pm 1$, in addition to any other quantum number n. As a result of V_{PV}, states of opposite parity become mixed:

$$|P,n\rangle \to |(P),n\rangle = |P,n\rangle + \sum_m \varepsilon_{mn}|-P,m\rangle\,, \tag{1.31a}$$

with

$$\varepsilon_{mn} = \frac{\langle -P,m|V_{\mathrm{PV}}|P,n\rangle}{E(P,n) - E(-P,m)}\,. \tag{1.31b}$$

The notion (P) in the perturbed state should indicate that this parity is now only nominal.

The matrix elements in (1.31b) can be estimated in an independent-particle model of the atom. Usually only mixing between $\mathrm{s}_{1/2}$ and $\mathrm{p}_{1/2}$ states need be considered because the delta function in (1.29) requires a non-zero value at the origin for the wave function or its gradient, respectively. In this case the admixture coefficients (1.31b) are typically of order $10^{-17}\,Z^3K_r$: one factor of Z comes from $Q_\mathrm{W}(\sim N \sim Z)$, the other two come from the value and the gradient of the atomic wave functions at the origin, and K_r is a relativistic correction factor for high-Z atoms ($K_r \sim 3$ for cesium and ~ 10 for bismuth). This factor shows the enormous advantage of using heavy atoms.

1.4.1 Experiments

We now consider an electromagnetic transition between two atomic states $|(P),n\rangle$ and $|(P),n'\rangle$ of the same nominal parity, i.e., in the absence of parity-violating effects magnetic dipole transitions. From (1.31) the transition amplitude now is

Table 1.3. Results of parity-violating effects in atoms

Transition	Quantity	Exp. Value	Theor. Value
Bi (648 nm)	$\mathrm{Im}\mathcal{E}_\mathrm{P}/\mathcal{M} \times 10^8$	-9.3 ± 1.15	-10.5 to -17
Bi (876 nm)	$\mathrm{Im}\mathcal{E}_\mathrm{P}/\mathcal{M} \times 10^8$	-10.4 ± 1.7	-8 to -13
Pb (1279 nm)	$\mathrm{Im}\mathcal{E}_\mathrm{P}/\mathcal{M} \times 10^8$	-9.9 ± 2.5	-11 to -14
Tl	$\mathcal{E}_\mathrm{P}/\beta$	-1.8 ± 0.6	-1.80 to -2.17
$(6\mathrm{P}_{1/2} - 7\mathrm{P}_{1/2})$	(mV/cm)	-1.73 ± 0.33	
Cs	$\mathcal{E}_\mathrm{P}/\beta$	-1.52 ± 0.18	-1.50 to -1.59
$(6\mathrm{S}_{1/2} - 7\mathrm{S}_{1/2})$	(mV/cm)	-1.65 ± 0.13	

$$\mathcal{A} = \mathcal{M} + \mathcal{E}_\mathrm{P} , \tag{1.32a}$$

with $\mathcal{M}$ being the zero-order M1 amplitude and $\mathcal{E}_\mathrm{P}$, given by

$$\mathcal{E}_\mathrm{P} = \sum_m [\varepsilon_{mn} \langle P, n' | H_\mathrm{elmag} | -P, m\rangle + \varepsilon_{n'm} \langle -P, m | H_\mathrm{elmag} | P, n\rangle] , \tag{1.32b}$$

is an electric dipole (E1) amplitude caused by parity violation. In a standard convention $\mathcal{M}$ is real and $\mathcal{E}_\mathrm{P}$ is purely imaginary. The existence of $\mathcal{E}_\mathrm{P}$ in (1.32a) means that reactions involving configurations related by mirror reflections have different rates. The size of these parity-violating asymmetries then is determined by the ratio

$$\Delta = \frac{\mathrm{Im}\,\mathcal{E}_\mathrm{P}}{\mathcal{M}} \sim 10^{-14}\, Z^3\, K_r , \tag{1.33}$$

which gives, for example, the relative difference between the cross sections of left and right circularly polarized photons.

Experiments have been proposed to detect the existence of $\mathcal{E}_\mathrm{P}$ in atomic hydrogen where the wave functions are precisely known. Unfortunately, the expected effects are extremely small – due to the lack of the enhancement factor $Z^3 K_r$ – and very difficult to observe. Parity-violating effects have, however, been observed in the heavy atoms bismuth, lead, thallium and cesium, where the effects are larger than in hydrogen but uncertainties in atomic theory make precise calculations of $\mathcal{E}_\mathrm{P}$ a difficult task. Two kinds of experiments have been performed: optical rotation in bismuth and lead and Stark-optical pumping of forbidden M1-transitions in thallium and cesium. In the first type the results are presented by the ratio Δ of (1.33), the second type yields the ratio of $\mathcal{E}_\mathrm{P}$ to β, the factor of proportionality between the Stark amplitude and the electric field.

In Table 1.3 (adapted from the review article by Bouchiat and Pottier [6]) the experimental results are compared to various theoretical calculations. The large spread in the theoretical numbers is due to the rather complicated structure of the atoms. An exception is cesium: here the atomic matrix elements are most reliably calculable since there is only one electron outside a closed shell. The agreement in this case is at the level of 10%.

In general the results demonstrate that parity conservation is violated in atoms at the level predicted by the Standard Model. All these investigations show that atomic physics will continue to contribute useful information to our knowledge of fundamental symmetries in nature.

1.5 References

1 Georgi H (1981) A unified theory of elementary particles and forces, Sci. Am. *244*(4): 40
2 Georgi H Glashow SL (1980) Unified theory of elementary particle forces, Phys. Today *33*(9): 30
3 't Hooft G (1980) Gauge theories of the forces between elementary particles, Sci. Am. *243*(6): 90
4 Fortson EN Lewis LL (1984) Atomic parity nonconservation experiments, Phys. Rep *113*: 289
5 Rich J Lloyd Owen D Spiro M (1987) Experimental particle physics without accelerators, Phys. Rep *151*: 239
6 Bouchiat MA Pottier L (1986) Optical experiments and weak interactions, Science *234*: 1203
7 Bouchiat MA Pottier L (1984) An atomic preference between left and right, Sci. Am. *250*(6): 76
8 Commins ED Bucksbaum PH (1980) The parity non-conserving electron-nucleon interaction, Ann. Rev. Nucl. Part. Sci. *30*: 1

2 Theories on the Origin of Biomolecular Homochirality

R. Janoschek

2.1 Introduction

From the discovery of dissymmetric crystals by Louis Pasteur in 1848, the conclusion was drawn that there exist dissymmetric molecular structures [1]. Their occurrence was explained by allpervasive and universal dissymmetric forces. Michael Faraday's discovery [2] that inactive materials such as glass show optical activity in a magnetic field, convinced Pasteur that the well-known classical polar fields are basically dissymmetric. However, all his related chemical experiments failed [3]. Pasteur's term *dissymétrie* was replaced later by the notion *chirality*, which was introduced by Kelvin, who adopted it from the familiar analogy of the morphological mirror-image relation between the left and the right hand [4].

Ever since the early days of biochemistry the question of the origin of biomolecular homochirality has been posed and is more topical than ever before. A subsequent question is concerned with the reason for the preference of L-amino acids and D-sugars in biochemical processes, in contrast to their mirror-image isomers.

The phenomenon of homochirality is also observed at the macroscopic level of living organisms although the connecting link to molecular chirality is not yet known. There is no a priori reason why a chiral object should be superior to its mirror-image. Yet the real world usually shows a propensity to prefer one kind of chirality over another. Human beings are generally not ambidextrous, and most people are right-handed. The majority of snail-shell spirals have a right-hand screw, however, certain species are predominantly left-handed. It is very uncommon that a species consists of the same numbers of left- and right-handed individuals, such as *Liguus poeyanus.* A preferred chirality is observed also for many types of plants. Bindweed (*Convolvulus arvensis*) winds to the right like the majority of helical plants, but honeysuckle (*Lonicera sempervirens*) grows as a left-handed helix. For further details on handedness in nature, a series of recent introductory reviews is recommended [5–10].

In this chapter a variety of kinetic models for the origin of biomolecular homochirality is described and discussed. As was established by these models, spontaneous asymmetric synthesis is an evident property of life. Any slight chiral excess is further amplified and leads to the disappearance of the mirror-image isomer with lower concentration. But what is the origin of such a chiral excess during the prebiotic period? Is there a consistent basis for the observed homochirality, or is it a mere matter of chance? A possible answer to this question was initiated by S. Glashow, A. Salam, and S. Weinberg who suggested a unified theory for the electromagnetic and weak interactions [11]. Their theory allows one to connect the parity-violating and dissymmetric weak interaction (represented by the weak neutral bosons in atomic nuclei) and the electromagnetic interaction, which causes the chemical bond. It seems to be evident that the whole universe is chiral on all scales, from the scale of elementary particles upward to the macroscopic scale of life. Therefore, unusual and interesting relations are to be expected.

2.2 Observability of Chiral Molecular Structures

The structure of chiral molecules is usually discussed by chemists in terms of three-dimensional molecular models in the framework of conventional stereochemistry. Two enantiomers, however, are never completely separated. They correspond to local minima on the energy hypersurface with a barrier in between (double-well potential). Quantum mechanics is in conflict with classical stereochemistry which was first recognized by Friedrich Hund [12]. The problem is the possibility of tunneling, which connects left- and right-handed structures. Thus, all stationary (time-independent) states of potentially chiral molecules would be achiral and therefore, optical activity should not exist. Hund proposed a solution to this problem by considering non-stationary states and the time scales for racemization by tunneling. Stationary states of achiral systems with positive (ψ_+) and negative (ψ_-) parity exhibit a small splitting $\Delta E_\pm$ which depends on the ratio $V/h\nu$ in the exponent of Hund's formula

$$\Delta E_\pm = \frac{1}{2\sqrt{\pi}} \sqrt{h\nu V}\, e^{-V/h\nu} , \tag{2.1}$$

where V is the height of the energy barrier and ν is the frequency for the vibration in a single well (Fig. 2.1). The time of racemization T based on the tunneling mechanism is connected with the energy splitting by

$$T = h/\Delta E_\pm . \tag{2.2}$$

In Table 2.1 significant data of double-well potentials and racemization-times are presented for few small systems. Tunneling rates are seen to be extremely sensitive to the potential energy function as well as to the kind of approximation of the wavefunction. Therefore, their calculation is an ambitious mathematical task. Hund's simple formula (2.1) fails completely as a proper description of tunneling rates. The level splitting $\Delta E_\pm$ can be deduced from spectra in the case of NH_3; for PH_3 the corresponding value is the result of a reasonable estimate [13]. The inversion splitting of CH_4 is estimated by means of the calculated energy barrier [14] and extrapolation.

Table 2.1. Wavenumber $\nu/c\,(\mathrm{cm}^{-1})$ for the vibration in a single well; energy barrier V (kcal/mol) of the double-well potential for molecular inversion; racemization-times: T (sec) from Eq. (2.1), and the most reliable values T_{rac} from spectra

	ν/c	V	$V/h\nu$	T	T_{rac}
NH_3	950	5	1.8	$5.3 \cdot 10^{-13}$	$2.1 \cdot 10^{-11}$
PH_3	991	37	13.1	$1.6 \cdot 10^{-8}$	$1.7 \cdot 10^{7}$
CH_4	1526	112	25.7	$3.4 \cdot 10^{-3}$	$\sim 10^{25}$

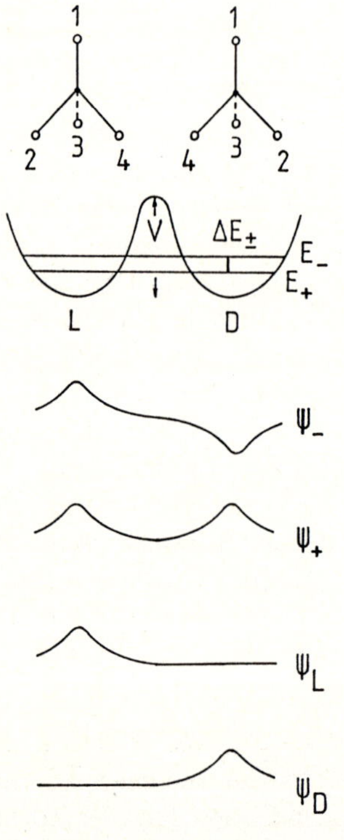

Fig. 2.1. Double-well potential for molecular inversion and achiral states, ψ_+, ψ_-. Chiral states ψ_L, ψ_D are the result of the transformation (2.3)

The data in Table 2.1 indicate no principal differences between optically active and inactive structures. If the lifetime T of a chiral structure in one of the wells is large compared with the time resolution of the spectroscopic experiment, then optical activity can be observed. Thus, optical activity occurs for phosphines, $PR^1R^2R^3$, but optically active ammines, $NR^1R^2R^3$, are unknown. The barrier of inversion for tetrahedral carbon is rather high. Among amino acids a "fast" racemization is believed to have been observed for aspartic acid [15]. D-aspartic acid has been shown to accumulate with age in human tooth enamel and lens at a rate of about $1.25 \times 10^{-3}\,\mathrm{yr}^{-1}$. However, a unimolecular process as the origin for this rate is still unproven. Most of the reaction occurs by thermal excitation, but not by tunneling, over a barrier of about 30 kcal/mol [16].

Conclusion. Hund's theory for the dynamics of chiral structures and the resulting phenomenon of optical activity is compatible with chemical experience, at least in a qualitative sense. Quantum mechanics provides achiral eigenfunctions, if the Hamiltonian for the nuclear motion is invariant under space reflection for the potentially enantiomeric system. Superposition of eigenfunctions ψ_+, ψ_- (Eq. (2.3)) leads to left- and right-handed enantiomers ψ_L, ψ_D, which can be interpreted as time dependent molecular states (see Fig. 2.1). The energies E_L and E_D are

$$\begin{aligned} \psi_L &= 1/\sqrt{2}(\psi_+ + \psi_-) \\ \psi_D &= 1/\sqrt{2}(\psi_+ - \psi_-) \end{aligned} \tag{2.3}$$

identical, $1/2(E_+ + E_-)$. The general importance of tunneling, apart from thermal activation, for racemization processes is still unclear. A more elaborate treatment of dynamics of chiral molecules was recently presented by Martin Quack [17].

2.3 Kinetic Models for Unstable Equilibrium

A simple model for L–D stereoselection rests on a chemical compound $L(D)$ which is a catalyst for its own production and an anti-catalyst for the production of its optical antimer $D(L)$. Besides autocatalysis also specific antagonism between L and D is considered. It is supposed that both enantiomers were present in similar concentrations during a prebiotic period. The model of Frank takes the following chemical reactions into account [18]:

$$\begin{aligned} &L + A \rightarrow 2L \qquad && k_1 \\ &D + A \rightarrow 2D \qquad && k_1 \\ &L + D \rightarrow A' \qquad && k_2 \end{aligned} \tag{2.4}$$

A and A' are achiral substances where the concentration of A is assumed not to vary with time. L and D are two optical isomers. The respective concentrations n_L and n_D are considered as functions of time which result from the coupled system of non-linear differential equations

$$\begin{aligned} dn_L/dt &= (k_1 - k_2 n_D)n_L \\ dn_D/dt &= (k_1 - k_2 n_L)n_D\ , \end{aligned} \tag{2.5}$$

where k_1 and k_2 are rate constants. Subtraction of Eqs. (2.5) makes the nonlinear terms vanish

$$d(n_L - n_D)/dt = k_1(n_L - n_D) \tag{2.6}$$

and leads to the solution

$$(n_L - n_D) = (n_L^0 - n_D^0)e^{k_1 t}\ , \tag{2.7}$$

where n_L^0 and n_D^0 are initial concentrations at $t = 0$. The excess of one enantiomer over the other increases exponentially if $n_L^0 \neq n_D^0$. Addition of Eqs. (2.5) leads to

$$d(n_L + n_D)/dt = k_1(n_L + n_D) - 2k_2 n_L n_D \,. \tag{2.8}$$

Thus, the sum $n_L + n_D$ has a slower relative rate of increase than the difference $n_L - n_D$. Eliminating dt from (2.5) yields

$$dn_L/dn_D = \{(k_1 - k_2 n_D)n_L\} \,/\, \{(k_1 - k_2 n_L)n_D\} \tag{2.9}$$

and hence

$$n_L/n_D = (n_L^0/n_D^0)\,\exp\left\{k_2(n_L - n_D - n_L^0 + n_D^0)/k_1\right\} . \tag{2.10}$$

Combining (2.7) and (2.10)

$$n_L/n_D = (n_L^0/n_D^0)\,\exp\left\{k_2(n_L^0 - n_D^0)(e^{k_1 t} - 1)/k_1\right\} . \tag{2.11}$$

Consequently the ratio n_L/n_D increases at a more than exponential rate if $n_L^0 > n_D^0$, and decreases correspondingly if $n_L^0 < n_D^0$. The general form of the solutions of Eqs. (2.5) are shown in Fig. 2.2. Equality of n_L and n_D represents unstable equilibrium which is caused by the terms representing specific mutual antagonism, i.e. the quadratic terms in Eqs. (2.5).

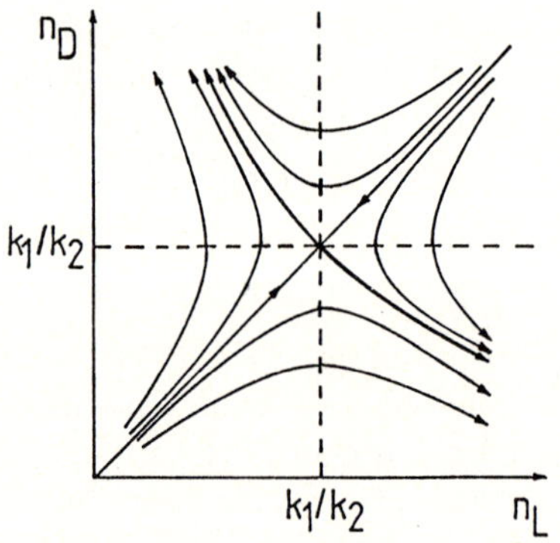

Fig. 2.2. Representation of the solutions of Eqs. (2.5). Every starting point not on the line $n_L = n_D$ leads to one of the asymptotes $n_L = 0$ or $n_D = 0$

Let us now consider unspecific antagonism which is an equally deleterious effect upon net reproduction rate. Equations (2.4) are extended by two equations so that

$$\begin{aligned}
L + A &\to 2L && k_1 \\
D + A &\to 2D && k_1 \\
L + D &\to A' && k_2 \\
L + L &\to A' && k_2 \\
D + D &\to A' && k_2 \,.
\end{aligned} \tag{2.12}$$

The corresponding system of differential equations is

$$\begin{aligned} dn_L/dt &= k_1 n_L - k_2 n_L n_D - k_2 n_L^2 \\ dn_D/dt &= k_1 n_D - k_2 n_L n_D - k_2 n_D^2 \, . \end{aligned} \tag{2.13}$$

Subtraction and addition yield

$$d(n_L - n_D)/dt = k_1(n_L - n_D) - k_2(n_L + n_D)(n_L - n_D) \tag{2.14}$$

and

$$d(n_L + n_D)/dt = k_1(n_L + n_D) - k_2(n_L + n_D)^2 \, , \tag{2.15}$$

respectively. Eliminating dt from these equations gives

$$d(n_L - n_D)/d(n_L + n_D) = (n_L - n_D)/(n_L + n_D) \, . \tag{2.16}$$

Integration of (2.15) and (2.16) leads to the final result

$$(n_L + n_D) = \text{const.}\, (n_L - n_D) = k_1/k_2 \, 1/(1 + \exp(-k_1 t)) \, . \tag{2.17}$$

With this equation, any initial disproportion is preserved, but not amplified.

The system of reactions (2.12) is now extended by two equations which represent specific antagonism. The effect of these extensions is that the rates for specific $(k_2 + k_3)$ and unspecific (k_2) antagonism differ.

$$\begin{array}{ll} L + A \rightarrow 2L & k_1 \\ D + A \rightarrow 2D & k_1 \\ L + D \rightarrow A' & k_2 \\ L + L \rightarrow A' & k_2 \\ D + D \rightarrow A' & k_2 \\ L + D \rightarrow A'' + D & k_3 \\ L + D \rightarrow A'' + L & k_3 \end{array} \tag{2.18}$$

Instead of (2.13) the following system of differential equations is obtained

$$\begin{aligned} dn_L/dt &= k_1 n_L - k_2 n_L n_D - k_2 n_L^2 - k_3 n_L n_D \\ dn_D/dt &= k_1 n_D - k_2 n_L n_D - k_2 n_D^2 - k_3 n_L n_D \end{aligned} \tag{2.19}$$

Subtraction and addition yield

$$d(n_L - n_D)/dt = k_1(n_L - n_D) - k_2(n_L + n_D)(n_L - n_D) \tag{2.20}$$

and

$$d(n_L + n_D)/dt = k_1(n_L + n_D) - k_2(n_L + n_D)^2 - 2k_3 n_L n_D \, , \tag{2.21}$$

respectively. Eliminating dt from these equations gives, in place of (2.16)

$$\begin{aligned} & d\left\{\ln(n_L + n_D)\right\} / d\left\{\ln(n_L - n_D)\right\} \\ & \quad = 1 - 2k_3 n_L n_D / \left\{(n_L + n_D)(k_1 - k_2(n_L + n_D))\right\} \end{aligned} \tag{2.22}$$

This expression is less than 1 in the case when $(n_L + n_D) < k_1/k_2$. Thus, the difference $n_L - n_D$ increases faster than the sum $n_L + n_D$ so that equality of n_L and n_D is again unstable.

In this section simple kinetic models of the Frank type were described where autocatalysis and specific antagonism of enantiomers are in competition. The latter leads to an unstable equilibrium for equal concentrations of enantiomers. This effect originates from non-linear terms in the systems of differential equations.

2.4 Kinetic Models with Intrinsic Asymmetry

All kinetic models in the preceding section are symmetric with respect to interchange of n_L and n_D. The source of an initial disproportion is assumed to be an external asymmetry such as circular polarisation of light. Since the discovery of parity violations in the weak interaction, connection to biochemical L–D stereoselection was recognized [19]. The energies of two enantiomers of the same compound are no longer identical and, therefore, a minute difference in the activation energies occurs. Consequently the two rate constants k_1 in (2.4) as well as k_3 in (2.18) are different. The following system of reactions will be examined [20, 21].

$$\begin{array}{ll} L + A \rightarrow 2L & k_{1L} \\ D + A \rightarrow 2D & k_{1D} \\ L + D \rightarrow A' & k_2 \\ L + D \rightarrow A'' + D & k'_{2L} \\ L + D \rightarrow A'' + L & k'_{2D} \end{array} \tag{2.23}$$

The rate equations describing these processes are

$$\begin{aligned} dn_L/dt &= k_{1L}n_L - k_{2L}n_L n_D \\ dn_D/dt &= k_{1D}n_D - k_{2D}n_L n_D \, , \end{aligned} \tag{2.24}$$

where $k_{2L} = k_2 + k'_{2L}$ and $k_{2D} = k_2 + k'_{2D}$. It will be assumed that the rate constants k_{1L} and k_{1D} as well as k_{2L} and k_{2D} are numerically very similar. Therefore,

$$\begin{aligned} k_{1L} &= (1 + \varepsilon_1)k_{1D} \\ k_{2L} &= (1 + \varepsilon_2)k_{2D} \, , \end{aligned}$$

where the increments ε_1, ε_2 are of the order of 10^{-13}. An iterative procedure, where the initial concentrations $n_L^0 = n_D^0 = n^0$ are involved, yields the first order approximation

$$\begin{aligned} n_L &= n^0 \exp(k_{1L}t) \left\{1 + (n^0 k_{2D}/k_{1D})(\exp(k_{1D}t) - 1)\right\}^{-1-\varepsilon_2} \\ n_D &= n^0 \exp(k_{1D}t) \left\{1 + (n^0 k_{2L}/k_{1L})(\exp(k_{1L}t) - 1)\right\}^{-1+\varepsilon_2} \, . \end{aligned} \tag{2.25}$$

For large values of t the asymptotic behaviour of (2.25) is

$$\begin{aligned} n_L &\sim (n^0)^{\varepsilon_2}(k_{1D}/k_{2D})^{1+\varepsilon_2}\exp\{(\varepsilon_1-\varepsilon_2)k_{1L}t\} \\ n_D &\sim (n^0)^{-\varepsilon_2}(k_{1L}/k_{2L})^{1-\varepsilon_2}\exp\{-(\varepsilon_1-\varepsilon_2)k_{1D}t\}\ . \end{aligned} \tag{2.26}$$

From (2.26) it is evident that for $\varepsilon_1-\varepsilon_2>0$ n_D decreases exponentially and finally vanishes, whereas n_L increases exponentially. For $\varepsilon_1-\varepsilon_2<0$ the behaviour of n_L and n_D is interchanged. The numerical values of the increments ε_1 and ε_2 are immaterial with respect to these results. According to (2.26) biochemical homochirality is a consequence of the difference $\varepsilon_1-\varepsilon_2$, i.e. the difference between differences of k_{1L}, k_{1D} and k_{2L}, k_{2D}.

An essential supposition in the models studied so far is that the initial production of chiral compounds can be treated as a rare event. The neglect of initial spontaneous generation of enantiomers in the kinetic models leads to an artificial starting point. The evolution of the chemical system, however, must have a beginning. Therefore, two processes complete the system (2.23) [22].

$$\begin{array}{ll} L+A\rightarrow 2L & k_{1L} \\ D+A\rightarrow 2D & k_{1D} \\ L+D\rightarrow A' & k_2 \\ L+D\rightarrow A''+D & k'_{2L} \\ L+D\rightarrow A''+L & k'_{2D} \\ A^0\rightarrow L & k_{0L} \\ A^0\rightarrow D & k_{0D} \end{array} \tag{2.27}$$

The achiral substrate A^0 is taken to be time-independent. The corresponding system of differential equations is

$$\begin{aligned} dn_L/dt &= k_{0L}+k_{1L}n_L-k_{2L}n_Dn_L \\ dn_D/dt &= k_{0D}+k_{1D}n_D-k_{2D}n_Ln_D\ . \end{aligned} \tag{2.28}$$

At the beginning the symmetric solution $(n(t=0)=n^0=n^0_L=n^0_D)$ for intrinsic symmetry $(k_{0L}=k_{0D}, k_{1L}=k_{1D}, k_{2L}=k_{2D})$ is considered. Equations (2.28) reduce to

$$dn/dt = k_0+k_1n-k_2n^2\ . \tag{2.29}$$

The symmetric solution has the form

$$n = k_1/(2k_2)+\alpha\,\mathrm{th}\,(\alpha k_2t+C) \tag{2.30}$$

where

$$\begin{aligned} \alpha &= \left\{(k_1/(2k_2))^2+k_0/k_2\right\}^{1/2} \\ C &= \mathrm{Arcth}\,\left\{(n^0-k_1/(2k_2))/\alpha\right\}. \end{aligned}$$

The solution (2.30) has an interesting property: The time origin t_0 can be defined from the natural condition $n=0$. It follows

$$t_0 = -\left\{\ln(k_1^2/(k_0 k_2)) + 2C\right\} / \left\{k_1 + 2k_0 k_2/k_1\right\} . \tag{2.31}$$

The evolution of enantiomers L and D has now a well defined starting point which can be expressed by means of rate constants and the concentration at an arbitrary point on the time scale.

Differences in k_0 and k_1 for L and D components will be introduced now. Without going into details the final approximate solutions are

$$\begin{aligned} n_L &\sim \exp(\Delta k_1 t - k_{0D} k_2 t / k_{1D}) \\ n_D &\sim \exp(-\Delta k_1 t - k_{0L} k_2 t / k_{1L}) , \end{aligned} \tag{2.32}$$

where $\Delta k_1 = k_{1L} - k_{1D}$. From (2.32) follows

$$n_L / n_D \sim \exp(2\Delta k_1 t + \Delta k_0 k_2 t) , \tag{2.33}$$

where $\Delta k_0 = k_{0L}/k_{1L} - k_{0D}/k_{1D}$. The sign of the exponent of (2.33) determines that component which will predominate later. It can be assumed that spontaneous chiral synthesis (k_0) is a rare event compared with autocatalysis (k_1). Suppose, for example, $\Delta k_1 > 0$, then an instability develops, once the concentrations n in (2.30) reach a value near the equilibrium concentrations $k_1/(2k_2)+\alpha$. This instability leads to a splitting of n_L and n_D with $n_L \to \infty$, $n_D \to 0$ as $t \to \infty$. The situation described is sketched in Fig. 2.3.

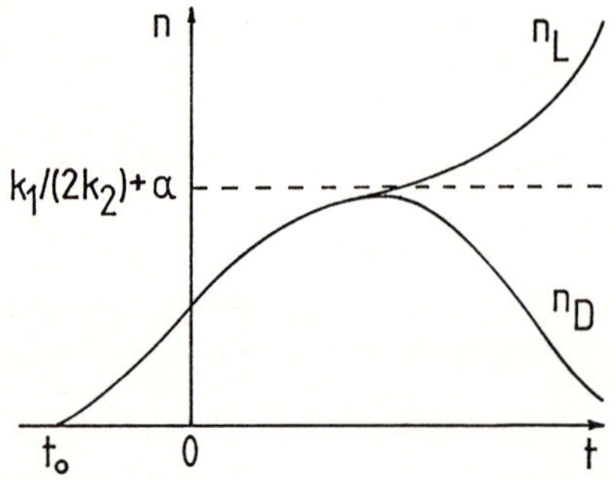

Fig. 2.3. Chiral evolution from a starting point t_0 (2.31) through a period of symmetry (2.30), and stereoselection through bifurcation near the equilibrium (2.32); ($t = 0$ is chosen arbitrarily)

2.5 Parity-Violating Energy Differences Between Enantiomers

The introduction of different reaction rates k_L and k_D for two enantiomers L and D in the last section rests on the hypothesis that E_L and E_D, other than assumed initially in (2.3), are different. The reason for this inequality can be found in the atomic nucleus, in particular in the parity-violating weak interaction which was outlined in the first chapter of this book. In this section calculations on the order of magnitude and sign of energy corrections for E_L and E_D will be described. A comprehensive treatment of such calculations can be found in literature [23]. In the following important features and recent results will be reviewed.

The energy operator H_{pv} for the parity-violating weak neutral current interaction is reducible to sums of proton–electron and neutron–electron terms

$$H_{pv} = H_{pv}^{pe} + H_{pv}^{ne} . \tag{2.34}$$

The potentials H_{pv}^{pe} and H_{pv}^{ne} both have a common dependence upon the scalar product of the momentum $\mathbf{p}_i$ and the Pauli spin matrix operator $\boldsymbol{\sigma}_i$ of electron i. Summing up all nuclear-electron pairs ai leads to an effective one-electron operator for the interaction

$$H_{pv} = -(\Gamma/2) \sum_a \sum_i Q_a \{\mathbf{p}_i \boldsymbol{\sigma}_i \, , \, \delta^3(\mathbf{r}_i - \mathbf{r}_a)\}_+ \, , \tag{2.35}$$

where $\{\ldots\}_+$ denotes an anticommutator. Γ is composed of the Fermi weak coupling constant G_F, the electron rest mass m_e, and the speed of light c,

$$\Gamma = G_F/(2\sqrt{2} m_e c) \, , \tag{2.36}$$

with $\Gamma = 5.732 \times 10^{-17}$ a.u. The charge density of electron i at the atomic nucleus a is represented by the Dirac delta function $\delta^3(\mathbf{r}_i - \mathbf{r}_a)$. The nucleus parameter Q_a is given by

$$Q_a = N_a - (1 - 4\sin^2\theta_w) Z_a \, , \tag{2.37}$$

where N_a is the neutron number and Z_a the proton number of the nucleus a. For the empirical value of the Weinberg angle θ_w, which relates the photon and the massive neutral boson Z^0, Q_a is represented by N_a.

The potential in (2.35) is purely imaginary because of the momentum operator $\mathbf{p}_i$ and thus the corresponding expectation value over real non-relativistic molecular wavefunctions vanishes. A non-zero value for E_{pv} of an enantiomer requires a spin–orbit coupling correction for the molecular wavefunction. The spin–orbit Hamiltonian H_{so} is given by

$$H_{so} = \sum_b \sum_j \xi(b,j) \mathbf{l}(b,j) \mathbf{s}(j) \, . \tag{2.38}$$

Here $\mathbf{s}(j)$ is the electron spin operator, and $\mathbf{l}(b,j)$ is the orbital angular momentum of electron j around nucleus b; $\xi(b,j)$ is the spin–orbit coupling parameter. The operator H_{so} connects triplet states $|\Psi_T\rangle$ to the uncorrected singlet ground state $|\Psi_S\rangle$ with energies E_T and E_S, respectively. The corrected singlet state $|\Psi_S'\rangle$ is

$$|\Psi_S'\rangle = |\Psi_S\rangle + \sum_T \langle\Psi_T|H_{so}|\Psi_S\rangle|\Psi_T\rangle/(E_S - E_T) \, . \tag{2.39}$$

To the first order of perturbation theory for H_{so}, the parity-violating energy correction of an enantiomer has the form

$$E_{pv} = 2\sum_T Re\{\langle\Psi_S|H_{pv}|\Psi_T\rangle\langle\Psi_T|H_{so}|\Psi_S\rangle/(E_S - E_T)\} \ . \tag{2.40}$$

The one-electron approximation for Ψ_S and Ψ_T allows us to represent (2.40) on the basis of the spin orbitals $|\psi_j m_s\rangle$. Factorization leads to

$$E_{pv} = 2\sum_j^{\text{occ}}\sum_k^{\text{vir}} Re\{\langle\psi_j|\mathbf{V}_{pv}|\psi_k\rangle\langle m_s|\mathbf{s}|m_s'\rangle \\ \langle\psi_k|\mathbf{V}_{so}|\psi_j\rangle\langle m_s'|\mathbf{s}|m_s\rangle/(\varepsilon_j - \varepsilon_k)\} \, . \tag{2.41}$$

The sums are taken over occupied and virtual molecular orbitals with the energies ε_j and ε_k, respectively, referred to the electronic ground state $|\Psi_S\rangle$. The spin-independent potentials in (2.41) are

$$\mathbf{V}_{pv} = -\Gamma\sum_a Q_a\left\{\mathbf{p}, \delta^3(\mathbf{r}-\mathbf{r}_a)\right\}_+ \tag{2.42}$$

and

$$\mathbf{V}_{so} = \sum_b \xi(b)\mathbf{l}(b) \ . \tag{2.43}$$

The separation of the spin in (2.41) reduces E_{pv} to the simple form

$$E_{pv} = \sum_j^{\text{occ}}\sum_k^{\text{vir}} P_{jk}/(\varepsilon_j - \varepsilon_k) \tag{2.44}$$

where

$$P_{jk} = Re\{\langle\psi_j|\mathbf{V}_{pv}|\psi_k\rangle\langle\psi_k|\mathbf{V}_{so}|\psi_j\rangle\} \ . \tag{2.45}$$

Space-inversion of the nuclear and electronic coordinates changes the sign of P_{jk} and, therefore, opposite directions for the energy corrections of L and D enantiomers will be obtained. This result is immediately evident from the vector operators $\mathbf{p}$ and $\mathbf{l}$ in (2.42) and (2.43) which are polar and axial, respectively. P_{jk} can be non-zero for many systems especially for molecules which are chiral in the classical sense.

The LCAO representation of molecular orbitals is especially suited to elicit matrix elements of $\mathbf{V}_{pv}$ since this operator is most effective at the atomic nucleus. Only one-center integrals are of importance such as

$$\langle ans|\left\{p_z, \delta^3(\mathbf{r}-\mathbf{r}_a)\right\}_+|an'p_z\rangle \sim R^a_{ns}(0)\cdot(dR^a_{n'p}(r)/dr)_{r=0} \ .$$

A corresponding integral for $\mathbf{V}_{so}$ which yields a non-vanishing contribution to (2.45) is

$$\langle bnp_y|l_z|bn'p_x\rangle \sim \int_0^\infty R^b_{np}r^2R^b_{n'p}dr \ .$$

Ab initio calculations of E_{pv} have been performed on small systems, for instance twisted ethylene (D_2 symmetry), hydrogen peroxide, and hydrogen disulphide. The majority of α-amino acids such as alanine, valine, serine, and aspartic acid adopt zwitterionic structures in aqueous solution and can be distinguished by their different side-groups *R*.

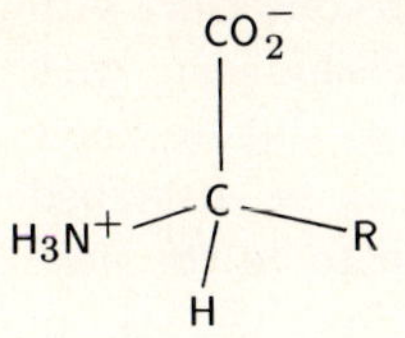

Scheme. *L*-amino acid, zwitterionic structure

In Table 2.2 ab initio calculated values for $\Delta E_{pv} = 2E_{pv}$ are listed, based on the preferred conformations in aqueous solution [9, 24].

Table 2.2. Calculated stability, $\Delta E_{pv}(10^{-17}$ kcal/mol), of naturally occurring amino acids and sugars compared to the respective naturally unpreferred enantiomers

Amino acid	R	ΔE_{pv}
L-alanine	CH_3	−2.24
L-valine	$CH(CH_3)_2$	−2.87
L-serine	CH_2OH	−1.05
L-aspartate anion	$CH_2CO_2^-$	−1.83
D-glyceraldehyde (hydrated)		−0.51
D-ribose	(C_2-*endo*)	−2.54
D-ribose	(C_3-*endo*)	−0.08

In each case the naturally occurring *L*-amino acid has a lower energy than the naturally unpreferred *D*-form. All the naturally occurring *D*-sugars are assumed to originate from the chiral *D*-glyceraldehyde. Calculations have shown that the naturally occurring *D*-enantiomer of the hydrated glyceraldehyde is more stable than the naturally unpreferred *L*-form. The C_2-C_3-*endo* ΔE_{pv} difference of *D*-ribose indicates a high sensitivity of this quantity with respect to the conformation.

The parity-violating energy difference ΔE_{pv} has been calculated also for a small fragment of a polypeptide chain. Polyglycine is more stable in a right-hand α-helix than in a left-hand one. This result is in accordance with the fact that the naturally occurring principal helical conformations of biopolymers are right-handed. Glycine (R = H) has no chiral center, and therefore, the stability of the right-hand α-helix is caused by the secondary structure.

The entire theory for biomolecular homochirality described so far rests upon the calculated data in Table 2.2 There is no doubt about the generally existing influence of the weak neutral current on the electronic structure as could be shown experimentally by the optical activity of heavy atoms. A frequently uttered doubt on the significance of ab initio calculations for a reliable estimate of relative stabilities in the order of 10^{-17} kcal/mol can be defeated as follows. Two enantiomers have not been calculated independently, but were treated in the framework of perturbation theory; the zero-order energies of L- and D-enantiomers are equal. Another objection is concerned with the question, whether the extremely low values for ΔE_{pv} are significant with respect to random fluctuations of concentrations n_L and n_D. But we know already from our kinetic models that unstable equilibria can be disturbed by the least reason. A corresponding theory will be presented in the next section. A further objection comes from experience in everyday chemistry in that the low values for ΔE_{pv} are not expected to be effective in kinetics. Daily chemistry, however, is concerned with small reaction volume and short reaction time. In contrast, a reaction volume of 1 km^3 and a reaction time of 10^4 yr, reasonable dimensions for evolution, could make ΔE_{pv} effective.

The phenomenon of ΔE_{pv} might lead to a fundamental change in the approach of chiral molecules. According to Hund's theory, which is in agreement with chemical experience so far, enantiomers L and D have a finite lifetime which is due to the tunneling splitting $\Delta E_{\pm}$ (Eq. (2.2)). However, Hund's theory is no longer valid when the condition

$$\Delta E_{\pm} \gg \Delta E_{pv} \tag{2.46}$$

is not fulfilled. The condition (2.46) could be violated with high energy barriers between L- and D-enantiomers as they occur in biomolecules. Then, L and D correspond to stationary states with wavefunctions ψ_L and ψ_D, respectively, in Fig. 2.1. This situation is sketched out in Fig. 2.4.

Experiments for a spectroscopic proof of ΔE_{pv} have been suggested, but not yet performed [17].

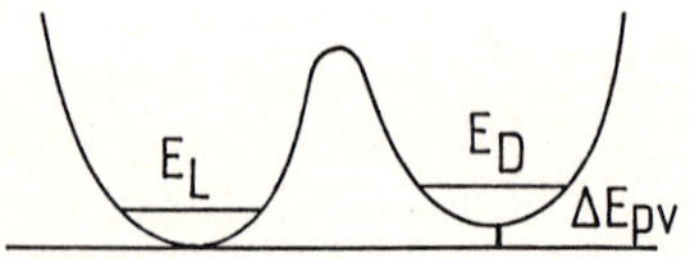

Fig. 2.4. Parity-violating energy difference ΔE_{pv} between enantiomers for the case $E_L < E_D$

2.6 Homochirality from Stochastic Equations

The kinetic models studied so far are constrained to an idealized homogeneous distribution of the reacting system. There is, however, no doubt that with a reaction volume as large as it was present during prebiotic biomolecular evolution, random fluctuations of the concentrations of chiral compounds have been ubiquitous. According to this assumption an important question arises: How large must a symmetry-breaking effect ΔE_{pv} be so that it can at least cause stereoselectivity, despite the presence of random fluctuations of the chiral dissymmetry $n_L - n_D$?

As an example the following model system of reactions will be considered

$$
\begin{aligned}
&A + A' \rightleftarrows L(D) && k_{\pm 1} \\
&A + A' + L(D) \rightleftarrows 2L(D) && k_{\pm 2} \\
&L + D \rightarrow A'' && k_3
\end{aligned}
\tag{2.47}
$$

where A, A', and A'' are achiral species. The first two processes are equilibrium reactions, whereas the third reaction describes the irreversible removal (specific antagonism) of enantiomers L and D. With a permanent supply of A and A' as well as the removal of L and D, the reacting system can be far from thermodynamic equilibrium which is a necessary condition for stereoselectivity. For convenience two new variables are introduced

$$
\begin{aligned}
\lambda &= n_A n_{A'} \,, \\
\alpha &= (n_L - n_D)/2
\end{aligned}
$$

where concentrations are denoted by n. The chiral dissymmetry α obeys the stochastic equation [25]

$$
d\alpha/dt = -U\alpha^3 + V(\lambda - \lambda_c)\alpha + Wg + C\eta f_2(t) + \sqrt{\varepsilon_1} f_1(t) \,. \tag{2.48}
$$

Without going into details of the kinetic constants U, V, W etc. the terms in (2.48) will be introduced as follows. The most simple form of the bifurcation equation consists of only the first two terms on the right-hand side. As λ evolves through the critical point λ_c the single steady state $\alpha = 0$ becomes unstable, and two stable steady states, $\alpha > 0$ and $\alpha < 0$, emerge symmetrically as is shown in Fig. 2.5. A detailed derivation of the bifurcation equation can be found in the literature [26]. The third term Wg with $g = \Delta E/kT$ is a measure for the different influence of the weak neutral current on the energy barriers of the reactions for L- and D-enantiomers. Here k is the Boltzmann constant and T the temperature. The last two terms describe two kinds of fluctuations which are of different origins. The first is related to an external chiral influence such as that from circularly polarized light. The second represents intrinsic thermodynamic fluctuations. The functions $f_1(t)$ and $f_2(t)$ are normalized gaussian white noise. For $g = 0$ chiral evolution is simply the result of mere chance, either $\alpha > 0$ or $\alpha < 0$, according to Fig. 2.5. With $g \neq 0$ the bifurcation at the critical point λ_c is avoided, but there are still two stable steady states branches for $\lambda > \lambda_c$. This situation is sketched out in Fig. 2.6 together with a calculated sample trajectory of α according to (2.48).

We should like make an attempt now to answer the above title question, whether the tiny values of g might be decisive for a selection of branches like that in Fig. 2.6, despite random fluctuations. The probability for the selection of branches obeys the Fokker–Planck equation associated with (2.48),

$$
\begin{aligned}
\partial P/\partial t = &- \partial/\partial\alpha(-U\alpha^3 + V(\lambda - \lambda_c)\alpha + Wg)P(\alpha, t) \\
&+ (\varepsilon/2)\partial^2/\partial\alpha^2 P(\alpha, t)
\end{aligned}
\tag{2.49}
$$

where the two parameters of the fluctuation terms in (2.48) are contracted to $\varepsilon = \varepsilon_1 + (C\eta)^2$. For $\lambda \ll \lambda_c$ $P(\alpha, t)$ is a gaussian which is strongly localized at

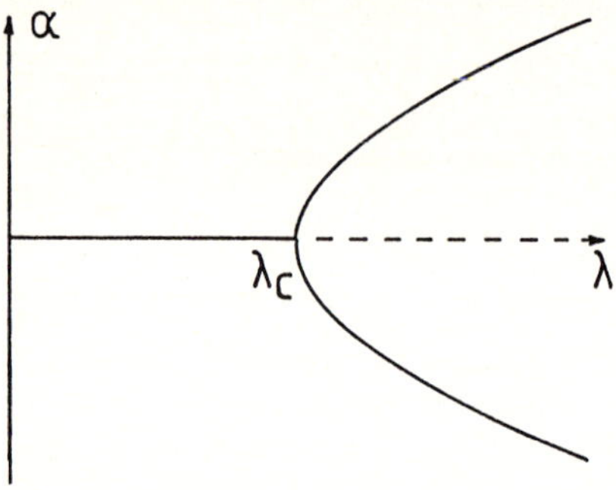

Fig. 2.5. Solution α of the bifurcation equation $d\alpha/dt = -U\alpha^3 + V(\lambda - \lambda_c)\alpha = 0$. *Solid/dashed lines* correspond to stable/unstable steady states; steady state means $d\alpha/dt = 0$

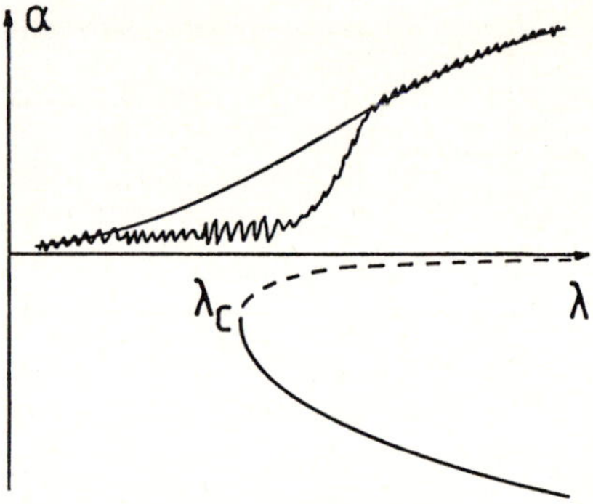

Fig. 2.6. A sample trajectory of α in Eq. (2.48) as λ is tuned through the critical point λ_c. *Solid/dashed lines:* stable/unstable steady states; fluctuations exaggerated

$\alpha = 0$. As λ evolves toward the critical point λ_c, $P(\alpha, t)$ becomes flat because of the diffusion term containing ε. $P(\alpha, t)$ is small for large α too, and hence the terms $-U\alpha^3$ and $V(\lambda - \lambda_c)\alpha$ in (2.49) can be neglected. Thus, the peak of the gaussian is shifted with a rate constant Wg. After a period of time t the peak is shifted by Wgt, while the increase of the width is $\sqrt{\varepsilon t}$. For sufficiently large t, Wgt can exceed $\sqrt{\varepsilon t}$, even for $Wg \ll \sqrt{\varepsilon}$. Accordingly the sign of Wg determines the chiral selection notwithstanding large fluctuations. These arguments are supported by the numerical solution of (2.49). For reasonable kinetic data the probability of branch selection $P_+(\alpha > 0 \text{ and } Wg > 0)$ turned out to be 98%, even though random chiral fluctuations are five orders of magnitude larger than the weak neutral current effect. This remarkable sensitivity of branch selection with respect to the parity-violating interaction can be understood only by means of the critical point λ_c.

Theories on homochiral evolution certainly offer a large field of interesting aspects in nuclear physics, molecular spectroscopy, and kinetics. One should always keep in mind, however, that any hypothesis on long time processes such as homochirality evolution will most probably remain experimentally unproven for ever.

Fruitful discussions with I. Gutman, R. Hegstrom, J. Kalcher, and M. Quack gave essential impact to this chapter and are gratefully acknowledged.

2.7 References

1 Pasteur L (1848) C R Hebd Séanc Acad Sci Paris *26*: 535
2 Faraday M (1846) Phil Mag *28*: 294
3 Pasteur L (1884) Bull Soc Chim France *41*: 215
4 Kelvin LD (1904) Baltimore Lectures Clay London
5 Mason SF (1984) New Scientist *101*:10
6 Mason SF (1984) Nature *311*: 19

7 Mason SF (1985) Nature *314*: 400
8 Janoschek R (1986) Naturwiss Rundsch *39*: 327
9 Tranter GE (1986) Nachr Chem Tech Lab *34*: 866
10 Hegstrom RA Kondepudi DK (1990) Scientific American 98
11 Latal H Chap. 1 in this book
12 Hund F (1927) Z Phys *43*: 805
13 Papousek D Aliev MR (1982) Molecular vibrational rotational spectra, Elsevier Amsterdam
14 Schleyer PvR Shavitt I Pepper MJM Janoschek R Quack M unpublished
15 Masters PM Bada JL Zigler JS (1977) Nature *268*: 71
16 Quack M Jans-Bürli S Molekulare Thermodynamik und Kinetik, Verlag der Fachvereine Zürich 1986
17 Quack M (1989) Angew Chem *101*: 588; Angew Chem Int Ed Engl *28*: 571
18 Frank FC (1953) Biochim Biophys Acta *11*: 459
19 Tennakone K (1984) Chem Phys Letters *105*: 444
20 Babović V Gutman I Jokić S (1987) Z Naturforsch *42a*: 1024
21 Gutman I Babović V Jokić S (1988) Chem Phys Letters *144*: 187
22 Babović V Gutman I Jokić S (1987) Collect of Scientific Papers of the Faculty of Science Kragujevac *8*: 51
23 Mason SF Tranter GE (1984) Mol Phys *53*: 1091
24 Tranter GE (1985) Mol Phys *56*: 825
25 Kondepudi DK Nelson GW (1985) Nature *314*: 438
26 Kondepudi DK Nelson GW (1984) Physica *125A*: 465

3 Chirality and Group Theory

G. Derflinger

3.1 Introduction

In 1961, Kauzmann, Clough, and Tobias [1] presented a semi-empirical concept for the description of chirality observations on molecules. Their Ansatz consists of a sum of terms, each of which corresponds to an interaction of certain parts of the molecule. By symmetry arguments alone, one can conclude which interactions can contribute to the chirality phenomenon and which cannot. Ruch and Schönhofer [2, 3] supplied the Kauzmann model with the group theoretical foundation and so created the theory of chirality functions. It is, however, not well-known that the first set-up to a chirality function is due to Ugi [4, see also 5, 6].

Within the approaches under discussion a molecule is assumed to consist of an achiral skeleton of given symmetry and ligands attached to the n skeletal sites. Based on this model pseudoscalar properties of molecules are described by chirality functions $\varphi(l_1, \ldots, l_n)$ which depend on ligand-specific parameters l_i. Obviously a chirality function has to fulfill the following requirements: its numerical value must be invariant under ligand permutations corresponding to proper rotations of the molecule and must change sign under improper rotations. Taking into account these restrictions Ruch and Schönhofer [2] construct chirality functions according to the principle of mathematical simplicity. Within the so-called first approximation method the polynomials of lowest degree in the ligand parameters which exhibit the required transformation behavior are used. The set-up according to the second approximation method, which – in its essential part – is quite similar to the model of Kauzmann et al., consists of a sum of terms depending on as few ligand parameters as possible. This means that, besides the symmetry requirements, no further physical arguments are used in constructing the chirality functions of [2]. Nevertheless, the application of these functions was extremely successful in the case of the skeleton symmetry D_{2d} [7–13]. In the case of other skeletal classes, however, the agreement with the experiments was less satisfactory [14–23] and some discrepancies have been reported [20-23].

In 1970 Ruch and Schönhofer [3] discovered a deficiency of their chirality functions of [2]. In order to remove this lack, which is not a lack in the opinion of other authors [24], they established the principle of qualitative completeness [3]. A qualitatively complete chirality function does not vanish identically for any non-racemic mixture of isomers, whatever the nature of the ligands may be. Besides the fact that qualitatively complete chirality functions consisting of more than one term, i.e. being not identical with functions of [2], have not yet been applied to the description of chirality phenomena there is a principal objection against the concept of qualitative completeness: One can construct non-racemic mixtures of non-isomers for which, independently of the nature of the ligands, the qualitatively complete chirality functions of [3] vanish identically [25]. There is no physical reason for this distinction between mixtures of isomers and mixtures of non-isomers. By the introduction of the principle of qualitative supercompleteness [26] this divergence has been removed. However, for a successful application the qualitatively supercomplete chirality functions seem to be too complicated. One would have to estimate too many ligand parameters. Nevertheless, as in [2, 3], some important qualitative results can be derived.

3.2 The Principle of Pairwise Interactions

A very interesting semi-empirical approach to the quantitative description of molar rotations of chiral molecules was suggested by Kauzmann, Clough and Tobias in 1961 [1]. This concept, which comprises a special type of cluster expansion, has been called Principle of Pairwise Interactions (PPI). It should be noted, however, that the authors do not restrict their considerations to pairwise interactions at all. Their general concept contains all possible contributions originating from pairwise, three-way, and higher interactions. When this theory was presented, no data for a convincing test were available and in consequence no further attention was paid to it.

Let us discuss the PPI on the example of a skeleton of symmetry D_{2d} containing four sites. The skeletons of allene and 2,2′-spirobiindane (cf. Scheme 3.1) are special cases of this type. We label the skeletal sites with 1,2,3,4 (in general 1,2,…,n). By u_{ij} we denote the contribution caused by the interaction between the ligand on site i and that on site j. The value of u_{ij} depends on the ligands attached to these sites. If we want to express this explicitly we write $u_{ij}(A, B)$, where A is the ligand on site i and B is the ligand on site j. Accordingly, u_{ijk} or $u_{ijk}(A, B, C)$, respectively, denote an effect caused by an interaction of three ligands, etc. For interactions in which the skeleton is also involved we write v_{iG}, $v_{ijG}, \ldots$ or $v_{iG}(A)$, $v_{ijG}(A, B), \ldots$, respectively. $v_{ijG}(A, B)$, for example, means the contribution due to the three-way interaction between the skeleton G, a ligand of sort A on site i and a ligand of sort B on site j. Thus, a chirality observation is assumed to result from a superposition of interactions, each interaction term being associated with some molecular fragment. Clearly, a fragment superimposable on its mirror image or, more general, invariant under an improper rotation can give rise to no chirality observation. Therefore, in the case of the D_{2d}-skeletons of Scheme 3.1, independently of the nature of the ligands, the following interaction terms are zero:

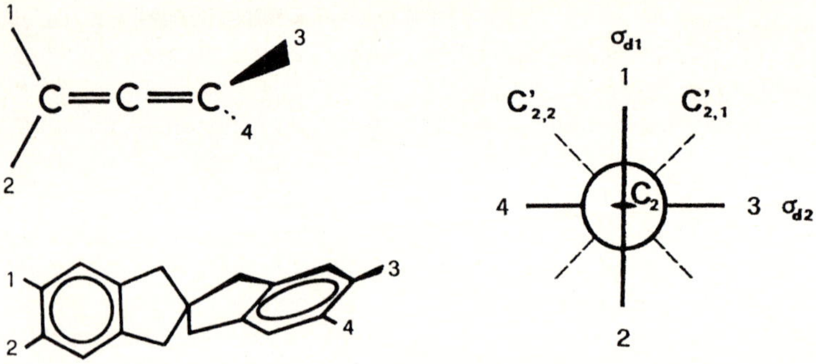

Scheme 3.1. Skeletons of allene and $2,2'$-spirobiindane with symmetry D_{2d}, numbering of the four skeletal sites, symmetry elements

$$v_{1G}, v_{2G}, v_{3G}, v_{4G}\ ; \quad u_{12}, u_{34}\ ; \quad u_{12G}, u_{34G}\ .$$

The general set-up φ after Kauzmann et al. [1] for a chirality observation on molecules of this skeletal class is then given by

$$\begin{aligned}\varphi = u_{13} + u_{14} + u_{23} + u_{24} + v_{13G} + v_{14G} + v_{23G} + v_{24G} + u_{123} + u_{124} \\ + u_{134} + u_{234} + v_{123G} + v_{124G} + v_{134G} + v_{234G} + u_{1234} + v_{1234G}\ .\end{aligned}$$

If only pairwise interactions are taken into account this is simplified to

$$\varphi = u_{13} + u_{14} + u_{23} + u_{24}\ .$$

In contrast to Kauzmann's treatment we collect the terms depending on the same ligands:

$$\begin{aligned} w_i &= u_i + v_{iG}\ , \quad (u_i \text{ formal})\ , \\ w_{ij} &= u_{ij} + v_{ijG}\ , \\ w_{ijk} &= u_{ijk} + v_{ijkG}\ , \\ &\dots \end{aligned}$$

This is justified by the fact that neither symmetry arguments, which naturally have to be based on the symmetry group of the skeleton, nor quantitative observations allow one to separate such a w-term. We call w_i or $w_i(A)$, respectively, a one-ligand effect, w_{ij} or $w_{ij}(A,B)$ a two-ligands effect, etc. In w-terms the simplest set-up for compounds having skeletal symmetry D_{2d} (Scheme 3.1) is

$$\varphi = w_{13} + w_{14} + w_{23} + w_{24}\ .$$

Terms due to molecular fragments which can be transformed into each other by a rotation are identical. Under an improper rotation of a fragment the corresponding term changes its sign. Thus, by symmetry we have

$$w_{13} = w_{24} = -w_{14} = -w_{23}\ .$$

Let us denote w_{13} by σ. Then

$$w_{13} = w_{24} = \sigma \ , \quad w_{14} = w_{23} = -\sigma \ .$$

For the molecule **4** (cf. Scheme 3.2) this leads to the set-up

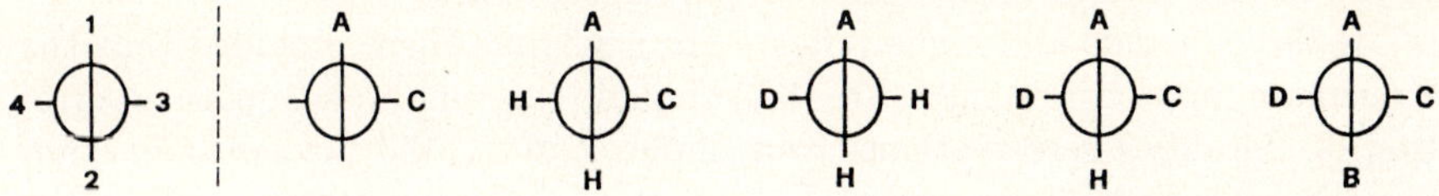

f **1** **2** **3** **4**

Scheme 3.2.

$$\varphi(A,B,C,D) = \sigma(A,C) + \sigma(B,D) - \sigma(A,D) - \sigma(B,C) \ . \tag{3.1}$$

The position of a ligand symbol in the argument vector of φ indicates to which site of the skeleton the corresponding ligand is attached. Interchanging ligands A and C in the fragment **f** is equivalent to a rotation. Therefore σ must be symmetric with respect to its two arguments:

$$\sigma(A,C) = \sigma(C,A) \ .$$

At first glance, Eq. (3.1) would motivate us to calculate the σ-terms for all possible pairs of ligands on the basis of observed data. However, due to the fact that σ is not unique, this is impossible in principle. Functions ω differing from σ by a constant K and terms depending only on one of the two ligands,

$$\omega(A,C) = \sigma(A,C) + K - f(A) - f(C) \ , \tag{3.2}$$

yield the same result when being composed analogously to (3.1):

$$\varphi(A,B,C,D) = \omega(A,C) + \omega(B,D) - \omega(A,D) - \omega(B,C) \ . \tag{3.3}$$

One can prove that this indeterminateness is a general property of set-ups which consist of terms depending on a minimum number of ligands, see also [2, 3]. This minimum number k is a function of the skeleton under consideration. In our example $k = 2$. It holds $k = n - o$ where n is the number of skeletal sites and o is the so-called chirality order which is the maximum number of ligands of the same kind in a chiral molecule [3].

The indeterminateness shown above was not recognized by Kauzmann et al. [1]. All one can do to remove it is to choose a reference ligand sort, so that every interaction term in which a reference ligand is involved vanishes. It is most convenient to select hydrogen as the reference ligand. If in (3.2) we set $K = \sigma(\mathrm{H},\mathrm{H})$, $f(A) = \sigma(A,\mathrm{H})$, $f(C) = \sigma(C,\mathrm{H})$ then $\omega(A,C)$,

$$\omega(A,C) = \sigma(A,C) + \sigma(\mathrm{H},\mathrm{H}) - \sigma(A,\mathrm{H}) - \sigma(C,\mathrm{H}) \ , \tag{3.4}$$

fulfills the required condition. As $\omega(A, C)$ of (3.4) may be identified with the chirality observation on the molecule **1** this can be easily understood: If we also replace A or C by H then the molecule becomes achiral and the term vanishes. We call ω the *reduced* interaction term. $\omega(A, C)$ corresponds to the chirality observation on that molecule which, besides the ligands occurring in the term, contains only ligands of the reference sort.

If three-ligands and higher effects are negligible then, provided that the Kauzmann approach is valid, Eq. (3.3) should represent an adequate description of chirality oberservations. Kauzmann et al. [1] did not consider compounds with skeletal symmetry D_{2d} at all. As already stated, also in the case of other skeletons they did not have enough data for a convincing test. Due to the work of Neudeck, Schlögl et al. (see [8–11]) data of more than a hundred derivatives of 2,2′-spirobiindane (cf. Scheme 3.1) are now available. This enables us to verify Eq. (3.3) in an excellent way. To do this we use the equation

$$\varphi(A, \mathrm{H}, C, D) = \varphi(A, \mathrm{H}, C, \mathrm{H}) + \varphi(A, \mathrm{H}, \mathrm{H}, D) \tag{3.5}$$

which, using H as reference sort, can be easily derived from (3.3). Equation (3.5) postulates that a chirality observation on a trisubstituted compound **3** is equal to the superposition of the corresponding observations on the disubstituted compounds **1** and **2**. Applying Eq. (3.5) to the molar rotations at $\lambda = 589$ nm the values for [M] shown in Table 3.1 have been obtained [13]. The agreement with the observed data (taken from [8, 9]) is in many cases very satisfactory. Some greater deviations might be explained by the influence of the alkyl group C on site 3 on the conformation of the substituent D on site 4. Within the theory of chirality functions this has been investigated for $D = \mathrm{CH_3CO}$ [12].

3.3 The Theory of Chirality Functions

Ruch and Schönhofer [2,3,27,28] supplied the Kauzmann model with the representation-theoretical foundation and called the resulting approach theory of chirality functions. Whereas within Kauzmann's model (cf. Eq. (3.3)) a chirality observation is seen as a function of the ligands themselves, the theory of chirality functions uses ligand-specific parameters. To each ligand sort one or more parameters are assigned. We shall discuss here only the case of one parameter per ligand. If we denote the parameter of the ligand attached to site i by l_i then analogously to (3.3) we can write

$$\varphi(L) = \omega(l_1, l_3) + \omega(l_2, l_4) - \omega(l_1, l_4) - \omega(l_2, l_3) , \tag{3.6}$$

where L means the argument vector $L = (l_1, l_2, l_3, l_4)$. $\varphi(L)$ is an example of a chirality function. In general a chirality function describes a pseudoscalar property of molecules of a given skeletal class as dependent on ligand-specific parameters. Taking into account the elementary requirements of chirality a chirality function $\varphi(L)$ must be invariant under permutations of ligands

Table 3.1. Calculated and experimental molar optical rotations ($\lambda = 589$ nm) for trisubstituted 2, 2′-spirobiindanes (**3** in Scheme 3.2)

Compounds	A	C	D	Calc.	Exp.
(1)	CH_3	C_2H_5	CH_3	+4.3	+5.5
(2)	CH_2OH	C_2H_5	CH_2OH	+0.1	+1.9
(3)	CHO	C_2H_5	CHO	−53.3	−54.8
(4)	CO_2H	C_2H_5	CO_2H	−33.9	−30.3
(5)	CO_2CH_3	CH_3	CO_2CH_3	−54.0	−53.6
(6)	CO_2CH_3	C_2H_5	CO_2CH_3	−43.4	−40.5
(7)	CN	C_2H_5	CN	−43.2	−46.3
(8)	CH_2OH	C_2H_5	CH_3	+4.0	+7.9
(9)	CHO	C_2H_5	CH_3	+12.2	+20.0
(10)	CO_2CH_3	C_2H_5	CH_3	+10.6	+16.0
(11)	CN	C_2H_5	CH_3	+12.1	+17.0
(12)	CH_3	C_2H_5	CH_2OH	+1.4	+0.9
(13)	CO_2CH_3	C_2H_5	CH_2OH	−0.6	+6.7
(14)	CO_2CH_3	C_2H_5	$CH = CH_2$	−62.8	−98.4
(15)	CH_3	C_2H_5	CHO	−17.1	−17.4
(16)	CH_3CO	CH_3	CHO	−55.8	−48.4
(17)	CO_2CH_3	CH_3	CHO	−62.4	−49.3
(18)	CO_2CH_3	C_2H_5	CHO	−51.8	−47.2
(19)	CH_3	C_2H_5	$CH = NN(CH_3)_2$	−53.7	−69.8
(20)	CO_2CH_3	CH_3	$CH = NN(CH_3)_2$	−161.3	−178.4
(21)	CO_2CH_3	C_2H_5	$CH = NN(CH_3)_2$	−150.7	−205.6
(22)	CHO	C_2H_5	CO_2CH_3	−48.1	−45.1
(23)	CH_3	C_2H_5	CH_3CO	−16.4	−11.6
(24)	C_2H_5	C_2H_5	CH_3CO	−23.6	−15.3
(25)	CH_2OH	C_2H_5	CH_3CO	−21.5	−12.8
(26)	CHO	C_2H_5	CH_3CO	−42.9	−32.2
(27)	CO_2H	C_2H_5	CH_3CO	−43.5	−28.1
(28)	CO_2CH_3	CH_3	CH_3CO	−57.6	−46.5
(29)	CO_2CH_3	C_2H_5	CH_3CO	−47.0	−27.9
(30)	CH_2OH	C_2H_5	$C(CH_3)(OCH_2)_2$	−6.6	−4.7
(31)	CHO	C_2H_5	$C(CH_3)(OCH_2)_2$	−14.5	−10.9
(32)	CO_2CH_3	C_2H_5	$C(CH_3)(OCH_2)_2$	−13.4	−10.6
(33)	CH_3	C_2H_5	CO_2CH_3	−15.0	−15.4
(34)	C_2H_5	CH_3	CO_2CH_3	−25.6	−19.5
(35)	CH_2OH	C_2H_5	CO_2CH_3	−23.6	−18.8
(36)	CH_3CO	CH_3	CO_2CH_3	−56.2	−42.8
(37)	CN	C_2H_5	CO_2CH_3	−40.6	−40.8
(38)	CH_3	C_2H_5	CN	−16.2	−15.5
(39)	CO_2CH_3	CH_3	CN	−53.9	−40.9
(40)	CO_2CH_3	C_2H_5	CN	−43.3	−41.8
(41)	CO_2H	CH_3	NH_2	−40.6	−39.3
(42)	CO_2H	C_2H_5	NH_2	−30.0	−46.6
(43)	CO_2H	CH_3	$NHCOCH_3$	−72.3	−71.2
(44)	CO_2H	C_2H_5	$NHCOCH_3$	−61.7	−55.2

corresponding to proper rotations of the molecule and must change sign under permutations corresponding to improper rotations. For the group-theoretical treatment we consider the symmetry group G of the skeleton. Each covering operation of G is equivalent to a certain permutation of the ligands. Thus G can be mapped homomorphically on a subgroup S of the symmetric group S_n of all $n!$ permutations of the n ligands. In the case of a skeletal class containing chiral molecules the homomorphism reduces to an isomorphism[1]. Let S_0 be that subgroup of S whose elements correspond to proper rotations of the skeleton. S_0 is of index two, by S^* we denote its coset, the elements of which correspond to improper rotations. Scheme 3.1 displays the numbering of the sites and the symmetry elements for our example. In Scheme 3.3, where the ligand permutations are given in cycle notation (see e.g. [29]), the homomorphism of $G = D_{2d}$ on S is shown. In the example

$$S_0 = \{e, (12)(34), (13)(24), (14)(23)\}\ , \quad S^* = \{(12), (34), (1324), (1423)\}\ .$$

The requirements a chirality function $\varphi(L)$ has to fulfill can be written as

$$\varphi(sL) = \begin{cases} \varphi(L) & \text{for } s \in S_0 \\ -\varphi(L) & \text{for } s \in S^*\ . \end{cases} \tag{3.7}$$

sL is the argument vector obtained by applying the ligand permutation s to L. The permutation s is referring to positions (i.e. ligand sites). By $O(s)$ we denote the permutation operators working on functions and defined by

$$O(s)\varphi(L) = \varphi(s^{-1}L)\ . \tag{3.8}$$

The $O(s)$ form groups $O(S_0)$, $O(S)$ or $O(S_n)$ which under the mapping $O(s) \leftrightarrow s^{-1}$ are isomorphic to S_0, S or S_n, respectively. Using (3.8) we get from Eq. (3.7)

$$O(s)\varphi(L) = \begin{cases} \varphi(L) & \text{for } s \in S_0 \\ -\varphi(L) & \text{for } s \in S^*\ . \end{cases} \tag{3.9}$$

This shows $\varphi(L)$ being a basis of a one-dimensional irreducible representation space of S or $O(S)$, respectively, the characters of which are

$$\chi_\chi(s) = \chi_\chi(O(s)) = \begin{cases} 1 & \text{for } s \in S_0 \\ -1 & \text{for } s \in S^*\ . \end{cases}$$

This representation, denoted by Γ_χ, is called the chirality representation. Clearly, Γ_χ is identical with one of the irreducible representations of the symmetry group G of the skeleton. In the case of a skeleton with symmetry D_{2d} this is the representation B_1. In Table 3.2 the characters of $\Gamma_\chi = B_1$ are recorded. Furthermore, the cycle structure of the ligand permutations corresponding to the symmetry operations is indicated by means of partition diagrams. To every cycle of a permutation, a row of the corresponding

[1] The benzene skeleton, $n = 6$, is an example in which the homomorphism is not an isomorphism.

partition diagram is assigned. The number of boxes in this row is equal to the length of the cycle. The rows are arranged in descending order. The projection operators associated with Γ_χ are

$$p_\chi = \frac{1}{|S|} \sum_{s \in S} \chi_\chi(s) s \leftrightarrow P_\chi = \frac{1}{|S|} \sum_{s \in S} \chi_\chi(s) O(s) \, . \tag{3.10}$$

Table 3.2. Characters of $\Gamma_\chi = B_1$

D_{2d}	E	C_2	$2C_2'$	$2\sigma_d$	$2S_4$
				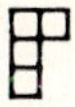	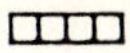
$\Gamma_\chi = B_1$	1	1	1	−1	−1

D_{2d} $\downarrow$ S	E $\downarrow$ e	C_2 $\downarrow$ (12)(34)	$C'_{2,1}$ $\downarrow$ (13)(24)	$C'_{2,2}$ $\downarrow$ (14)(23)	σ_{d1} $\downarrow$ (34)	σ_{d2} $\downarrow$ (12)	S_4 $\downarrow$ (1324)	S_4^- $\downarrow$ (1423)

Scheme 3.3

A function $\varphi(L)$ is a chirality function if, and only if,

$$P_\chi \varphi(L) = \varphi(L) \, . \tag{3.11}$$

If for any function $\psi(L)$ the projection $P_\chi \psi(L)$ does not vanish identically then $P_\chi \psi(L)$ is a chirality function. This means that chirality functions can be constructed by the well known projection operator technique of group representation theory (see e.g. [30]). We show this on the example of the skeleton of Scheme 3.1 with symmetry D_{2d}. For $\psi(L)$ we set the function $\varrho(l_1, l_3)$ which does not depend explicitly on l_2 and l_4 and can be interpreted as an interaction term in Kauzmann's sense. With respect to Eq. (3.10) and Scheme 3.3 we get

$$\begin{aligned} P_\chi \varrho(l_1, l_3) = \tfrac{1}{8} \, [&\varrho(l_1, l_3) + \varrho(l_2, l_4) + \varrho(l_3, l_1) + \varrho(l_4, l_2) \\ &- \varrho(l_2, l_3) - \varrho(l_1, l_4) - \varrho(l_3, l_2) - \varrho(l_4, l_1)] \, . \end{aligned} \tag{3.12}$$

Collecting the terms that depend on the same ligands leads to the chirality function

$$\varphi(L) = P_\chi \varrho(l_1, l_3) = \omega(l_1, l_3) + \omega(l_2, l_4) - \omega(l_1, l_4) - \omega(l_2, l_3) \, , \tag{3.13}$$

where

$$\omega(l_i, l_j) = \omega(l_j, l_i) = \tfrac{1}{8}[\varrho(l_i, l_j) + \varrho(l_j, l_i)] \ .$$

The function of (3.13) is identical with that of (3.6). It also follows that ω is symmetric with respect to its two arguments. Thus, the projection by P_χ can be understood as the representation-theoretical formalization of Kauzmann's proceeding.

3.4 The Approximation Methods

Within the condition (3.11), which follows from symmetry, Ruch and Schönhofer [2] suggested two methods, the so-called first and second approximation methods, for constructing chirality functions according to the principle of mathematical simplicity. This means that besides the qualitative conditions implied by symmetry requirements no further physical arguments are used in setting up these chirality functions. Nevertheless, in the case of skeletons of symmetry D_{2d} these functions permit a very excellent description of chirality phenomena [7–13]. In the case of other classes, however, their application was less satisfactory [14–23].

As the second approximation method is more general let us discuss it first. Within this method a chirality function is set up as a superposition of terms depending on as few ligand parameters as possible. We explain this on our example of a skeleton with symmetry D_{2d} (Scheme 3.1). Let D denote the representation induced by the four one-ligand terms

$$\varrho^{(1)}(l_i) \ , \quad i = 1, 2, 3, 4 \ ,$$

where $\varrho^{(1)}$ is a certain function. The characters of D in the permutation group S, which is isomorphic to D_{2d}, are given by the number of terms, i.e. ligands, invariant under the corresponding group element. The two-ligands terms

$$\varrho^{(2)}(l_i, l_j) \ , \quad i, j = 1, 2, 3, 4$$

span the direct product $D^2 = D \times D$, the characters of which are the squares of the characters of D, etc. Table 3.3 shows the characters of the chirality representation Γ_χ and of the representations D^m, $m = 1, 2, 3, 4$. Furthermore the last column contains the multiplicities τ_m of Γ_χ in D^m. The τ_m can be got in the usual way by the standard procedure of subduction [30–32]. Thus

$$\tau_m = \frac{1}{|S|} \sum_{s \in S} \chi_m(s) \chi_\chi(s) \ , \tag{3.14}$$

where $\chi_m(s)$ is the character of s in D^m. By $\tau_1 = 0$ the fact is expressed that no chirality function can be built up from one-ligand terms. This is in agreement with the result which one gets following Kauzmann [1]. As $\tau_2 = 1$ the set-up according to the second approximation method consists of two-ligands terms. As has been shown (Eq. (3.13)) one gets the corresponding chirality function $\varphi(L)$ by applying P_χ to some term, e.g. $\varrho^{(2)}(l_1, l_3)$:

Table 3.3. Characters of Γ_χ and D^m

D_{2d}	E	C_2	$2C_2'$	$2\sigma_d$	$2S_4$	
S	(1)	(12)(34)	(13)(24) (14)(23)	(12) (34)	(1324) (1423)	τ_m
Γ_χ	1	1	1	−1	−1	
D	4	0	0	2	0	0
D^2	16	0	0	4	0	1
D^3	64	0	0	8	0	6
D^4	256	0	0	16	0	28

$$\varphi(L) = P_\chi \varrho^{(2)}(l_1, l_3) = \omega^{(2)}(l_1, l_3) + \omega^{(2)}(l_2, l_4) - \omega^{(2)}(l_1, l_4) - \omega^{(2)}(l_2, l_3) \,. \tag{3.15}$$

(By using the superscripts we state here explicitly that we are concerned with two-ligands terms.) The fact that interaction terms like $\omega^{(2)}(l_1, l_2)$ vanish by symmetry is expressed by $P_\chi \varrho^{(2)}(l_1, l_2) = 0$. We denote the smallest m for which τ_m is greater than zero by k. As already mentioned $k = n - o$, where o is the chirality order [3]. In general a chirality function after the second method is an element of a τ_k-dimensional function space which consists of the Γ_χ-components of D^k. The corresponding set-up is then given by

$$\varphi(L) = \sum_{r=1}^{\tau_k} P_\chi \varrho_r^{(k)}(l_{i_{r1}}, l_{i_{r2}}, \ldots, l_{i_{rk}}) \,. \tag{3.16}$$

The τ_k terms $\varrho_r^{(k)}$ are generally different functions depending on non-equivalent k-tuples of ligands.

Within the first approximation method Ruch and Schönhofer [2], taking into account their principle of mathematical simplicity, set up the polynomials of lowest degree exhibiting the required transformation behaviour. In simple cases, e.g. when k is 1 or 2, the chirality polynomial of lowest degree can be got directly from the chirality function (3.16) after the second method or one of its components. $\omega^{(1)}(l_i)$ and $\omega^{(2)}(l_i, l_j)$ have to be specialized according to

$$\omega^{(1)}(l_i) = l_i \,, \tag{3.17}$$

$$\omega^{(2)}(l_i, l_j) = l_i l_j \quad \text{if } \omega^{(2)} \text{ is symmetric,} \tag{3.18}$$

$$\omega^{(2)}(l_i, l_j) = l_i l_j (l_i - l_j) \quad \text{if } \omega^{(2)} \text{ is antimetric.} \tag{3.19}$$

Equation (3.17) is trivial: The ligand parameter is identified with the one-ligand term, the first and second approximation methods become identical. In Eqs. (3.18, 19) $l_i l_j$ and $l_i l_j (l_i - l_j)$, respectively, are the simplest polynomials which consist of terms depending on two ligands and show the transformation property of the corresponding $\omega^{(2)}(l_i, l_j)$. Applying Eq. (3.18) to the chirality function (3.15) for the D_{2d} skeleton gives

$$\varphi(L) = l_1 l_3 + l_2 l_4 - l_1 l_4 - l_2 l_3 = (l_1 - l_2)(l_3 - l_4) \ . \qquad (3.20)$$

$\varphi(L)$ depends only on differences of ligand parameters. This is a general property of the chirality polynomials of lowest degree [2]. Therefore one can set the parameter of a reference ligand sort (usually H) equal to zero. This is in agreement with the fact, demonstrated on the example D_{2d}, that an $\omega^{(k)}$-term vanishes if a ligand of the reference sort is involved. For each argument l_i of $\omega^{(k)}$ appears as a factor in the polynomial which is set for $\omega^{(k)}$, see (3.18, 19).

In Sect. 3.2 using the example of the D_{2d}-skeleton, $\omega^{(2)}(l_i, l_j)$ have been introduced as reduced interaction terms. $\omega^{(2)}(l_1, l_3)$, $\omega^{(2)}(l_2, l_4)$, $-\omega^{(2)}(l_1, l_4)$, $-\omega^{(2)}(l_2, l_3)$ describe the individual two-ligands effects. As has been shown, $\omega^{(2)}(l_1, l_3)$, for example, can be identified with the chirality observation on a molecule containing ligands of the reference sort on sites 2 and 4. Within the scope of the first approximation method for $\omega^{(2)}(l_1, l_3)$ the product $l_1 l_3$ is set:

$$\omega^{(2)}(l_1, l_3) = l_1 l_3 \ . \qquad (3.21)$$

Therefore, according to this method, for the chirality observations φ_1, φ_5, φ_6 on the molecules **1, 5, 6** of Scheme 3.4

$$\varphi_1 = ac \ , \quad \varphi_5 = a^2 \ , \quad \varphi_6 = c^2 \ , \qquad (3.22)$$

should hold, where a and c are the parameters of the ligands A and C, respectively. Thus, the principle of mathematical simplicity, implying the use of the lowest-degree polynomials within the first approximation method, leads to the conclusion that a chirality observation on the hetero-disubstituted compound **1** should be equal to the geometric mean of the observations on the corresponding homo-disubstituted compounds **5** and **6**:

$$\varphi_1 = \sqrt{\varphi_5 \varphi_6} \ . \qquad (3.23)$$

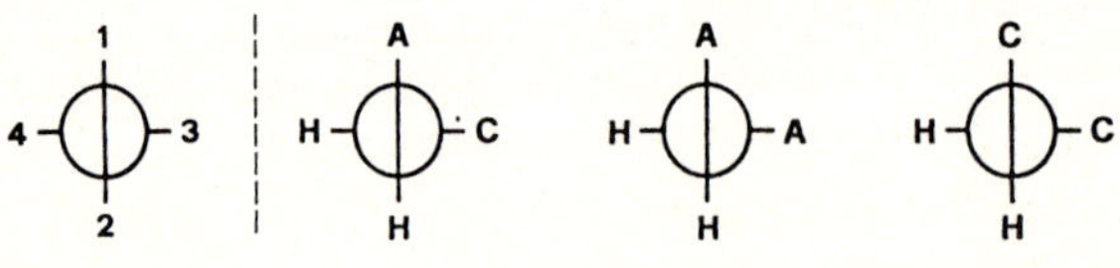

1 **5** **6** **Scheme 3.4.**

Table 3.4 shows that the molar optical rotations of disubstituted spirobi-indanes (see Scheme 3.4) fulfill Eq. (3.23) in an excellent way. The parameters a and c of the ligands A and C, respectively, are calculated as the square-roots of the rotations of the corresponding homo-disubstituted compounds which are recorded in the diagonal of the table (cf. Eq. (3.22)). In the case

Table 3.4. Molar optical rotations ($\lambda = 589$ nm) for homo- and hetero-disubstituted 2,2′-spirobiindanes (**5,6,1** in Scheme 3.4)

	C	CH_3	C_2H_5	CH_2OH	CHO	CH_3CO	COOH	$COOCH_3$	CN	OCH_3
	c	3.38	4.39	4.27	9.91	9.52	8.54	9.20	9.34	4.43
A	*a*									
CH_3	3.38	11.4								
C_2H_5	4.39	15.7 14.8	19.3							
CH_2OH	4.27	14.3 14.4	18.3 18.7	18.2						
CHO	9.91	32.8 33.5	45.0 43.6	39.5 42.3	98.3					
CH_3CO	9.52	32.1 32.1	42.9 41.8	39.8 40.6	87.9 94.4	90.6				
COOH	8.54	– 28.8	39.1 37.5	38.6 36.4	– 84.7	82.6 81.3	73.0			
$COOCH_3$	9.20	30.7 31.1	41.3 40.4	41.9 39.3	93.1 91.2	88.3 87.6	– 78.6	84.7		
CN	9.34	31.9 31.5	44.0 41.0	– 39.8	– 92.6	– 88.9	– 79.8	84.6 85.9	87.2	
OCH_3	4.43	15.8 14.9	– 19.4	– 18.9	– 43.9	– 42.1	– 37.8	– 40.7	– 41.3	19.6

a, c: Ligand parameters of A and C, resp. (square roots of diagonal elements); diagonal and upper entries in off-diagonal part: experimental molar rotations; lower entries in off-diagonal part: calculated molar rotations.

of the hetero-disubstituted compounds the upper entries give the observed rotations (taken from [8, 9]), the lower ones are the values calculated according to $\varphi_1 = ac$. Since, besides symmetry, only the argument of mathematical simplicity has been used for deriving Eq. (3.20) of which the Eqs. (3.22) are special cases, the physical reason for the highly satisfactory agreement of (3.20) and especially (3.22) with data observed on allenes [7] and spirobiindanes [8–13] should be investigated independently. Haase and Ruch [33] tried this by applying a quantum chemical treatment. But their arguments do not seem to be very convincing.

From the view of this paper two facts are essential for the validity of the chirality polynomial (3.20):

1. The Kauzmann model holds for the allene and spirobiindane derivatives under consideration and within this model two-ligands terms $\omega^{(2)}(l_i, l_j)$ are sufficient for describing the experimental data.
2. For $\omega^{(2)}(l_i, l_j)$ one can set the product $l_i l_j$.

There is a strong argument for point 2, though it is also mainly a mathematical one: Suppose the symmetric interaction term $\omega^{(2)}(l_1, l_3)$ be expanded into a power series,

$$\omega^{(2)}(l_1, l_3) = \alpha + \beta l_1 + \beta l_3 + \gamma l_1^2 + \delta l_1 l_3 + \gamma l_3^2 + \dots,$$

and apply the projection operator P_χ for generating a chirality function $\varphi(L)$. As P_χ annihilates all terms depending on only one ligand we get

$$\varphi(L) = P_\chi \omega^{(2)}(l_1, l_3) = \delta P_\chi(l_1 l_3) + \ldots = \frac{\delta}{4}(l_1 - l_2)(l_3 - l_4) + \ldots$$

This means that $l_1 l_3$ is the first term in the expansion which is not cancelled out by symmetry. Apart from a numerical factor, which can be eliminated by a suitable scaling of the ligand parameters, $P_\chi(l_1 l_3)$ is identical with the chirality polynomial (3.20).

3.5 Determining the Lowest-Degree Chirality Polynomials

Rules have been given for constructing the chirality polynomial of lowest degree for a given skeleton [3, 28, 34–37], which require the characters of the symmetric group S_n. Here we describe a method which does not need these characters. (The complete proof will be given elsewhere.) To each ligand assortment one can assign a partition $(\gamma_1, \gamma_2, \ldots, \gamma_c)$ of n represented by a partition diagram [3]. The length γ_i of the i-th row is equal to the number of ligands of sort i. The rows are arranged in such a way that $\gamma_1 \geq \gamma_2 \geq \ldots \geq \gamma_c$, where c, $c \leq n$, is the number of ligand sorts. For example, the partition (3,2,2,1,1) with the diagram

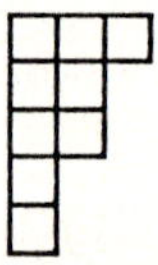

represents an assortment $A, A, A, B, B, C, C, D, E$. The degree g of a partition is defined as

$$g = \sum_{i=1}^{c} (i-1)\gamma_i \tag{3.24}$$

[34]. Analogously we define the degree of a molecule as the degree of the partition implied by its ligand assortment. A partition is called active if chiral molecules can be constructed from the corresponding ligand assortments.

For constructing the lowest-degree chirality polynomial one has to proceed as follows:

1. Find all active partitions of lowest degree and for each of these partitions obtain the number of different chiral molecules. This can be done either by inspection or by the method given in Sect. 3.7. In many cases of practical interest there is only one active partition of lowest degree. (The most simple skeleton for which more than one active partition of lowest degree exists is the octahedron, $n = 6$. These are the partitions (3,1,1,1) and (2,2,2), both with

degree 6.) Furthermore, in many cases one can build only one chiral molecule from an assortment corresponding to an active lowest-degree partition.

2. With each chiral molecule of lowest degree a component of the chirality function after the first approximation method is associated. For getting such a component consider the corresponding molecule and let

$$j_{i1}, j_{i2}, \ldots, j_{i\gamma_i}\,, \quad i = 1, \ldots, c\,, \tag{3.25}$$

be the skeletal sites to which the ligands of the i-th sort are attached. The component is then got as

$$P_\chi \prod_{i=2}^{c} \prod_{\alpha=1}^{\gamma_i} l_{j_{i\alpha}}^{i-1}\,. \tag{3.26}$$

Clearly, the lowest-degree chirality polynomial consists of as many components (3.26) as there are chiral molecules of lowest degree.

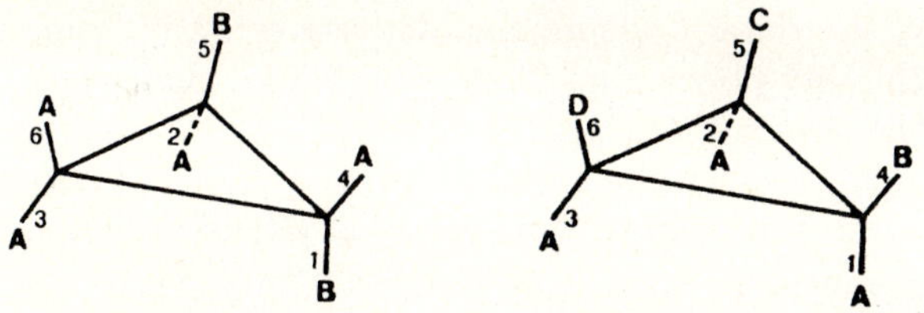

10 **11** **Scheme 3.5.**

We illustrate the presented method on the cyclopropane skeleton with symmetry D_{3h} and $n = 6$, see Scheme 3.5. (5,1) is the only partition of degree 1. It is not active. There is also only one partition of degree 2, namely (4,2), this partition is active. From a ligand assortment corresponding to this partition one can construct only one chiral molecule, which – for A, A, A, A, B, B – is shown in Scheme 3.5 (compound **10**). Thus, the lowest-degree chirality polynomial $\varphi(L)$ consists of only one component. The arrangement (3.25) of the labels of the skeletal sites can be illustrated by filling them into the partition diagram

2	3	4	6
1	5		

The first row contains the ligand sites to which a ligand A is attached, in the second row the ligand sites are recorded which are fitted with a ligand B. (The ranking within the rows has no relevance.) As $c = 2$, $\nu_2 = 2$, $j_{21} = 1$, $j_{22} = 5$ the lowest-degree chirality polynomial $\varphi(L)$ is given by

$$\varphi(L) = P_\chi(l_1 l_5)\,. \tag{3.27}$$

Using $D_{3h} = C_{1h} \times D_3$ and a decomposition of D_3 into cosets we get

$$P_\chi = \tfrac{1}{12}(e - (14)(25)(36))(e + (14)(26)(35))(e + (123)(456) + (132)(465)) .$$

Inserting this into (3.27) gives

$$\begin{aligned}\varphi(L) &= l_1 l_5 + l_2 l_6 + l_3 l_4 - l_2 l_4 - l_3 l_5 - l_1 l_6 \\ &= (l_1 - l_2)(l_4 - l_6) - (l_1 - l_3)(l_4 - l_5) .\end{aligned} \tag{3.28}$$

3.6 Qualitative Completeness and Supercompleteness

The chirality functions after the first and second approximation methods, based on the principle of mathematical simplicity, give rise to some criticism. One point has been expressed by Ruch and Schönhofer [3]: It may happen that these functions, independently of the nature of the ligands, vanish identically for a chiral molecule or a non-racemic mixture of isomers.

Consider, for example, the chiral cyclopropane derivatives **11** (Scheme 3.5). As $l_1 = l_2 = l_3$ the chirality polynomial (3.28) vanishes for these compounds independently of the nature of the ligands A, B, C, D. The reason for this lack of the function (3.28) is quite obvious: The terms $l_1 l_5, l_2 l_6, \ldots, l_1 l_6$ represent interactions between two ligands *trans*-placed at adjacent carbon atoms. As the corresponding terms $\omega_t^{(2)}(l_i, l_j)$ are symmetric they can be approximated by $l_i l_j$, cf. Eq. (3.18). From a physical viewpoint, however, it is evident that, compared with these interactions, the interactions $\omega_c^{(2)}(l_i, l_j)$, $(i,j) = (1,2), (2,3), \ldots$ between *cis*-placed ligands are not negligible at all. But, as these terms are antimetric, one needs a polynomial of degree three for their representation (Eq. (3.19)). This leads to these interaction terms not being included in the lowest-degree poynomial (3.28), which is of degree two. This example shows that the principle of mathematical simplicity is ill-founded. A chirality function which also contains *cis*-interaction terms would not vanish identically for **11**. Concerning the lowest-degree chirality polynomial (3.20) for the D_{2d}-skeleton there is no chiral molecule for which this function vanishes independently of the nature of the ligands. But one can construct non-racemic mixtures of isomers for which this vanishing occurs. Ruch and Schönhofer [3] gave the example shown in Scheme 3.6. This lack of the chirality functions after the first and second approximation procedures, which is not a lack in the opinion of other authors [24], inspired Ruch and Schönhofer [3] to establish the principle of qualitative completeness. A chirality function is called qualitatively complete if it does not vanish identically for any non-racemic mixture of isomers whatever the nature of the ligands may be.

Isomers may be thought of as being formed by permuting the ligands of a given molecule. This leads to a treatment within the frame of the symmetric group S_n of all $n!$ ligand permutations. The permutation group S,

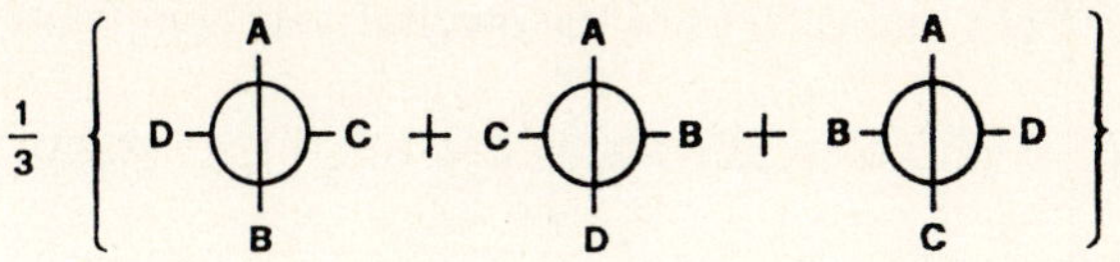

Scheme 3.6.

which is isomorphic to the group G of the covering operations of the skeleton, is a subgroup of S_n. A chirality function is qualitatively complete if it induces every irreducible representation Γ_r of S_n as many times as Γ_r contains the one-dimensional chirality representation Γ_χ of S [3]. Nevertheless, as Dugundji, Marquarding, and Ugi [24] have demonstrated the conditions for qualitative completeness can be stated without using representation theory: Let L_k, $k = 1, 2, \ldots, g$, be the ligand vector of one enantiomer of the enantiomeric pair k. It holds that a chirality function $\varphi(L)$ is qualitatively complete if, and only if, the functions

$$\varphi(L_1), \varphi(L_2), \ldots, \varphi(L_g)$$

are linearly independent. This result is directly reasonable because of every isomeric mixture M being a linear combination

$$M = a_1 L_1 + a_2 L_2 + \ldots + a_g L_g \ .$$

Thereby a negative concentration a_k is to be interpreted as the positive concentration $-a_k$ of the antipode of L_k. The chirality function

$$\varphi(M) = a_1 \varphi(L_1) + \ldots + a_g \varphi(L_g)$$

of an isomeric mixture M cannot be identically annihilated if the $\varphi(L_k)$, $k = 1, \ldots, g$, are linearly independent.

Using their principle of mathematical simplicity Ruch and Schönhofer construct qualitatively complete chirality functions. These functions consist of components each of which is associated with a Γ_r containing Γ_χ. Within the first approximation method for every such component polynomials of lowest degree are set. In the set-up according to the second approximation method each component is represented by a superposition of terms depending on as few ligands as possible. This combining of the principle of qualitative completeness with the principle of mathematical simplicity often leads to paradox results which are against physical insight [22]. Consider, for example, the skeleton of [2.2]metacyclophane shown in Scheme 3.7, its symmetry group is C_{2h}, $n = 4$. From the irreducible representations Γ_r, $r = 1, \ldots, 5$, of S_4 to which the partition diagrams of Scheme 3.8 are assigned Γ_2 and Γ_4 contain the chirality representation[2]. In both cases the multiplicity is one. The

[2] Within the scope of this paper a partition diagram may have three different meanings: 1. It may indicate the cycle structure of a permutation. As far as the symmetric group S_n is concerned all permutations associated with the same diagram form a conjugate class. 2. There is a one-to-one correspondence between the partition diagrams consisting of n boxes and the irreducible representations of the S_n. 3. A partition diagram may represent a ligand assortment.

corresponding components $\varphi^{(2)}(L)$ and $\varphi^{(4)}(L)$ of the qualitatively complete chirality function

$$\varphi(L) = \varphi^{(2)}(L) + \varphi^{(4)}(L) , \tag{3.29}$$

derived in [38], are given by

$$\varphi^{(2)}(L) = \omega^{(2)}(l_1) - \omega^{(2)}(l_2) - \omega^{(2)}(l_3) + \omega^{(2)}(l_4) , \tag{3.30}$$

$$\varphi^{(4)}(L) = \omega^{(4)}(l_1, l_2) + \omega^{(4)}(l_4, l_3) + \omega^{(4)}(l_3, l_1) + \omega^{(4)}(l_2, l_4) , \tag{3.31}$$

where

$$\omega^{(4)}(l_1, l_2) = -\omega^{(4)}(l_2, l_1) .$$

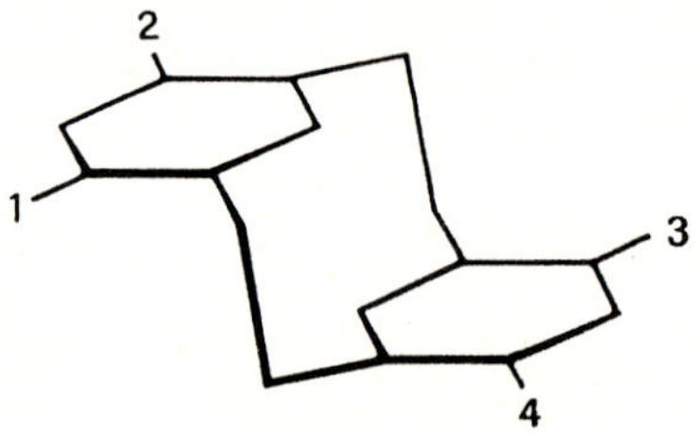

Scheme 3.7.

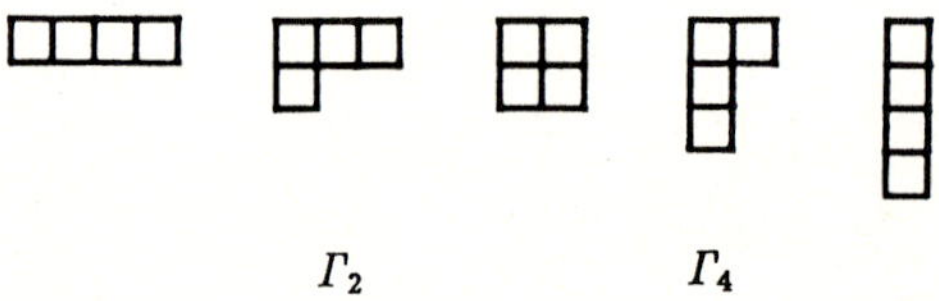

Scheme 3.8.

The qualitatively complete chirality function $\varphi(L)$ of Eqs. (3.29–31) contains no terms for the pair effects caused by ligand pairs attached to sites 1 and 4, or 2 and 3, respectively. This is not reasonable from the point of view of physics. One would expect that the 1,4 effect is of about the same importance as the other pair effects for which terms are included in $\varphi(L)$. Furthermore the 1,2 and 1,3 pair effects which correspond to non-equivalent pairs of ligand sites are described by the same function $\omega^{(4)}$. This simplification is unjustifiable.

Objections against the principle of qualitative completeness which are even more serious have been raised in [25]: One can construct non-racemic mixtures of non-isomers for which, independently of the nature of the ligands, the qualitatively complete chirality functions after [3] vanish identically. Thus, on the one hand, the principle of qualitative completeness forbids systematical zeroes for non-racemic mixtures of isomers but, on the other hand, it allows such zeroes for non-racemic mixtures of non-isomers. There is no physical reason for this distinction between isomeric and non-isomeric mixtures. In order to remove this lack the principle of qualitative supercompleteness has been established [26]. A qualitatively supercomplete chirality function does not vanish identically for any non-racemic mixture, whatever the nature of the ligands may be. However, for a successful application the qualitatively supercomplete chirality functions seem to be too complicated.

One would have to estimate too many ligand parameters. In this context it should also be mentioned that qualitatively complete chirality functions consisting of more than one component, i.e. being not identical with functions of [2], have not yet been applied to the description of chirality phenomena.

The author (see [25, 26]) does not share Ruch and Schönhofer's [3, 27] opinion that the decomposition of chirality phenomena according to the irreducible representations of the symmetric group S_n is of physical relevance. In this context the work of Meinköhn [39, 40] is of interest: By means of the algebraic invariant theory [41] another classification of chirality phenomena is derived which is not related to the irreducible representations of the S_n. Meinköhn shows that any analytic chirality function decomposes as a linear combination in the elementary functions of a suitable module basis which is finite. The linear combination coefficients are rational integral functions in the ligand parameters which are totally symmetric with respect to the permutations of S. In the case of the skeleton of Scheme 3.7 with symmetry C_{2h}, for example, the module basis is given by the two chirality polynomials

$$l_1 - l_2 - l_3 + l_4 \,,$$
$$l_1 l_4 - l_2 l_3 \,.$$

On the other hand, within the concept of qualitative completeness, the lowest-degree polynomials inducing Γ_2 and Γ_4, respectively, of S_4, (Scheme 3.8) are

$$l_1 - l_2 - l_3 + l_4 \,,$$
$$(l_1 - l_4)(l_2 - l_3)(l_1 - l_2 - l_3 + l_4) \,.$$

(Note that $(l_1 - l_4)(l_2 - l_3)$ is totally symmetric with respect to S.) These polynomials can be got from (3.30) and (3.31) by using (3.17) and (3.19), respectively. Summarizing one can state that the concept of qualitative completeness and the application of the theory of invariants lead to two different classifications of chirality phenomena which are very interesting from the mathematical point of view. However, a physical relevance has been proved in neither case. King avoids to discuss qualitative completeness in his review papers [34, 37]. A very formal development of the theory of chirality functions with emphasis on qualitative completeness has been given by Dress [42].

Dugundji, Marquarding, and Ugi [24] have developed the concept of hyperchirality. A hyperchiral family is the set of all distinct permutation isomers which have in common the same group of ligand permutations representing the rotations, and have also in common that coset of ligand permutations which represents the conversion of each isomer into its respective enantiomer. Frequently a hyperchiral family consists of more than just one molecule and its enantiomer. A hyperchiral family with x pairs of enantiomers contains $x(x-1)$ pairs of isomers, which – according to Dugundji et al. [24] – in equal amounts should behave like racemates, i.e. any chirality observation should yield zero for such "pseudo-racemates". Thus, in contradiction to the principle of qualitative completeness, the concept of hyperchirality postulates the existence of non-racemic mixtures of isomers for which a chirality observation independently of the nature of the ligands yields zero. The different views of the concepts of qualitative completeness, qualitative supercompleteness, and hyperchirality, with respect to the possibilty of such systematical zero points are summarized by Table 3.5. Obviously the concept of qualitative completeness is the only of the three standpoints which is inconsistent in itself. Hässelbarth [43] and Mead [44] have criticized the concept of hyperchirality in a non objective way, see also [45–47]. A short summary has been given by King [37].

Table 3.5. Systematical zeroes are allowed for non-racemic mixtures of ...

		... non-isomers	
		no	yes
... isomers	no	qualitative supercompleteness	qualitative completeness
	yes		hyperchirality

3.7 Counting Enantiomeric Pairs

The enumeration of isomers is a fascinating problem to which attention has been paid since the end of the last century. For a long time the final breakthrough in this field has been ascribed to Pólya. His paper [48], entitled "Kombinatorische Anzahlbestimmungen für Gruppen, Graphen und chemische Verbindungen", appeared in 1937. Unfortunately the work of Redfield [49] who, ten years before, had anticipated most of Pólya's results was overlooked up to the early 1960s [50]. Nevertheless, as Pólya's paper is much more elaborate in detail, it was worthwile translating it into English [51] even after Redfield's work had been discovered.

For enumerating the enantiomeric pairs for a given skeleton with a given ligand assortment we develop a method which is related to Redfield's enumeration theorem [49, 52], see also [53]. Let A be a permutation group working on a set X. A implies a decomposition of X into equivalence classes which are called orbits. Two elements $x, y \in X$ are said to belong to the same orbit if there is an $a \in A$ so that $y = ax$. Most of the enumerative methods are based on the so-called Burnside lemma. This lemma states that the number $o(A; X)$ of orbits is given by

$$o(A;X) = \frac{1}{|A|} \sum_{a \in A} j_1(a) \tag{3.32}$$

where $j_k(a)$ means the number of cycles of order k in the permutation a. For the proof see for instance [29, 54, 55, 56]. Eq. (3.32) can also be written as

$$o(A;X) = \frac{1}{|A|} \sum_{a \in A} \sum_{x \in X} \delta_{ax,x} \tag{3.33}$$

where $\delta_{x,y}$ is the Kronecker delta.

Now we number the skeletal sites as well as the ligands and consider a molecule as a bijection, i.e. a one-to-one mapping, l of the set of the n ligand numbers on the set of the n site numbers. We write this mapping as a matrix consisting of two rows:

$$l = \begin{pmatrix} 1 & \dots & i & \dots & n \\ l(1) & \dots & l(i) & \dots & l(n) \end{pmatrix} . \tag{3.34}$$

The upper row contains the site numbers, in the lower row the ligand numbers are recorded. Equation (3.34) means that ligand $l(i)$ is attached to site i, $i = 1, \dots, n$. As a short-hand we will use

$$l = \begin{pmatrix} i \\ l(i) \end{pmatrix} . \tag{3.35}$$

l has been called an ordered molecule [3, 57]. A permutation π of the columns leaves l unchanged. This is expressed by

$$l = \begin{pmatrix} i \\ l(i) \end{pmatrix} = \begin{pmatrix} \pi(i) \\ l(\pi(i)) \end{pmatrix} . \tag{3.36}$$

l may also be interpreted as a permutation, namely as that permutation which replaces i by $l(i), i = 1, \dots, n$. Use of this will be made in the following.

Let S now be the group of those site permutations which are equivalent to covering operations of the skeleton. S operates on the first row of l. Analogously let T, which acts on the second row of l, be the group of those ligand permutations which only permute ligands within the same sort. Obviously T is the direct product

$$T = T_{\gamma_1} \times \dots \times T_{\gamma_i} \times \dots \times T_{\gamma_c} \tag{3.37}$$

of the symmetric groups T_{γ_i}, where γ_i is the number of ligands of sort $i, i = 1, \dots, c$, and is the length of the i-th row of the partition diagram corresponding to the given ligand assortment. Let now operate simultaneously a permutation s on the upper row of l and a permutation t on the lower row of l. Using (3.36) with $\pi = s^{-1}$ we get

$$(s,t)l = \begin{pmatrix} s(i) \\ t(l(i)) \end{pmatrix} = \begin{pmatrix} i \\ t(l(s^{-1}(i))) \end{pmatrix} . \tag{3.38}$$

Equation (3.38) shows that applying s to the site numbers is equivalent to applying s^{-1} to the ligand numbers. Interpreting the double rows as permutations, $l \in S_n$, (3.38) may be written as

$$(s,t)l = tls^{-1} . \tag{3.39}$$

If $s \in S$ and $t \in T$ then l and $(s,t)l = tls^{-1}$ represent the same molecule. On the other hand, for a given pair l_1, l_2 of bijections an (s,t), $s \in S$, $t \in T$, transforming l_1 into l_2 according to

$$l_2 = (s,t)l_1$$

can be found if, and only if, l_1 and l_2 correspond to the same molecule. It follows that every isomer is represented by an orbit of the power group S^T which consists of the elements (s,t), $s \in S$, $t \in T$, and acts on the set of the $n!$ bijections l of Eq. (3.34) which can also be regarded as permutations, $l \in S_n$. The number z of isomers is therefore equal to the number of orbits of S^T which works on S_n according to (3.39). Using (3.33) one gets

$$z = o(S^T; S_n) = \frac{1}{|S|.|T|} \sum_{s \in S} \sum_{t \in T} \sum_{l \in S_n} \delta_{l,(s,t)l} \cdot \tag{3.40}$$

$l = (s,t)l$ or $l = tls^{-1}$, respectively, are equivalent to $lsl^{-1} = t$. There follows

$$\delta_{l,(s,t)l} = \delta_{lsl^{-1},t}$$

which can be different from zero only if s and t belong to the same conjugate class C_j of the S_n. Such a C_j is formed by all permutations having the same cycle structure, which is represented by a partition j of n. In this context it is convenient to introduce a different notation for partitions: We write j as

$$j = (1^{j_1}, 2^{j_2}, \ldots, k^{j_k}, \ldots) \tag{3.41}$$

where j_k indicates how often the term k occurs in the decomposition of n. If j refers to a permutation then j_k is the number of cycles of length k. Equation (3.40) can be slightly modified to

$$z = \frac{1}{|S|.|T|} \sum_{j} \sum_{s \in S \cap C_j} \sum_{t \in T \cap C_j} \sum_{l \in S_n} \delta_{lsl^{-1},t} \cdot \tag{3.42}$$

For $s \in C_j$, $t \in C_j$ the sum over l

$$\sum_{l \in S_n} \delta_{lsl^{-1},t}$$

is independent of s and t. To show this let s, s', t, t' be arbitrary elements of C_j. There exist permutations $\pi, \varrho \in S_n$ so that $s' = \pi s \pi^{-1}$ and $t' = \varrho^{-1} t \varrho$. Therefore

$$\sum_{l \in S_n} \delta_{ls'l^{-1},t'} = \sum_{l \in S_n} \delta_{l\pi s\pi^{-1}l^{-1},\varrho^{-1}t\varrho} = \sum_{l \in S_n} \delta_{(\varrho l\pi)s(\varrho l\pi)^{-1},t} \cdot$$

In the same way as l, $\varrho l \pi$ runs over all elements of the S_n. From this the independence of the sum of s and t follows. In (3.42) we let now s as well as t run over C_j instead of over $S \cap C_j$ or $T \cap C_j$, respectively. Because of the proved independence this extension of the domain can be easily compensated for by a correction factor. This factor is given by the product of the ratios σ_j/λ_j and τ_j/λ_j where $\sigma_j = |S \cap C_j|$, $\tau_j = |T \cap C_j|$ and $\lambda_j = |C_j|$ specify how many permutations with partition j there are in the groups S, T, and S_n, respectively. In addition we change the summation sequence and so get from (3.42)

$$z = \frac{1}{|S|.|T|} \sum_j \frac{\sigma_j \tau_j}{\lambda_j^2} \sum_{l \in S_n} \sum_{t \in C_j} \sum_{s \in C_j} \delta_{lsl^{-1},t} \, . \tag{3.43}$$

In the sum over s in (3.43) lsl^{-1} runs over all λ_j elements of C_j and meets t exactly once. Therefore this sum is equal to one. From this

$$z = \frac{n!}{|S|.|T|} \sum_j \frac{\sigma_j \tau_j}{\lambda_j} \tag{3.44}$$

follows for the number of isomers. An enantiomeric pair is counted as one compound by (3.44). For S contains also the site permutations which correspond to improper rotations of the skeleton and convert a chiral molecule into its antipode. Thus z is the sum of the number z_a of achiral compounds and the number z_e of enantiomeric pairs,

$$z = z_a + z_e \, . \tag{3.45}$$

If we replace S by the group S_0 of those site permutations which correspond to proper rotations of the skeleton then a pair of antipodes is counted as two compounds. Analogously to (3.44) there holds

$$z_a + 2z_e = z + z_e = \frac{n!}{|S_o|.|T|} \sum_j \frac{\sigma_j^0 \tau_j}{\lambda_j} \tag{3.46}$$

where $\sigma_j^0 = |S_0 \cap C_j|$. By S^* we denote the coset of S_0 in S. Taking into account $|S_0| = \frac{1}{2}|S|$ and $\sigma_j = \sigma_j^0 + \sigma_j^*$, where $\sigma_j^* = |S^* \cap C_j|$, we get from (3.44–46) for the number z_e of enantiomeric pairs

$$z_e = \frac{n!}{|S|.|T|} \sum_j \frac{(\sigma_j^o - \sigma_j^*)\tau_j}{\lambda_j} \, . \tag{3.47}$$

The number λ_j of elements of the conjugate class C_j of the S_n is given by

$$\lambda_j = \frac{n!}{\prod j_k! k^{j_k}} \, ,$$

see for instance [29]. Using this equation, (3.47) may be rewritten as

$$z_e = \frac{1}{|S|.|T|} \sum_j (\sigma_j^o - \sigma_j^*)\tau_j \prod_k j_k! k^{j_k} \, . \tag{3.48}$$

In (3.47, 48) the sum has to be taken over all partitions j of n. But in most cases of interest there are only relatively few partitions j for which σ_j as well as τ_j are different from zero.

The number $\tau_j = |T \cap C_j|$ of elements of T with cycle structure j can be got by means of Redfield's group reduction function [49] which was called cycle index (German: Zyklenzeiger) by Pólya [48] and plays a fundamental role in enumeration theory. The group reduction function (grf) Z of a permutation group A working on a set of n objects is defined as

$$Z(A; p_1, p_2, \ldots, p_n) = \frac{1}{|A|} \sum_j \alpha_j \prod_k p_k^{j_k} \tag{3.49}$$

where $\alpha_j = |A \cap C_j|$ and $p_1, p_2, \ldots, p_n$ are variables. As the grf of a direct product group is the product of the grfs of the single factors [49] we get with respect to (3.37) and (3.49)

$$Z(T; p_1, \ldots, p_n) = \frac{1}{|T|} \sum_j \tau_j \prod_k p_k^{j_k} = \prod_i Z(T_{\gamma_i}; p_1, \ldots, p_{\gamma_i}) \, . \tag{3.50}$$

This offers an easy way for calculating the τ_j.

Table 3.6. Data for calculating the number of enantiomeric pairs for an assortment (2,2,1,1) of ligands on an octahedral skeleton, $|S| = 48$, $|T| = 4$

Partition j	σ_j^o	σ_j^*	τ_j	$\prod j_k! k^{j_k}$
(1^6)	$1(E)$	0	1	720
$(1^4, 2)$	0	$3(3\sigma_h)$	2	48
$(1^2, 2^2)$	$3(3C_2)$	$6(6\sigma_d)$	1	16

The elements of O_h from which σ_j^o, or σ_j^*, respectively, arise are specified within parentheses.

As an example let us calculate the number z_e of enantiomeric pairs which can be built by distributing a ligand assortment corresponding to the partition (2,2,1,1) over the sites of an octahedral skeleton (point symmetry group O_h). For the grf of the direct product group

$$T = T_2 \times T_2 \times T_1 \times T_1 \ ,$$

which is of order 4, we get

$$Z(T; p_1, \ldots, p_6) = \left[\tfrac{1}{2}(p_1^2 + p_2)\right]^2 p_1^2 = \tfrac{1}{4}(p_1^6 + 2p_1^4 p_2 + p_1^2 p_2^2) \, .$$

From this

$$\tau_{(1^6)} = 1 \ , \quad \tau_{(1^4,2)} = 2 \ , \quad \tau_{(1^2,2^2)} = 1 \tag{3.51}$$

follow. All other τ_j's are zero. The data necessary for calculating z_e for our example by means of Eq. (3.48) are compiled in Table 3.6. Note that we need σ_j^0 and σ_j^* only for those partitions j for which τ_j is different from zero. Inserting into (3.48) gives

$$z_e = \frac{1}{48 \times 4}(1 \times 1 \times 720 - 3 \times 2 \times 48 - 3 \times 1 \times 16) = 2\,.$$

This can be easily verified by inspection.

3.8 References

1 Kauzmann W Clough FB Tobias I (1961) Tetrahedron *13:* 57
2 Ruch E Schönhofer A (1968) Theoret Chim Acta *10:* 91
3 Ruch E Schönhofer A (1970) Theoret Chim Acta *19:* 225
4 Ugi I (1965) Z Naturforsch *20b:* 405
5 Ruch E Ugi I (1966) Theoret Chim Acta *4:* 287
6 Ruch E Schönhofer A Ugi I (1967) Theoret Chim Acta *7:* 420
7 Ruch E Runge W Kresze G (1973) Angew Chem *85:* 10; (1973) Angew Chem, Int Ed Engl *12:* 20
8 Neudeck H Schlögl K (1977) Chem Ber *110:* 2624
9 Neudeck H Richter B Schlögl K (1979) Mh Chem *110:* 931
10 Neudeck H Schlögl K (1981) Mh Chem *112:* 801
11 Neudeck H Schlögl K Tscheplak H (1985) Mh Chem *116:* 789
12 Haslinger E Neudeck H Robien W (1981) Mh Chem *112:* 405
13 Langer E Lehner H Derflinger (1984) J Chem Soc Perkin Trans II *1984:* 201
14 Richter WJ Richter B Ruch E (1973) Angew Chem *85:* 21; (1973) Angew Chem, Int Ed Engl *12:* 30
15 Richter WJ (1978) Z Naturforsch Teil B *33:* 165
16 Richter WJ Heggemeier H Krabbe HJ Korte EH Schrader B (1980) Ber Bunsenges Phys Chem *84:* 200
17 Richter WJ (1980) Theoret Chim Acta *58:* 9
18 Rapić V Schlögl K Steinitz B (1977) Mh Chem *108:* 767
19 Richter WJ Richter B (1976) Isr J Chem *15:* 57
20 Keller H Krieger Ch Langer E Lehner H Derflinger G (1977) Ann Chem *1977:* 1296
21 Keller H Krieger Ch Langer E Lehner H Derflinger G (1978) Tetrahedron *34:* 871
22 Keller H Langer E Lehner H Derflinger G (1978) Theoret Chim Acta *49:* 93
23 Langer E Lehner H (1979) Mh Chem *110:* 1003
24 Dugundji J Marquarding D Ugi I (1976) Chem Scripta *9:* 74
25 Derflinger G Keller H (1978) Theoret Chim Acta *49:* 101
26 Derflinger G Keller H (1980) Theoret Chim Acta *56:* 1
27 Ruch E (1972) Acc Chem Res *5:* 49
28 Mead CA (1974) Top Curr Chem *49:* 1
29 Berge C (1971) Principles of combinatorics. Academic New York
30 Hamermesh M (1964) Group theory and its application to physical problems. Addison-Wesley Reading MA
31 Altmann SL (1977) Induced representations in crystals and molecules. Academic New York
32 Cotton FA (1971) Chemical applications of group theory. Wiley-Interscience New York
33 Haase D Ruch E (1973) Theoret Chim Acta *29:* 247
34 King RB (1982) Theoret Chim Acta *63:* 103
35 King RB (1987) J Math Chem *1:* 15
36 King RB (1987) J Math Chem *1:* 45
37 King RB (1988) J Math Chem *2:* 89

38 Keller H Krieger Ch Langer E Lehner H Derflinger G (1977) J Mol Struct *40:* 279
39 Meinköhn D (1978) Theoret Chim Acta *47:* 67
40 Meinköhn D (1980) J Chem Phys *72:* 1968
41 Schur I (1968) In: Grunsky H (ed) Vorlesungen über Invariantentheorie. Springer Berlin Heidelberg New York
42 Dress W (1979) Lecture Notes in Chemistry *12:* 215
43 Hässelbarth W (1976) Chem Scripta *10:* 97
44 Mead CA (1976) Chem Scripta *10:* 101
45 Dugundji J Marquarding D Ugi I (1977) Chem Scripta *11:* 17
46 Mead CA (1977) Chem Scripta *11:* 145
47 Hässelbarth W (1977) Chem Scripta *11:* 148
48 Pólya G (1937) Acta Math *68:* 145
49 Redfield JH (1927) Amer J Math *49:* 433
50 Harary F Palmer EM (1967) Amer J Math *89:* 373
51 Pólya G Read RC (1987) Combinatorial enumeration of groups, graphs, and chemical compounds. Springer Berlin Heidelberg New York
52 James G Kerber A (1981) The representation theory of the symmetric group. Addison-Wesley Reading MA
53 Ruch E Hässelbarth W Richter B (1970) Theoret Chim Acta *19:* 288
54 Harary F (1969) Graph Theory. Addison-Wesley Reading MA
55 Harary F Palmer EM (1973) Graphical enumeration. Academic New York
56 Kerber A (1979) Lecture Notes in Chemistry *12:* 1
57 Dugundji J Gillespie P Marquarding D Ugi I Ramirez F (1976) In: Balaban AT (ed) Chemical applications of graph theory. Academic London

4 Helicity of Molecules Different Definitions and Application to Circular Dichroism

G. Snatzke

4.1 Introduction

Any object which is not superposable onto its mirror image was called "chiral" at the end of last century by Lord Kelvin [1], who derived the term from the Greek word $\chi\varepsilon\iota\varrho$ for hand. In particular, molecules are called chiral if they have this mentioned property, and it has been well known since the middle of last century [2] that in the non-ordered state (gases, liquids, amorphous solids) *optical activity* can be measured only if the molecules are chiral. *Chirality* is thus a molecular property, whereas optical activity is a bulk property of a substance, i.e. in all practical cases can be measured only for a large ensemble of molecules. It is thus wrong to speak of a "chiral substance", as it is incorrect to speak of "optically active molecules"!

A special form of chirality is *Helicity*, which implies the coupling of a translation with a concomitant rotation. The radius of such a *helix* may be constant all over, or it may change along the direction of propagation ("conical helix" etc.). Every chemist will cite the well known *α-helix*-conformation of a protein [3] as a typical example of a helix, but there exist many others, as e.g. the house of a snail (the edible or Roman snail *Helix pomata* is the source of the name), a bean stalk, the stem of a vetch (*Vicia*), or again on the molecular level the famous DNA double helix [4] and the molecule of heptahelicene [5]. A helix must be distinctly differentiated from a spiral, which, in the strict sense, is a curve in one plane, whose radius vector changes length and direction with time. Mechanical watches contain "the" typical example of a spiral, viz. the balance wheel of the movement, which has, of course, disappeared from modern digital watches. A somewhat less known example is one of the old forms of the Chinese characters for cloud, 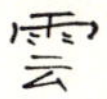 , which calligraphers sometimes write as  .

Whereas the spiral is an achiral object (its mirror image e.g. with respect to its own plane is congruent to itself) the helix is not, and one differentiates between *right-handed* and *left-handed* helices. The definition of *Helicity* is unequivocal; when proceeding along the direction of its axis (direction of translation) a helix can wind around this axis in one of two opposite senses, and *per definitionem* a right-handed helix is characterized in the following way. The thumb of our right hand should be aligned with the axis of the given helix; if the bent fingers of this right hand then follow the rotation of the helix, this is called right-handed, otherwise it is a left-handed one.

From this definition for a helix as a coupled motion of translation and rotation follows (and the reader should try this out with his right and his left hand!) that the sense of helicity is independent of which side we are looking from. A well-known practical example: a right-hand bolt will fit into its appropriate nut equally well from both sides, but we cannot screw it into a left-handed nut. Another very typical example for helicity is given by the way women in India wear their Saris: the several meters-long cloth is always wound around the body as a right-handed helix. Helicity also characterizes any propeller; only two-bladed ones are mentioned here, since for nomenclature purposes more complicated propellers can always be broken down into sets of two-bladed ones.

The Cahn-Ingold-Prelog system [6] gives us rules for the unequivocal notation of chiral and helical molecules ("CIP" code); in the latter case a right-handed helix is designated by the prefix *P*-, a left-handed one by *M*- (these names were derived from "plus" and "minus").

Whereas the sense of helicity can always be assigned unequivocally if at least one full turn of the helix is present, this is no longer the case if only part of a full turn of a molecular helix can be identified, as e.g. in the case of conjugated dienes, enones, or other short parts of "molecular screws". Such molecular helices are never smooth and into any short part of it we can easily fit either a right-handed or a left-handed screw! Several conventions ("working rules") have, therefore, been proposed, and some of them call a given helical arrangement in such a molecular helix, say, right-handed, others, however, left-handed. In the following it is tried to summarize the most often used definitions of the "sense of helicity", and their applications to "molecular propellers". The inherent ambiguity can be overcome only if in each paper that contains notations like *P*- and *M*-helicity the author's preferred use is defined in the text.

4.2 The Ideal Finite Helix

A helix, which might be cylindrical, elliptical, conical, etc. is obtained when a translation and a rotational movement are coupled in some systematic way. For our purposes it suffices to discuss only the cylindrical helix (a "screw"). Figure 4.1 shows the projection of one with more than 2 turns. In a right-handed Cartesian system (X, Y, Z) the Z-axis usually coincides with the axis $\mathbf{A}$ of the helix, and at $z = 0$ the vector $\mathbf{r}$, perpendicular to Z, may point in the X-direction. This vector now experiences a translation in the Z-direction, while it is concomitantly rotated around this same axis in such a way, that the angle ϕ, measured from the (X, Z)-plane, is proportional to this translation:

$$\phi = \pm(2\pi z)/\lambda \quad \text{(Rad)} . \tag{4.1}$$

It is counted positive if it is rotated in a right-handed Cartesian system from X- toward Y- through an angle of 90° (Fig. 4.1a). By this the tip of the radius vector describes a cylindrical regular helix, which is right-handed with the (+)-sign, left-handed with the (−)-sign in Eq. (4.1). Such a helix belongs to point group C_2, with its C_2-axis being perpendicular to the helix axis **A**, and cutting the helix at its midpoint. Figure 4.1b gives a projection along this C_2-axis (from the periphery towards the axis), and in Fig. 4.1c a tangent **T** is drawn together with the axis **A** (projection along $-\mathbf{r}$ at any point of the helix). This angle between the projections of **T** and **A** is constant throughout the whole screw.

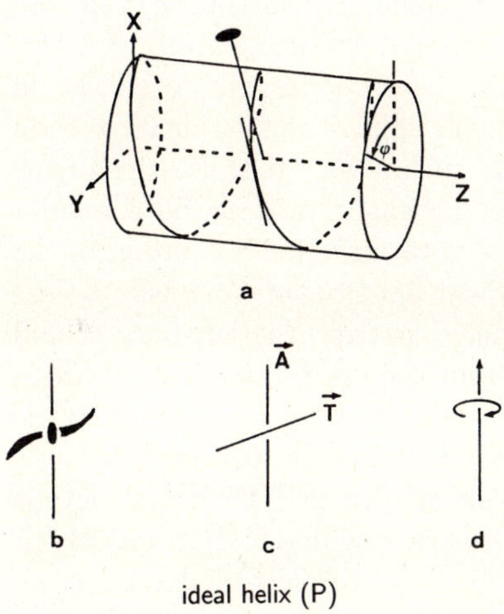

Fig. 4.1. (*a*) Dimetric projection of a cylindrical *P*-helix in a right-handed Cartesian system, C_2-axis indicated. (*b,c*) the same, viewed along the C_2-axis. (*d*) Shorthand notation for a *P*-helix

In looking at such a cylindrical helix one gets the feeling that there is "more" symmetry in it than just this C_2-axis, but this is not the case as long as we discuss *point groups*, as is usual in stereochemistry. We would have to resort to one of the additional symmetry elements defined for *space groups*, viz. the screw axis **A** in order to get the full symmetry properties of such a helix.

Obviously such a helix is a chiral object, and its mirror image is obtained when the (−)-sign is used instead of the (+)-sign in (4.1). This definition of handedness is, of course, independent of the length of the radius vector and the pitch of the helix, and it is unambiguous. All what we have to know is the arrangement of **T** vs **A** according to Fig. 4.1c. This right-handed helix of Fig. 4.1a is often symbolically written as the axial vector shown in Fig. 4.1d.

It cannot be repeated enough, that the helicity of a screw or helix is the same, independent of the direction in which one looks upon it. The moment one decouples the two combined movements, the translation and the rotation, however, this is no longer the case: in a given plane parallel to the (X, Y)-plane at z_1 the angle ϕ increases clockwise for the right-handed helix, when viewed in the $(-Z)$-direction ("against the translation"), but anticlockwise, when viewed in the $(+Z)$-direction ("with the translation").

Many physical properties of matter can be described by helices; only a single one, viz. "optical rotation", will be used here as an example [7]. Light, which is an electromagnetic radiation, can be characterized by several parameters: In this context we choose its speed c, its wavelength λ, and its frequency ν, which are related by the simple equation

$$c = \lambda \cdot \nu \, . \tag{4.2}$$

The frequency ν is independent of the medium, whereas c and λ change with it, and are largest in vacuum. If the electric field vector $\mathbf{E}$ of this electromagnetic field oscillates only in one plane ("plane of polarization") we call such a light beam "linearly-" or "plane-polarized". Mathematically, but also physically, any linearly polarized light beam can be thought to consist of the superposition of two "circularly polarized" light beams of identical radii but opposite helicities. These are defined in such a way, that the tips of the electric field vectors along the direction of propagation describe a regular helix, right-handed or left-handed. "Optical rotation", i.e. a rotation of the plane of polarization of this linearly polarized light beam can easily be explained by assuming, that in "optically active matter" the left-handed and the right-handed light ray travel with different speeds. In wavelength ranges of absorption both these rays are also absorbed to different extents, and the difference of absorption coefficients is called "Circular Dichroism (CD)".

Two special cases follow easily from these definitions: if the wavelength is of infinite length then the helix degenerates into a long line (only translation). At the other extreme with λ approaching 0, we have only rotation, but no translation at all, and the helix degenerates into a circle.

If the magnitude of $\mathbf{r}$ oscillates between two extrema with a phase difference $\delta\phi$ of $\pi/2$ during such a combined movement, then the cross section of the auxiliary cylinder which one can wrap around the helix axis becomes an ellipse. The tip of the vector $\mathbf{r}$ describes then what is called an "elliptically polarized light beam". Its "ellipticity" is defined as the arctan of the ratio of the minor (b) to the major axis (a) of this ellipse,

$$\phi = \arctan(b/a) \, . \tag{4.3}$$

In ranges of absorption of light the intensity diminishes, i.e. the radius vector $\mathbf{r}$ becomes smaller and smaller as the ray travels through such a medium: the helix becomes conical, i.e. the mentioned auxiliary cylinder is a cone.

4.3 Real Molecules or Parts of them, Fractions of a Helix

Since bonds can be envisaged as straight lines between two atoms, a contiguous train of bonds of, say, a polypeptide α-helix may be *approximated* by a regular helix. This is analogous to the approximation of a circle by a regular polygon, and by this the exact correlation as given in (4.1) is lost. As mentioned before, the sense of helicity is unambiguously defined the moment we know the directions of **A** and **T**. **A**, and therefore also **T**, can always be obtained from a helical train of bonds, as long at least one full turn of the helix is present. With polypeptides we need, therefore, at least four amino acid units, since on the average 3.6 amino acid residues make one full turn. Also for a molecule like heptahelicene (**1**) the axis can unequivocally be defined.

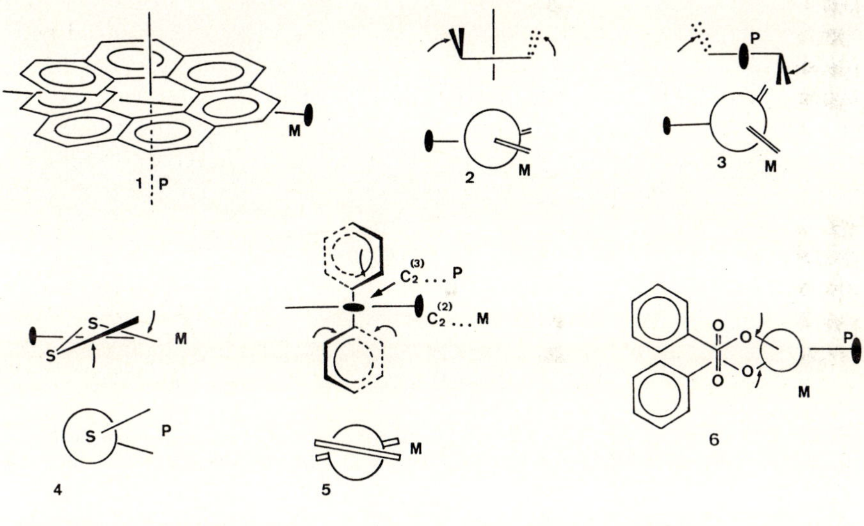

Scheme 4.1. **1**: Heptahelicene. M-Helicity along the C_2-axis, P-helicity along the helix-axis. **2**: Non-coplanar *cisoid* conjugated diene. Most frequently used stereodescriptors: P-helicity along C_2-axis, M-helicity because of negative torsional angle along middle bond. **3**: Non-coplanar *transoid* conjugated diene. Notation as for **2**. **4**: Non-coplanar disulphide. M-helicity along C_2-axis, P-helicity along S-S-bond. **5**: Twisted biphenyl. P-helicity along $C_2^{(2)}$, M-helicities along $C_2^{(3)}$ and the long axis ($C_2^{(1)}$). **6**: Dibenzoate of a 1,2-diol with negative torsional angle. M-helicity along the (O)-C-C(-O) bond according to CIP-rules, P-helicity along C_2-axis, also according to the CIP-rules

In many other molecules with which we associate usually helicity only less than one full turn of such a helix is present, although we usually find (for the idealized molecule) a C_2-axis (cf. the examples **2** through **8**). This C_2-axis is perpendicular to the axis **A** of the helix, and would we deal with an ideal regular helix then we could find out the direction of **A** as shown in Fig. 4.1c. For the rough approximation as given by a train of chemical bonds this is, however, no longer always possible, and we can often fit in an infinite number of axes corresponding to helices of different radii and pitches (the magnitudes of the latter are inverse to the length of the radii). As a consequence, we can approximate a given helical train of bonds even by either a right- or a left-handed helix at will! Figure 4.2 shows that also two ideal helices of opposite senses can share *approximately* a common piece of curve, if their axes are orthogonal. In Fig. 4.3 two helices of opposite sense are fitted onto a non-coplanar butadiene. Only two possible axes, $\mathbf{A}_\mathrm{L}$ and $\mathbf{A}_\mathrm{R}$, perpendicular to each other, are drawn, and the helix fitted along $\mathbf{A}_\mathrm{L}$ is left-handed, that along $\mathbf{A}_\mathrm{R}$ right-handed. In the first case the pitch is small, but the rotation around $\mathbf{A}_\mathrm{L}$ is large, whereas in the second helix the pitch is large and the rotation around $\mathbf{A}_\mathrm{R}$ is small. In a semiquantitative way one may say that for each such choice of pairs of perpendicular axes $\mathbf{A}_\mathrm{L}$ and $\mathbf{A}_\mathrm{R}$ the (formal) product of "translation" times "rotation" remains constant in magnitude, whereas the handedness of one helix is opposite to the other one.

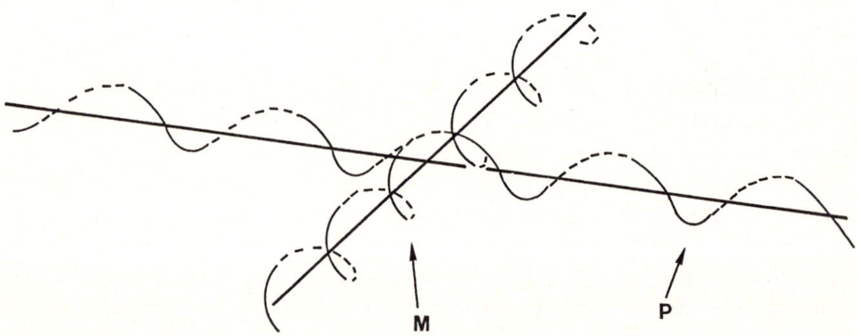

Fig. 4.2. To a short part of a threedimensional curve a P- as well as an M-helix fits well. The axes of the two helices are orthogonal to each other

This explains (at least in part) why the correlation between the sign of the Cotton-effect of the first $\pi \to \pi^*$-transition of a conjugated non-coplanar diene and the sense of the helicity of this molecular moiety is not a very safe one [7]. CD is produced when an electron is excited "on a helical path" in the electron cloud of the chromophore, and along two perpendicular axes (orthogonal to the C_2-axis) these helices always have opposite senses! Other factors (e.g. the presence of axial allylic or homoallylic bonds) may then become determining for the Cotton-effect [8].

In spite of this principal impossibility to find in a "natural" or "symmetry-determined" way the direction of the axis **A**, which would be needed to specify unequivocally the sense of helicity according to the right-hand-rule, any of

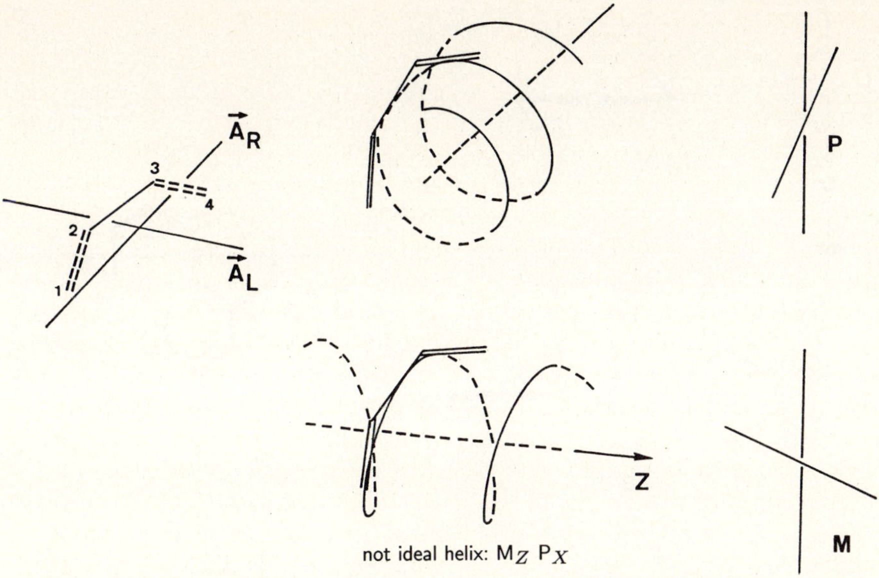

Fig. 4.3. *Middle*: Best fitting of two helices along a non-coplanar train of three consecutive bonds (*left*), and arising helicity (*right*) from direction of its axis and tangent

the structures (**2**) to (**8**) describes a unique type of helicity. Consequently, for each of these types one should be able to assign to each enantiomer one, *and only one*, descriptor in an unequivocal way. Such descriptors in use are (D/L), (P/M), $(+,-)$, etc., and they allow a classification into one of only two categories, which we may then identify or associate with *positive* or *negative sense of helicity*. In this context should be mentioned the famous paradigm introduced by Ruch for this classification: we may take 100 different shoes of all colours or fashions, but in each case we can unequivocally put each of them either in the box of "Left shoes", or in another one with "Right shoes". Not so, however, for potatoes: we can easily identify these as potatoes, but it is impossible to divide them in a similar fashion into two categories – there exists an infinite number of shapes without any such ordering principle. A real example for this situation is a (still hypothetical) metal complex with the structure of a tetragonal pyramid, in which the transition metal sits at the top, and four different ligands occupy the four corners of the basis. The exchange of two of these ligands does not lead to its enantiomer, as in the tetrahedral case with the C-atom in the middle. Only after a second such ligand exchange is the mirror of the original pyramid obtained.

Practically all proposed methods for the specification of the sense of the helicity of a given molecule make use of the chiral arrangement of two bonds which are correlated to each other by the application of the C_2-operation. Such pairs of bonds are marked in formulae **2** to **8** by arrows and called in the following "marked bonds".

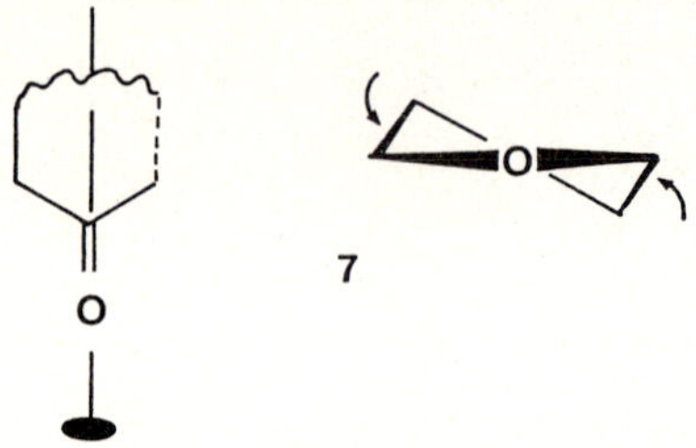

7

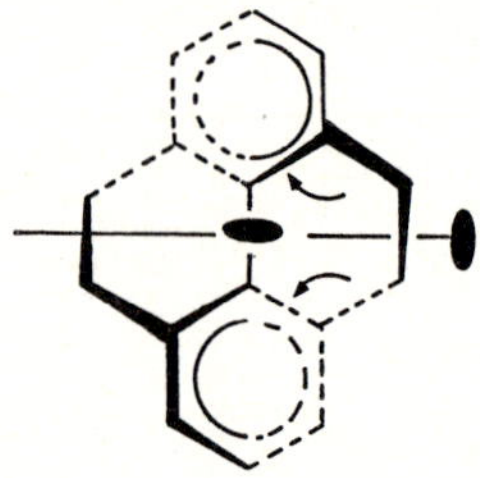

9

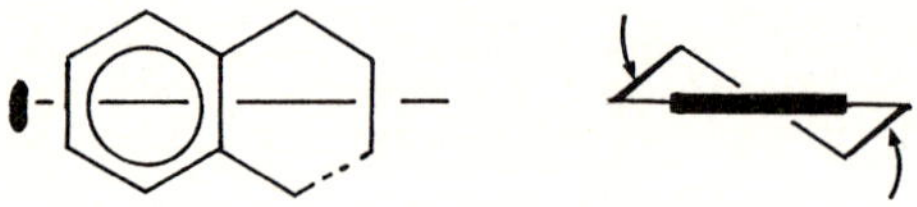

8

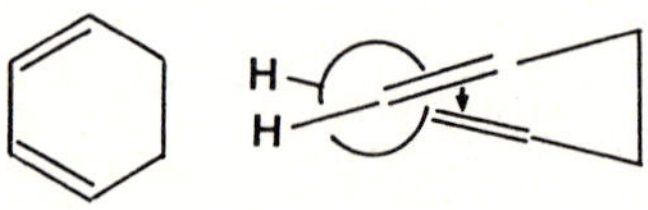

10

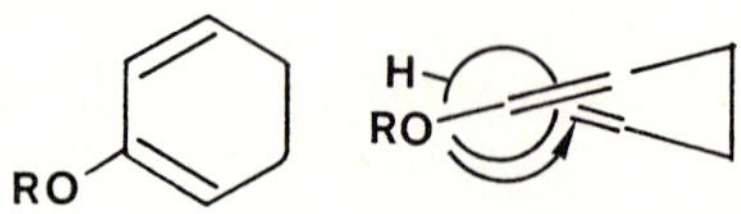

11

CIP – P

CIP – M

Scheme 4.2. 7, 8: *P*-Helicity along C_2-axis. **9**: Like **5**, but also chiral for 90° twist. **10**: Positive torsional angle according to CIP-rules. **11**: Same sense of helicity as for **10**, but negative torsional angle according to CIP-rules

4.4 Rules

4.4.1 The Torsional-Angle-Rule (CIP)

For **2**, **3**, **4** (lower projection), **5**, and **6** the marked bonds are connected by only one bond, so the torsional angle around this middle bond can be specified by the *Klyne-Prelog* [9] or the *Cahn-Ingold-Prelog* [6] *rule.* For **7** and **8** more than one bond may lie between these two marked bonds, but if we connect one of the possible pairs of endpoints which are correlated by the C_2 operation by a line, then this line can serve as an auxiliary bond for our purpose of classification. If this torsional angle is positive, the arrangement is called P, otherwise the descriptor is M. In this way and without any further assumptions an unequivocal classification is possible except for **5**, and this rule is given in row 1 of Table 4.1

5 belongs to point group D_2, so either C_2^2 or C_2^3 may be chosen as the axis perpendicular to the connecting middle bond. Is it C_2^2 then the torsional angle is acute and negative, is it C_2^3 then this torsional angle is oblique and positive. This unsubsituted **5** is, of course, of no practical importance,

Table 4.1. Stereodescriptors for **2** to **9** according to different systems of nomenclature

molecule rule	**2**	**3**	**4**	**5**	**6**	**7**	**8**	**9**
4.4.1, 4.4.5	M	M	P	–[b]	M	M	M	M
4.4.2	Δ	Λ	Λ	Δ	Δ	Δ	Δ	Δ
4.4.3[a] (a)	P	P	M	–[b]	P	P	P	P
4.4.3[b] (b)	M	M	P	–[b]	M	M	M	M
4.4.4	(–)	(–)	(+)	–[b]	(–)	(–)	(–)	(–)

[a] Achiral for torsional angles of $\pm 90°$.
[b] Additional definition required (e.g.: acute torsional angle preferred over oblique one).

since we are not able to isolate the two enantiomers (this should be possible, however, at very low temperatures); **9** is, however, of the same symmetry and could be resolved. Nevertheless, **5** and **9** differ in one important instance: for a torsional angle of $\pm 90°$ around the bond connecting the two phenyls **5** acquires D_{2d}-symmetry and becomes thus achiral, whereas **9**, for these torsional angles, does not change its D_2-symmetry and remains chiral. By the additional bridges of **9** the ambiguity of marking a bond in the "lower" ring is resolved: according to the CIP-rules one has to proceed from one o-position to the other along the additional shortest path, and this is through this same cyclohexadiene ring.

Although it is tempting to identify *P* and *M* with the two senses of helicity this cannot be done, because of the inherent property of the CIP-nomenclature, that mere replacement of one atom by another may change the descriptor without touching the geometry. For example, the 1,3-cyclohexadiene **10** and the homochirally analogous enol acetate **11** have opposite descriptors (*P*-**10**, *M*-**11**) for the same "geometric ring-helicity". To circumvent this problem one may e.g. specify the torsional angle by an attribute like "within the ring", or "for the train of bonds U-V-W-X". As long as one does not pretend to use the CIP-notation this would be a "legal" procedure. *It is unacceptable, however, although unfortunately sometimes found in the literature, that an author creates "modified CIP-rules" at will for each special case.*

4.4.2 The IUPAC-Axis-Tangent Rule

This rule was originally proposed by Sargeson [10] for the characterization of configurational as well as conformational chirality of octahedral complexes with two or three bidentate ligands. From the way it is defined it is applicable only to situations where **A** and **T** lie in two parallel planes which are perpendicular to the line connecting the midpoints of **A** and **T**, so that for an angle of $\pm 90°$ between the projections of **A** and **T** onto such a plane an achiral arrangement results (C_{2v}- or D_{2d}-symmetry).

The rule says: make one of the two marked bonds to the axis **A**, the other to a tangent **T** of a helix, then the sense of helicity is unequivocally determined by application of the right-hand-rule (cf. Fig. 4.1c). It is easy to prove that it does not matter, which of the two marked bonds is made **A**, and which **T**; both choices lead to identical results. In this IUPAC-rule capital Δ and Λ are used to classify the configurational, small δ and λ the conformational helicity. The pair Δ and δ refers to right-handed, Λ and λ to left-handed helices.

One could try to extend this rule to other helical fragments of molecules, which are not so restricted in the relative arrangement of **A** and **T**, and applying this notation (using deliberately capital descriptors) to the marked bonds, the given examples have to be named as shown in the second row of Table 4.1.

A closer inspection reveals, however, that this Δ, Λ-system, while serving perfectly the purpose for which it was developed, fails with our examples (except **5**) for a torsional angle of $\pm 90°$. In that case the helix as defined by the pair **A**/**T** degenerates into an achiral circle, whereas **2**–**4** and **6**–**9** are, of course, still chiral with these torsional angles. One may further note (cf. the pair **2**/**3**) that the sense of helicity changes when the torsional angle changes between acute and oblique. Since usually for organic molecules torsional angles cannot be specified that exactly, in a range of, say, $\pm|80\ldots100|°$, one may use an incorrect model and ascribe thus a wrong descriptor to the still well-defined helicity. On the other hand, the biphenyl helicity of **5** is described correctly, and it does not matter, which of the two marked bonds in the "lower" ring are used: −**5** in the confirmation as given is always −**5** (and becomes indeed achiral for a perpendicular arrangement of the two rings!).

4.4.3 The Two-Tangent-Rule

The two marked bonds are considered as two different tangents $\mathbf{T}_1$ and $\mathbf{T}_2$ to a helix. This specifies then also immediately the axis **A** as the sum vector of $\mathbf{T}_1$ and $\mathbf{T}_2$, but presents us at the same time the difficulty that for this purpose one has to give polar directions to $\mathbf{T}_1$ and $\mathbf{T}_2$, and there exist two (pairs of) choices, leading to opposite sense of helicity! The situation is depicted in Fig. 4.4. The choice a) (with A-symmetry for the C_2-operation) places **A** along the C_2-axis, which, however, can never be any helix axis of the approximate helical train of bonds, and leads to a right-handed helicity for the arrangement chosen, whereas b) (with B-symmetry for the C_2-operation) selects from the manifold of possible directions of the axes a single one. For this the sense of helicity of **2** is negative. Although b) would be the more "natural" way to specify the axis of the "molecular helix", because only here do we go consecutively from one end of this unit to the other, most authors who used this convention have, in general, tacitly decided for modification a), in complete perversion of the original idea of "train of bonds following approximately a helix". Furthermore, the P/M-notation has been borrowed from the CIP-nomenclature, P standing for the right-handed ("Plus"), M for the left-handed ("Minus") helix as defined by the procedure outlined in Fig. 4.4a. One should be aware that this use of P/M has nothing to do with the original definition of these two descriptors in the CIP-system; the only excuse for this "nomenclature-stealing" was, to avoid the creation of still another pair of new descriptors.

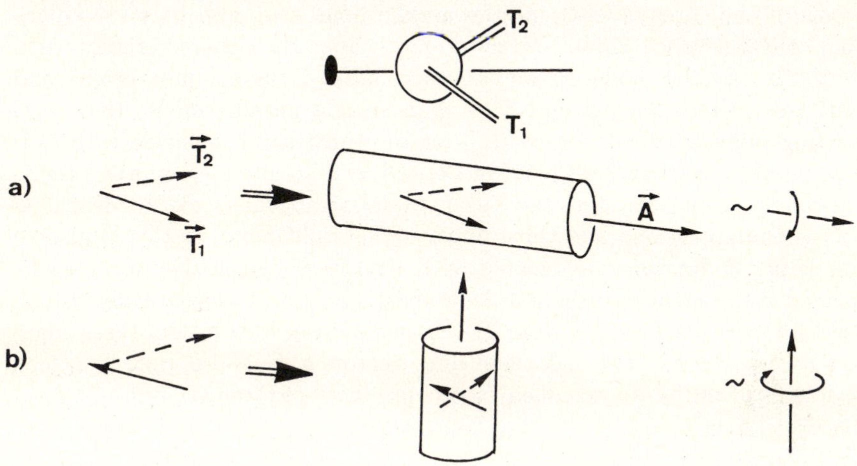

Fig. 4.4. (*a*) A-mode coupling of two tangents T_1 and T_2, representing the two formal double bonds of a non-coplanar butadiene moiety; leads to P-helicity. (*b*) B-mode coupling of same moiety leads to M-helicity

P- and M- in this sense is identical with positive and negative sense of helicity, because they describe skeleton-helicity independent of substituents, and the third line of Table 4.1 lists the appropriate descriptors for the molecules **2** to **9**. For **5** there exists again the ambiguity that the descriptor depends on the (deliberate) choice of either C_2^2 or C_2^3 as the determining axis. An additional convention may resolve this, e.g. such, that one chooses those two of the three marked bonds for which the torsional angle is acute. This would, of course, have to be specified explicitly in any paper.

Rules 4.4.1 and 4.4.3 are fully equivalent; there is no trouble with a torsional angle of $\pm 90°$, and no change of the descriptor is caused by the change from an acute to an oblique torsional angle of same sign (as long as one uses rule 4.4.1 in the modified form: torsional angle around a connecting real or auxiliary bond "within the ring"). It is indeed a great pity that version a) of rule 4.4.3 has hitherto mostly been used instead of version b), so that in general rules 4.4.1 and 4.4.3 give opposite descriptors. A redefinition of rule 4.4.3 into version b) would, however, create more confusion than it would help!

4.4.4 The Spade-Product Rule

Three non-coplanar vectors form either a right-handed or a left-handed co-ordinate system (usually not Cartesian, i.e. in general not orthogonal). The handedness of this co-ordinate system, constructed in accordance with rules which are well established in vector analysis, can serve as descriptor not only for the helical train of three bonds, but even for such molecular fragments which are devoid of any C_2-axis. The procedure is as follows: name the four atoms of the three-bond-train (the middle bond may also be an auxiliary one) consecutively 1-2-3-4. Vectors do not change their characteristics when shifted in parallel mode, so we can make atom 1 the common origin, and shift the vectors $2 \rightarrow 3$ and $3 \rightarrow 4$ in such a parallel mode, that their starting points (2 for the second, 3 for the third bond) coincide with 1. If now the three vectors $1 \rightarrow 2$, 1(from 2)$\rightarrow$ 3, and 1(from 3)$\rightarrow$ 4 form a right-handed co-ordinate system (i.e. their spacial arrangement can be described by the thumb, second and third finger of the right hand, in this sequence) this is *per definitionem* associated with a positive sign, if they form a left-handed system, this corresponds to a negative sign. Mathematically stated, the three vectors $1 \rightarrow 2$, 1(from 2)$\rightarrow$ 3, and 1(from 3)$\rightarrow$ 4 form three edges of a parallelepiped. If we call these three vectors **a**, **b**, and **c**, then its signed volume is given by the so-called "spade product" of these vectors:

$$V = a \cdot [b \times c] = b \cdot [c \times a] = c \cdot [a \times b] = \begin{vmatrix} a_x & a_y & a_z \\ b_x & b_y & b_z \\ c_x & c_y & c_z \end{vmatrix} . \tag{4.3}$$

This same sign is now given to the helicity of the molecular bond train. Application of this rule, which is the procedure of choice whenever one wants to determine the sense of helicity (or chirality) from co-ordinates as e.g. obtained from X-ray diffraction patterns, is shown in Fig. 4.5. As with rules 4.4.1 and 4.4.3, no ambiguity appears except for **5**, where again we would have to define further specifications about the preference of one of the two equivalent C_2-axes. The respective descriptors are summarized in line 4 of Table 4.1

At first glance it seems disturbing that with torsional angles of $\pm 90°$ for the then achiral molecule **5** one also gets a non-zero spade product, and by this a "handedness" for the co-ordinate system, when one of the two C_2-axes had been selected. Note, however, that for 90° the difference between an acute and an oblique angle is lost, so we get degeneracy: both of the possible spade-products have the same magnitude, but opposite signs. We can equally well construct two co-ordinate systems which are mirror images to each other. This is, however, the situation as found in any racemate, and anyway no stereochemist would try to assign a handedness to an achiral molecule!

Another discrepancy seems to be that the spade product leads, for any given geometry, to one and only one sign, whereas it has been shown above, that a helical fragment can be approximated by either a right-handed or a left-handed helix. This latter fact arises, however, only because one can use

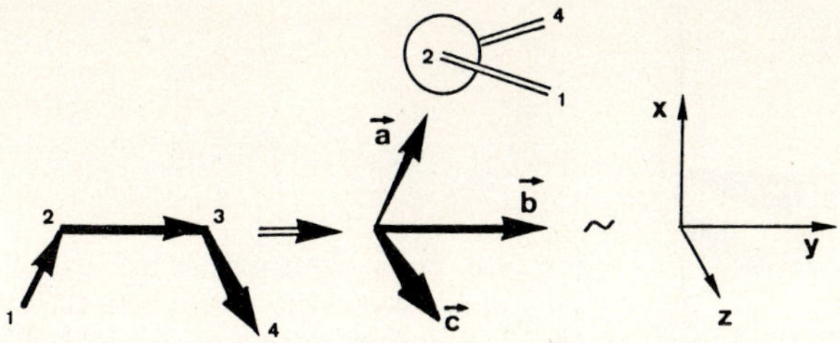

Fig. 4.5. Same diene units as in Figs. 4.3 and 4.4, represented by a (usually non-Cartesian) co-ordinate system (x, y, z): the spade product defines the sign of the helix

different axes for a co-ordinate system around the real, not "curve-smoothed" molecule, whereas the spade product is unequivocally defined by the molecular moiety 1-2-3-4.

The calculation of the spade product is exemplified by using "atomic co-ordinates" for the four atoms of **2**, which are not at all realistic, but they show the principle. Let the atoms **1** to **4** in Fig. 4.3 have the following co-ordinates:

1: $(3, -1, 0)$
2: $(2, 0, 2)$
3: $(-2, 0, 2)$ and
4: $(-3, 1, 0)$, then our three vectors have the components
a $\equiv 1 \rightarrow 2 : (-1, 1, 2)$
b $\equiv 2 \rightarrow 3 : (-4, 0, 0)$ and
c $\equiv 3 \rightarrow 4 : (-1, 1, -2)$. This leads to the spade product

$$V = \begin{vmatrix} -1 & 1 & 2 \\ -4 & 0 & 0 \\ -1 & 1 & -2 \end{vmatrix} = -(-4).(-4) = -16 < 0 \,.$$

From Fig. 4.5 one sees that $\mathbf{x}, \mathbf{y}, \mathbf{z}$ do indeed form a left-handed co-ordinate system, in agreement with this negative sign of the spade product V. The magnitude of the spade product may even be taken as a semiquantitative measure of helicity or chirality. Whenever it becomes very small one should, of course, also consider that this "exact" procedure is inaccurate for the real molecule.

4.4.5 The Spiral-Staircase-Rule

Apart from a screw, a "spiral" staircase is also a common representation of a helix; if both bond angles are made 90° (without changing the helicity) in our three-bond-train, then the middle one can be taken as the axis of such a staircase, the first and third bond represent two consecutive steps. Also this visualization of helicity has been occasionally used, and it is exemplified in Fig. 4.6. The descriptors are summarized also in line 1 of Table 4.1, since they are identical with those of rule 4.4.1 (P and M have been used for right-handed and left-handed staircases). In essence, this rule determines the sign

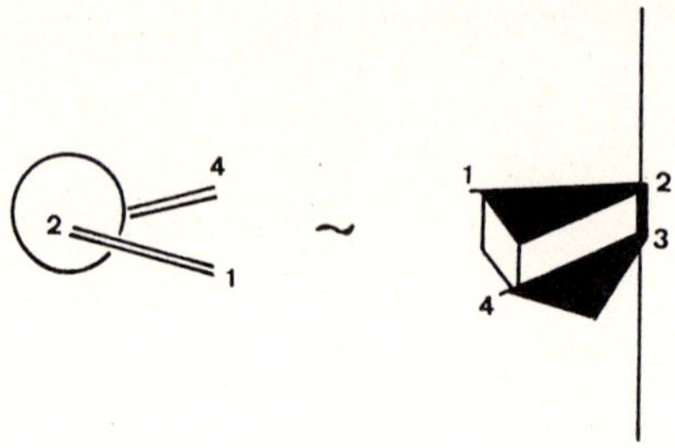

Fig. 4.6. Same diene unit as in Figs. 4.3, 4.4 and 4.5, represented by a helical (but usually called spiral!) staircase with *M*-helicity

of the torsional angle (1-)2-3(-4) in a similar way as it does the Klyne-Prelog convention, but in a more picturesque way. A bonus is, that the descriptor is independent of substitution. Molecule **5** poses the same problems as with most of the other rules, and necessitates additional conventions.

4.5 Some Applications

In Sect. 4.2 Circular Dichroism (CD) was used to explain "Helicity", and in this chapter some practical applications will be given. Since $\Delta\varepsilon$ is the "natural" magnitude for the characterization of the phenomenon, this will also be used in the following (molar ellipticity values are obtained from $\Delta\varepsilon$ by merely multiplying by 3300). CD can be measured only if the substance which is investigated absorbs light. According to Moscowitz [11] we can differentiate between such molecules in which the chromophore is chiral itself ("Class I") and those, for which it is achiral, but the rest of the molecule is chiral ("Class II"; the achiral chromophore is chirally perturbed by its chiral environment. Since any perturbation is transferred to the chromophore both through the bonds and directly through space even chiral solvent molecules – if all of the same absolute configuration – give rise to a measurement of an effect, but since its sign depends on the optical activity of the solvent and not on that of the dissolved compound this "solvent-induced CD" is not discussed here).

It is appropriate to subdivide this Class II into IIA and IIB. The latter subclass includes all those molecules for which the nearest chiral perturbation is at least two bonds away from the chromophore, and only in this case one can build up a "Sector Rule", of which the "Octant Rule" may be the most familiar one. In all other cases, however, "Helicity (or Chirality) Rules" have to be used to correlate molecular "absolute" geometry with the sign of the CD.

If the chromophore is built into a ring then it is usually relatively easy to assign a given chiral molecule to one of these classes, but open-chain molecules may fall into any one of these categories depending on their preferred conformation under the conditions of the measurement. Furthermore, there may be found "Exciton Interaction" (Davydov splitting) [12] whenever two (or more) chromophores (chiral or achiral!) with strong electrical transition moments interact in a chiral way. Two CD bands of opposite signs are obtained, which in principle should have the same magnitude, and their $|\Delta\varepsilon|$-values may be several hundred units in size. A few of each of these applications will now be given.

In simple cases Qualitative MO-theory is able to describe the orbitals involved in such a way that by the application of a few "recipes" the sign of some individual CD-bands can be predicted from the inherent twist of the chromophore [13]. These "recipes", which can be derived from the theory of CD and are easier presented within the LCAO-MO – rather than the VB – formalism, read as follows:

Inherently chiral chromophores (Class IA). 1) Identify the MOs Ψ_i and Ψ_j, which are involved in the transition, and draw them (only crude approximations needed; the hatching of the first lobe is deliberate, the other hatchings follow from the form of the MO) for the chiral chromophore;

2) Multiply them formally with each other, equal signs leading to a (+)-, opposite to a (−)-sign;

3) Ivert these signs (nothing mysterious! There is no physical reason for this, but it is necessary with the choice of definitions of the direction of transition moments and of $\Delta\varepsilon = \varepsilon_L - \varepsilon_R$ as also used here);

4) Determine the centre of gravity for the positive and the negative charges separately. The vector from the positive towards the negative charge represents then the direction of the electric transition moment vector $\boldsymbol{\mu}$, and even semiquantitatively its magnitude;

5) Determine the rotation of charge during the excitation (i.e. along the direction of $\boldsymbol{\mu}$) and apply to it the right-hand-rule: the thumb gives the direction of the magnetic transition moment $\mathbf{m}$;

6) Is the angle so obtained between these two vectors an acute one (extreme: both are parallel) then the CD is positive, is it obtuse (extreme: both antiparallel) the CD is negative.

Corollary (Class IIA). For Class IIA similar recipes hold, but since then either $\boldsymbol{\mu}$ or $\mathbf{m}$ (in rare cases even both) is $\equiv 0$ this missing transition moment has to be stolen from another nearby transition. QMO theory is often able to give you this moment.

4.5.1 Practical Applications

Class I: Three typical chromophores will be discussed, viz. that of vinyl ethers (both Ψ_i and Ψ_j are chiral), of non-coplanar conjugated enones (Ψ_i is achiral), and the non-coplanar disulfide chromophore (Ψ_j is chiral).

I.1: Vinyl ethers [14]. In a vinyl ether the C=C-π-MOs are conjugated with that one of the two lone pairs on oxygen which has its axis perpendicular to the C-O-C plane. In any pyranosyl-glycal this chromophore is present, and it is inherently twisted [a survey of published X-ray structures revealed for one case a torsional angle of approx. 15° along the (C-)O-C(=C) bond with a nearly planar C=C system, but many others, for which this angle is only

7°, and then the C=C-double bond is also twisted by that same angle (and with the same sign of the twist)].

Step 1: The chromophoric system (three centres, four electrons) is (nearly) isoelectronic with that of an allyl anion, the first transition is thus from $\pi_2 \rightarrow \pi_3^*$ (Fig. 4.7a). *Step 2:* The formal multiplication is done in Fig. 4.7b, Fig. 4.7c shows the same after the sign inversion (*Step 3*). With the signs chosen the positive transition charge is built up around the oxygen atom, the negative one around C(2); $\boldsymbol{\mu}$ points, therefore, from the ring-O towards C(2), and by following the "transition charge" along this vector one notices that electron cloud is rotated anticlockwise. According to the right-hand-rule this corresponds to a magnetic transition moment vector pointing antiparallel to $\boldsymbol{\mu}$, so a negative sign for the CD is predicted. Now indeed all D-glycals show a negative CD between 210 to 190 nm, in full agreement with the prediction from the "recipes".

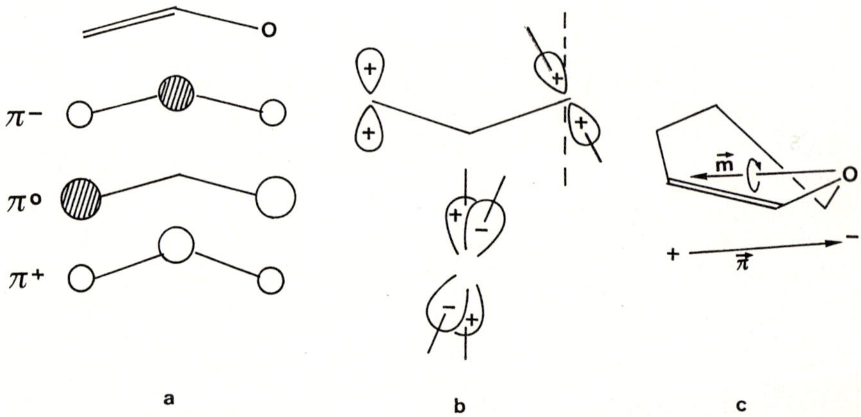

Fig. 4.7. The vinylether chromophore of a D-hexose. (*a*) MO-Scheme for isoelectronic allyl-anion. First and second MO doubly filled, third MO LUMO. (*b*) Formal multiplication scheme of HOMO · LUMO. (*c*) $\boldsymbol{\mu}$ and **m** for D-glucal

It should be kept in mind that in general we follow the opposite way: we measure the CD, get for a given band a positive or negative sign, and from this we want to determine the absolute configuration. Say, we obtain a negative Cotton effect for a substance, which could be identified as glucal, but it should be unknown whether of D- or L-configuration. Can we determine this immediately from the measured CD? Remember, for the determination of the correlation between the stereochemistry and the CD we used *only* the geometry of the chromophore, but not explicitely any of the D- or L-configurations of the centres of chirality. The same correlation would have been obtained if we would have taken L-glucal, but then in a conformation where all the ring substituents are in axial arrangement. If present, this conformation would, however, immediately flip over to the other chairlike one, and then the critical torsional angle would become positive. By this the CD also would become positive, which obviously disagrees with the postulated measurement.

It should be noted that this is the general situation in nearly all cases of flexible molecules; we need a second measurement (may be UV, NMR, etc.) in order to be able to determine both the preferred conformation and the absolute configuration! This is comparable to the mathematical problem of the determination of two unknowns. In order to do so we need two equations not one.

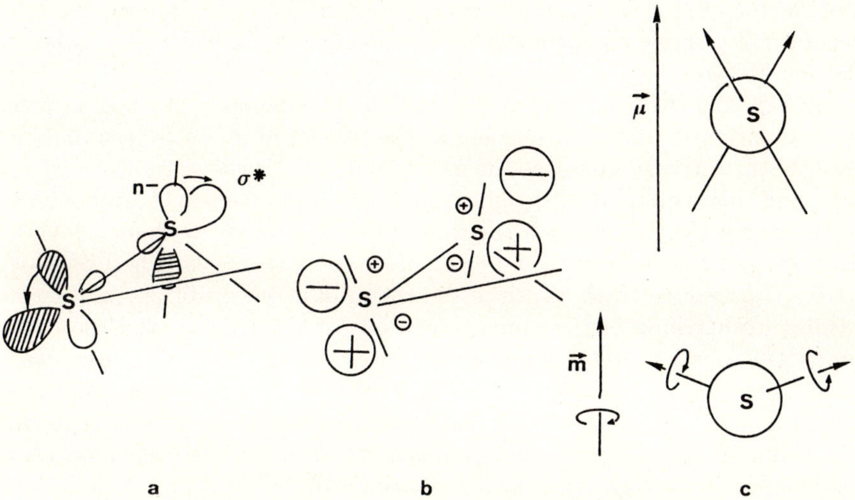

Fig. 4.8. The disulfide chromophore of 2,3-dithia-5α-cholestane. (*a*) The n^- and σ^*-MOs of the S-S–chromophore. (*b*) The transitional charges. (*c*) *Top:* the two individual electric transitional moments and their sum vector. *Bottom:* The two individual magnetic transitional moments and their sum vector

I.2: Chiral Disulfides [14]. The sense of helicity of a *cisoid* non-coplanar disulfide is that given in formula **4**. The first band in the UV- and CD-spectrum has the origin $n^- \rightarrow \sigma^*$, in which n^- stands for the antibonding combination of those non-bonding orbitals at the two sulphur atoms which are perpendicular to the respective C-S-S plane (HOMO), and σ^* for the S-S antibond (LUMO). Figure 4.8a shows these MOs for both halves of the full chromophore, Fig. 4.8b the signs of the respective products (note that the "inner" lobes of the σ^*-antibond are appreciably smaller than the "outer" ones), and Fig. 4.8c the sign pattern after the sign inversion. The dominating two electric transition moments point both up (with our deliberate choice of the first orbital hatching), so the overall $\boldsymbol{\mu}$ must also point upwards and is perpendicular to the C_2-axis of the chromophore. Following then the rotation of charge from the appr. vertically arranged n-lobes into the horizontally arranged σ^*-lobes of identical sign we obtain two magnetic moments, of which the one in front points upwards right, that at the back upwards left. Their sum vector is thus parallel to $\boldsymbol{\mu}$, and a positive CD is predicted for this positive torsion angle as defined in **4** (lower picture). In all hitherto investigated cases this prediction has proved to be true.

I.3: Conjugated *transoid* Enones [13]. The HOMO → LUMO transition is of $n \rightarrow n_3^*$ parentage, and this time the HOMO is achiral, but the LUMO is chiral. In a coplanar system n and π are orthogonal; this property is lost, however, when the C=C-C=O moiety is bent, so n and π may mix. Such a mixing is the more effective the smaller is the MO energy difference. It will thus be strongest for π_2 and n, and since the nonbonding MO is the HOMO, π_2 will be stabilized by this interaction, and n destabilized. The new MOs after mixing will thus be: $\pi_2 \rightarrow \pi_2 + \delta_1 \cdot n$, and $n \rightarrow n - \delta_2 \cdot \pi_2$, the δ_i's being small positive numbers (MOs not normalized). Of these the second is the new HOMO.

In Fig. 4.9a the front lobe of the "old" n is deliberately hatched; in first approximation it will be orthogonal to the p-lobes of π_2 on oxygen and on the C of the carbonyl group, so the upper lobe at C_α is the nearest part of π_2 to n, and this should be not-hatched, since this interaction is of antibonding type. Having thus determined the sign of one lobe of π_2 we know all signs from the usual sign pattern, and these are given in Fig. 4.9a. The LUMO is practically identical with π_3^*, since the energy difference to n is quite large (deliberate hatching of the upper p-lobe on oxygen for π_3^*), and the MO-product is shown in Fig. 4.9b, after sign inversion in Fig. 4.9c. In usual way we obtain then $\boldsymbol{\mu}$ from C(3) towards O; if the charge rotation from the modified n-MO into the LUMO is followed with the right hand then **m** points approximately from O towards C(1). The two transition moments are thus nearly opposite to each other, so a *negative* Cotton effect is predicted. There are ample examples which follow this rule, and the few exceptions which are known are substituted at C(3) by a polar group in such a way that the bond C(3)-X is approx. perpendicular to the plane of the C=C-double bond, which allows maximum interaction of this bond with the π-system.

One should note that in case I.1 the charge rotation was distributed equally along the whole $\boldsymbol{\mu}$, whereas here this seems concentrated on one end of μ. Both these extreme cases arise from our simplifications and are in the end equivalent.

Class II.A: In all cases of class II no component of **m** exists in the direction of $\boldsymbol{\mu}$, so one of the two transitional moment vectors has to be "stolen" from another transition (with chromophores of high symmetry even both transitional moments may become $\equiv 0$). In simpler cases this can also be handled by the QMO-theory, here I refrain, however, from going through all the arguments and would rather only mention a few rules for illustration.

II.A.1. Twisted cyclopentanones and cyclohexanones [7]. Let us take the standard projection as given in formula **7** on the right side. The torsional angles on both sides of the carbonyl, viz. those in the (O=)C-C(-C) moieties, are positive, so the chiral perturbation of the chromophore has the same sense from both sides. If this torsional angle is positive, as in formula **7**, then the

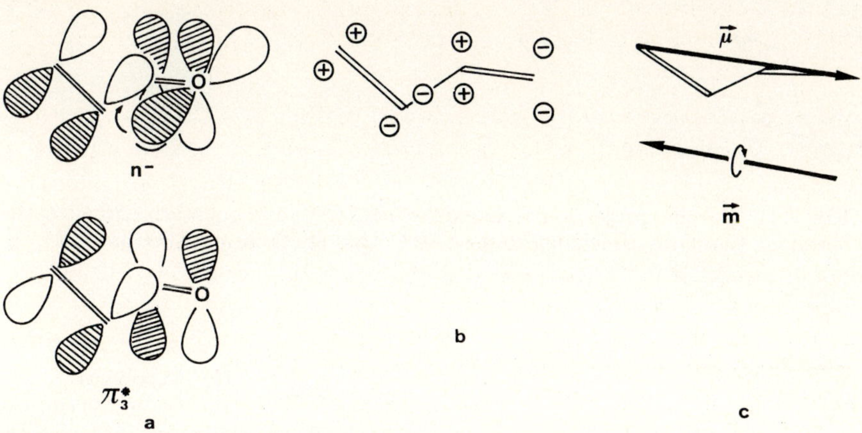

Fig. 4.9. The non-coplanar conjugated enone chromophore. (*a*) The involved MOs. The deliberately front-hatched n-MO induces the non-hatched sign of the p-orbital at C_β. From this the signs of the other π_2-p-lobes follow from their usual distribution. (*b*) Formal multiplication of lobes and sign inversion (transitional charge distribution). (*c*) $\boldsymbol{\mu}$ and **m**

$n \to n^*$ CD of the carbonyl is also positive $(+4\ldots+7)$. The magnitude of the CD is actually a good indicator that something unusual is present, because the CD of saturated ketones with chair conformation does usually not exceed 3.4.

II.A.2. Tetralins and tetrahydroisoquinolines [7]. If that conformation is present which is given in formula **8** (right) for a tetralin or a tetrahydroisoquinoline then the CD within the α-band is again positive, but this time it depends strongly on the substitution pattern of the aromatic system. One has to estimate the direction of the overall $\boldsymbol{\mu}$; if this falls into a sector $\pm 30°$ from the original C_2-axis then the rule has not to be modified. Is the deviation larger, however, then anyway the C_2-axis is not anymore there at all, and the simple picture may fail. Let us e.g. consider estradiol (**12**, R = H), its 3-O-acetate (**12**, R = Ac), and its 3-O-methyl ether (**12**, R = Me). The CD within the α-band is negative, as it is for the methyl ether, for the 3-O-acetate on the contrary it is positive. Such sign-changes led some chemists to believe that the α-band CD is not very valuable for the structure determination at all, but if one takes into account for such rules sign-changes which come from the change of the symmetry of the chromophore then these CD-signs are on the contrary of very great practical value.

II.A.3. α-axially [7] and β-equatorially [15] substituted cyclohexanones with chair conformation. The carbonyl chromophore is severely perturbed by the mentioned group; is this Cl, Br, I, SR, rhodanide or cyanate then $\Delta\varepsilon$ becomes very large, and it is positive for the absolute configuration depicted in Fig. 4.10. Again two different informations are obtained from one single CD-spectrum: the unusually large magnitude of $\Delta\varepsilon$ indicates the presence

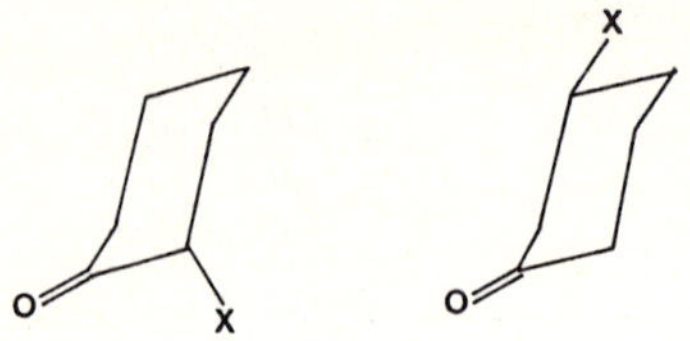

Fig. 4.10. Parallel projection of an axial α-halo (*left*) and equatorial β-halo cyclohexanone (*right*), both leading to positive Cotton effects around 300 nm

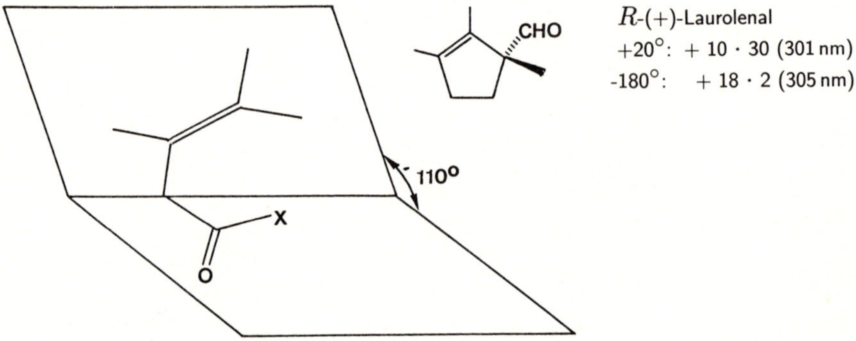

Fig. 4.11. Necessary conformation for strong $n \rightarrow \pi^*$-UV and positive CD-bands of a β,γ-unsaturated oxo compound (aldehyde, ketone, acid derivative) and example of a corresponding aldehyde [*R*-(+)-Laurolenal, two different temperatures]

of such a special arrangement, the sign leads to its absolute configuration. This rule is as safe as an X-ray diffraction using Bijvoet's method for the determination of the absolute configuration.

II.A.4. Another rule which never seems to fail is that for certain β,γ-unsaturated ketones, if their geometry is as given in Fig. 4.11 (or its mirror image) [7]. It has already earlier been noted that such ketones may have an $n \rightarrow \pi^*$ UV-absorption with a molar absorption coefficient of up to 10^3. The prerequisite is a geometry as shown in Fig. 4.11, i.e., the p_1 AO of the double bond comes close to the p-AO on C of the carbonyl, so the π-MOs of both systems mix but do not really undergo (homo-)conjugation. By this the overall symmetry of the C=O-chromophore (C_{2v}) is reduced, so the corresponding transition gains some allowedness. By the same mechanism also the CD-values grow, and since the UV is proportional to μ^2, the CD to μ, $\Delta\varepsilon$ should appr. be proportional to $\sqrt{\varepsilon}$, and this had indeed been found experimentally, $|\Delta\varepsilon|$ becoming as large as 35. Also this rule is "100% safe". An example of such an aldehyde [16] (instead of a ketone) is also given in Fig. 4.11. Acids and their derivatives with the correct geometry show a very strong $n \rightarrow \pi^*$ band, too, which appears then below 220 nm.

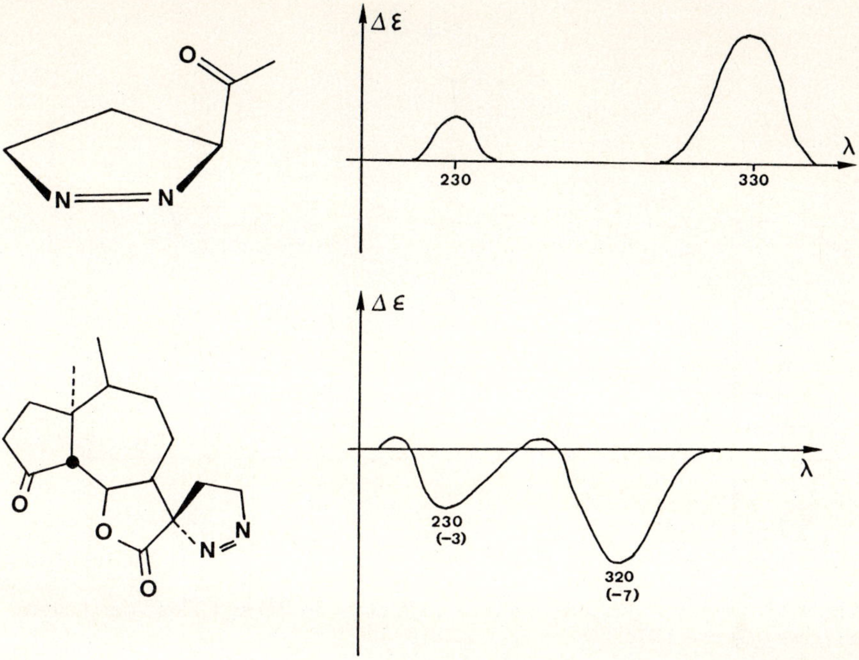

Fig. 4.12. Absolute configuration and necessary conformation for the very intense positive $n \to \pi^*$ Cotton effects of an α-pyrazolino oxocompound, and an example (damsin, positive CD between the two negative pyrazoline Cotton effects is from the cyclopentanone chromophore)

II.A.5. A similar perturbation leads to an equally safe rule for certain pyrazolino ketones or acids (Fig. 4.12), this time, however, not within the keto but within the N=N-absorption bands around 330 and 230 nm [17]. Again the $|\Delta\varepsilon|$-values may become as high as 30, and both mentioned Cotton effects have the same sign, which depends on the absolute configuration of the chromophore as shown in Fig. 4.12. It may seem as if this were a rather fancy chromophore, but nature provides us with many sesquiterpenoids, which contain the α-methylene lactone moiety, and this adds easily diazomethane to give the wanted derivatives. For the α-methylene lactone chromophore itself also a rule exists, but it is somewhat complicated since the CD-signs depend on the position of this unit at a larger ring, and on its configuration, too. An example of the first mentioned rule is given in Fig. 4.12.

Class II.B.: II.B.1. "The" prototype of such a chromophore is the cyclohexanone, in which the sixmembered ring can adopt the chair conformation [7]. Only in such a case, when the chromophore and the ring into which it is built are both achiral, can the sector rules be valid; for cyclohexanones this is an octant rule. The chromophore has two symmetry planes, which become then nodal planes of the sector rule. By this a quadrant rule seems appropriate, but also the nodal spheres of the involved MOs have to be considered.

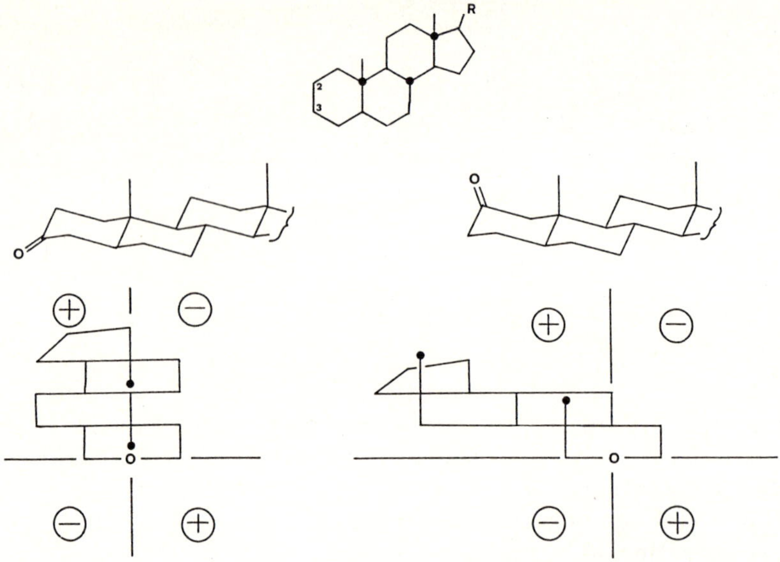

Fig. 4.13. Octant projections of 3-keto- (*left*, $\Delta\varepsilon = +1.27$) and 2-keto-5$\alpha$-cholestane (*right*, $\Delta\varepsilon = +2.97$)

For the n-MO this does not lead to any additional plane, but the π^* has one more nodal sphere, whose position and curvature is not known for sure. So a plane is used instead, making the quadrant rule into an octant rule. It is the custom to use a standard projection from O towards the C of the carbonyl group and give the signs for the "rear" sectors, since it is very rare that groups are attached in a "front" octant. The rule and its application to cholestan-2-one and cholestan-3-one is shown in Fig. 4.13. Although it is not the most frequently applied rule it still is the best known for historical reasons, and many chemists apply it to cases where it cannot be used at all!

II.B.2. Hemisphere Rule for Sulfoxides [7]. A sulfoxide molecule is chiral, since the two S-C bonds form one plane, and the S-O bond deviates from this plane appreciably. Only one single nodal plane is involved, and thus a "hemisphere rule" should be applicable. An example is given in Fig. 4.14.

For any other symmetric chromophore the appropriate sector rule can be found in the same way: make the nodal spheres to planes of the sector rule and you have a good start for such a rule. The signs of the individual sectors may be found by inspection of one correct example, from which one can extrapolate.

Exciton Interaction [12]. The prerequisite to apply this formalism, which was introduced into science by the Russian physicist Davydov, is the presence of two strong electric transition moment vectors which are chirally arranged to each other (it is irrelevant whether these chromophores are themselves achiral or chiral). Benzoates or other similar aromatic ester functions serve well our purposes, and it became the custom to speak of the "Exciton

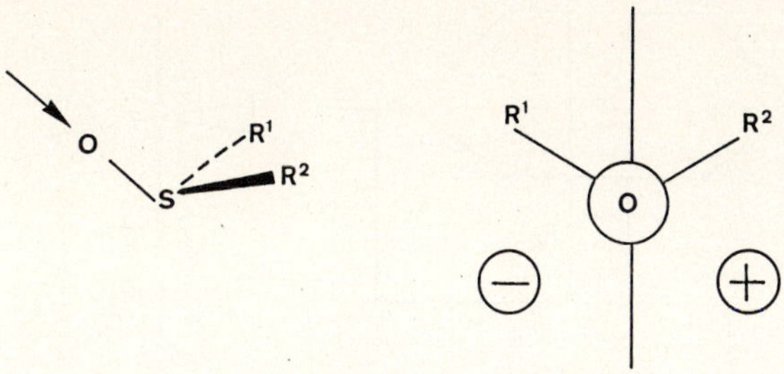

Fig. 4.14. Parallel projection of a chiral sulfoxide ($R^1 > R^2$) which leads to a negative CD around 210 nm

Chirality Method", although it is not a new method, but the application of Davydov's theory to the special case of two benzoates (or similar units). The rule is derived here for a $2\alpha,3\beta$-dihydroxy-5α-cholestane dibenzoate (cf. formula **6**), but it may be generalized and used in other, similar cases, even if the two benzoate units are farther apart by more than only one bond.

Benzoates show a strong B_{1u}-band at 230 nm ($\varepsilon = 14\,000$, by some authors called "charge transfer band"), which is polarized appr. along the long axis of the benzoate chromophore ($\boldsymbol{\mu}$ nearly along the line from the p-position to the midpoint of the two oxygen atoms of the ester grouping). The preferred conformation of a benzoate of a secondary alcohol is so that the O=C-bond is syn-periplanar to the hydrogen which is geminal to the ester grouping, as indicated by many X-ray structures. The arrangement of the two benzoates is, therefore, that shown in formula **6**.

Taking into account the mentioned facts and drawing these electric transition moments we have two choices for their directions: both may point to the right (Fig. 4.15 left), or down (Fig. 4.15 right). (The remaining two combinations are equivalent to the two just mentioned.) Their sum vectors point either horizontally or vertically, and we can interpret these sum vectors as axes of two cylinders, the individual vectors as tangents to it. In the first case the two tangents describe a right-handed helix, so the magnetic transition vector also points to the right, whereas in the second combination the two tangents define a left-handed helix around this downwards pointing sum vector. The first combination corresponds thus to a positive CD, the second to a negative one. Where will they be found in the spectrum?

Thinking of the physical phenomenon described by $\boldsymbol{\mu}$ we can interpret it as a shift of positive charge from one end to the other (convention as above), so the transition charges have also been indicated in the figure by $+/-$. Using Coulomb's law (in its simplest form: the smallest of the four possible distances will give rise to the biggest contribution to the interaction energy) and looking onto the projection it is obvious that the overall interaction of the left combination is repulsive, whereas that on the right is attractive. This leads to a splitting of the excited state into two levels, i.e. we get two UV- and CD-bands. The latter have opposite signs, and we read off the diagram that

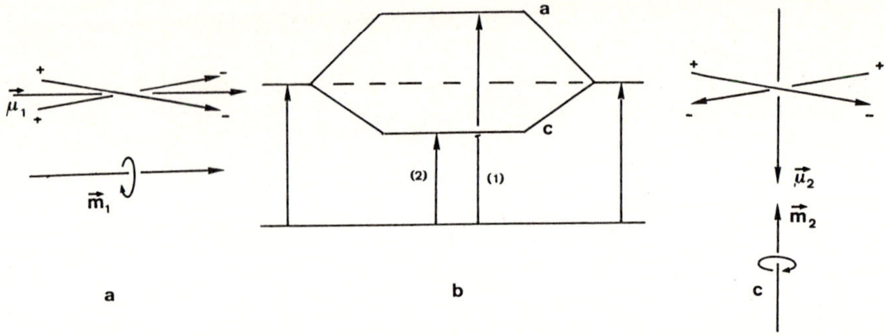

Fig. 4.15. Exciton interaction of two (equal) chromophores with strong electric transitional moments. (*a*) One possible combination of the two μ-s (both pointing to the right), (*c*) the other possibility (both pointing downwards. The other two combinations, both pointing left and both pointing upwards lead to identical combinations of μ and **m**). (*b*) The MO-scheme: only the excited state is split into two, the ground state is not. Combination (*a*) corresponds to the higher, (*c*) to the lower energy

the *P*-helix corresponds to the higher-, the *M*-helix to the lower-lying excited state: we obtain what is called a negative "CD-couplet" with negative wing at longer, positive at shorter wavelength than λ_{max} of a monobenzoate. The CD-amplitude may become several hundreds in such a case, and in literature many applications of this rule can be found.

In general it is always possible to find out without any doubt for any such coupling the sign, but it may sometimes become difficult to determine for sure the relative energies of the two combinations. Furthermore, it is good practice to discuss two CD-bands of opposite sign only then as a CD-couplet if the $\Delta\varepsilon$-values are larger than those found for the respective chromophore alone. Furthermore, with benzoates usually the second wing at shorter wavelengths is not any more well detectable because of the stronger absorption, so other esters (*p*-dimethylamino benzoates etc.) have been proposed. On this basis Nakanishi and his colleagues [18] have developed a procedure which allows the determination of the branches of sugars in an oligosaccharido terpenoid or steroid glycoside with less than 1 mg of material!

Application of CD to non-absorbing substances. Since CD appears only within absorption bands one has to introduce a chromophore into the molecule for its application, and the benzoates of diols, which were mentioned above, are one example for this principle. Cotton applied this principle in the last century by using transition metal complexes of chiral diols, diamines etc. I will cite only two examples from our research group to demonstrate this application.

Dinuclear complexes like $[Mo_2(OAc)_4]$ exchange, in solution, quickly their acetate ligands against other acids, but also against diols, aminols, phosphanes etc., just to name a few groups [19]. If we have a di-secondary *threo*-1,2-diol (on cyclohexane with e/e-, e/a- or a/e-conformation or in an aliphatic chain) then this diol-moiety can span the Mo-Mo distance in the complex best if it is in a gauche conformation. The (O-)C-C(-O) torsional angle is either defined in the cyclohexane case by the absolute configuration, with flexible alphatic diols this conformation prevails where the gem-hydrogens point "inwards" in the complex, the other two groups "outwards". This defines for any diol unequivocally the conformation in the bound case, and the torsional angle of the glycol moiety has always the same sign as the Cotton effect of this in-situ complex around 300 nm. No special preparation or purification is necessary, the glycol is just added to the stock solution of $[Mo_2(OAc)_4]$ in DMSO, and one can immediately afterwards measure the CD and determine so very quickly and safely the absolute configuration of the glycol unit. As an example the two diols 2α-, 3β- and $2\beta, 3\beta$-cholestanediols are used. In Fig. 4.16 the two CD-curves are given; although the a/e $2\beta, 3\beta$-diol is much more hindered than the e/e $2\alpha, 3\beta$-diol, under same conditions the two induced CD-curves are practically mirror images of each other.

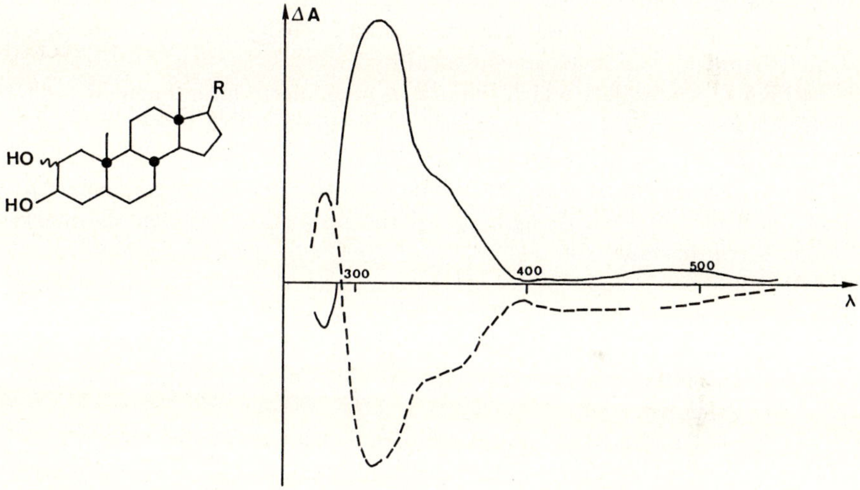

Fig. 4.16. The CD-curves of in situ complexes between $[Mo_2(OAc)_4]$ and 2β- (——) and 2α-hydroxy-5α-cholestan-3β-ol (- - - - -), in DMSO solution

Quite recently we [20] found that the corresponding $[Rh_2(O_2CCF_3)_4]$ complex accepts also monoalcolhols, olefins, epoxides and even ethers, and several rules could be developed. As an example is cited (+)-*p*-menth-1-ene (Fig. 4.17), and the CD-band which can be used for the correlation is around 350 nm. It is positive for the mentioned olefin, in agreement with the prediction from the rule. The configuration of a large number of non-absorbing compounds may now also be investigated by CD via this complex.

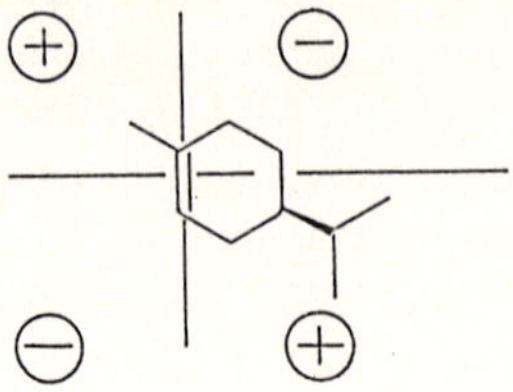

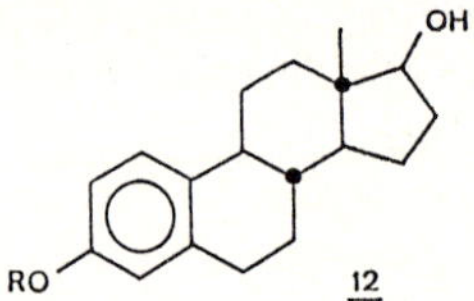

12

Fig. 4.17. Projection from the axial position towards the Rh_2-unit of the (+)-*p*-menth-1-ene/[$Rh_2(O_2CCF_3)_4$]-complex, leading to a positive sign of the CD-band around 350 nm

4.6 Summary

1) If the position of the axis of the helix is known for sure (e.g. protein α-helix with at least 4 amino acid residues; single or multiple-stranded oligonucleoside helix with known mode of base stacking; helicenes with at least one full turn) the right-hand-rule will give the sense of helicity [right-handed, (+), or P, left-handed, (−), or M] unequivocally.

2) For D_2-molecules, which acquire D_{2d}-symmetry for a torsional angle of $\pm 90°$ between marked bonds, octahedral complexes with bidentate ligands, etc., the Δ/Λ (or δ, λ-) convention can serve for the unequivocal characterization of the helicity. It may be extended to other molecular fragments, too, fails, however, for such if the angle of the projection between the marked bonds is approaching $\pm 90°$.

3) Rules which make use of the torsional angle around a bond connecting the two marked bonds for characterization of helicity, or such, which are equivalent to these (e.g. the Two-Tangent-Rule 4.4.3) are unambiguous except for molecules with D_2-symmetry, in which case additional conventions have to be defined.

4) Spade-Product-Rule: it defines unequivocally a (+)- or (−)-sign for any helical or otherwise chiral arrangement, which can be characterized by three non-coplanar bonds. For molecules with D_2-symmetry additional definitions are required.

5) Whatever rule one applies, it is essential that one specifies it in each publication.

4.7 References

1 Lord Kelvin (1904) The Baltimore lectures on molecular dynamics and the wave theory of light, p 436. Clay & Sons London
2 Pasteur L (1922) Œuvres T 1, p 327. Mason Paris
3 Schulz GE Schirmer RH (1979) Principles of protein structure. Springer Berlin Heidelberg New York
4 Saenger W (1984) Principles of Nucleic Acid Structure. Springer New York Berlin Heidelberg Tokyo
5 Martin RH (1982) Angew Chem *94:* 614
6 Prelog V Helmchen G (1982) Angew Chem *94:* 614
7 Snatzke F Snatzke G (1980) In: Kienitz H Bock R Fresenius W Huber W Tölg G (eds) Analytiker-Taschenbuch Bd 1, p 217. Springer Berlin Heidelberg New York. When a rule is described there then rather that reference will be cited than the original literature
8 cf Nishio M Hirota M (1989) Tetrahedron *45:* 7201
9 Klyne W Prelog V (1960) Experientia *16:* 521
10 cf. Block BP Powell WH Fernelius WC (1990) Inorganic Chemical Nomenclature, ACS Professional Reference Book, p 148/9. ACS Washington DC
11 Moffitt W Moscowitz A (1959) J Chem Phys *30:* 648
12 Harada N Nakanishi K (1983) Circular Dichroic Spectroscopy – Exciton Coupling in Organic Stereochemistry. University Science Books Mill Valley
13 Snatzke G (1978) In: Mason SF Optical Activity and Chiral Discrimination, pp 25 ff and 43 ff. D Reidel Publ Coy Dordrecht
14 Snatzke G (1979) Angew Chem *91:* 380
15 Snatzke G Eckhardt G (1970) Tetrahedron *26:* 1143
16 Snatzke G Schaffner K (1968) Tetrahedron *51:* 986
17 Snatzke G (1969) Riechst., Aromen, Körperpfl. *19:* 98
18 Wiesler WT Berova N Ojika M Meyers HV Chang M Zhou P Lo L-C Niwa M Takeda R Nakanishi K (1990) Helv Chim Acta *73:* 509
19 Frelek J Perkowska A Snatzke G Tima M Wagner U Wolff HP (1983) Spectroscopy Interntl J *2:* 274
20 Gerards M Snatzke G (1990) Tetrahedron: Asymmetry *1:* 221

5 Anomalous Dispersion of X-Rays and the Determination of the Handedness of Chiral Molecules

C. Kratky

5.1 Introduction

Most biological and many other molecules are *chiral*, i.e. they cannot be superimposed on their mirror image. The two mirror images of a chiral molecule are called *enantiomers*. Chiral molecules show the phenomenon of *optical activity*, i.e. a solution of one enantiomer rotates the plane of polarized light. A 1 : 1 mixture of both enantiomers is called a *racemate*; racemates are not optically active, because the optical activities of the two *antipodes* cancel each other, since the two enantiomers rotate the plane of polarized light by the same amount, but in opposite directions.

Until about 1950 there was no physical or chemical method available to determine the *absolute configuration* of a chiral molecule, i.e. to decide which of the two possible enantiomeric structures of an optically active molecule corresponds to the dextrorotatory and which to the levorotatory isomer. Configurations were assigned relative to a standard, glyceraldehyde, which was originally chosen by Emil Fischer in 1891 [1] for the purpose of correlating the configurations of carbohydrates but has also been related to many other classes of compounds, including amino acids, terpenes and steroids, as well as a variety of many other biochemically important substances.

Dextrorotatory glyceraldehyde was arbitrarily assigned the configuration shown in Fig. 5.1, and it was named D-(+)-glyceraldehyde. The levorotatory enantiomer was correspondingly termed L-(−)-glyceraldehyde. The sign in parentheses refers to the experimentally observable sense of rotation (which in itself does not allow an unambiguous assignment of the handedness of the molecule), while the capital letters D and L denote the absolute configuration (relative to the one of glyceraldehyde).

The type of correlations leading to the assignment of relative configurations for most known biologically active molecules is outlined in Fig. 5.2; these correlations were based on chemical transformations of known stereochemical consequences. An imperfect nomenclature system was devised for families of asymmetric compounds (like sugars and amino acids), which designated them as D or L according to the configurational similarity of one asymmetric carbon with D or L glyceraldehyde. Thus, as outlined in Fig. 5.2, it was shown that all the α-amino acids from proteins are L-amino acids.

There were several shortcomings to this approach:

– At the time the choice of absolute configuration for glyceraldehyde was

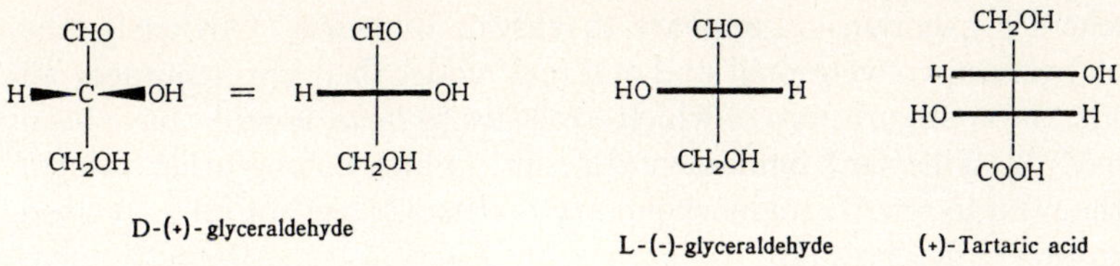

Fig. 5.1. Fischer projections of D- and L-glyceraldehyde and (+)-tartaric acid. In the Fischer convention it is understood that substituents at the right and left of a tetrahedral carbon atom are above the plane of the paper, the substituents above and below a tetrahedral carbon atom are under the plane of the paper

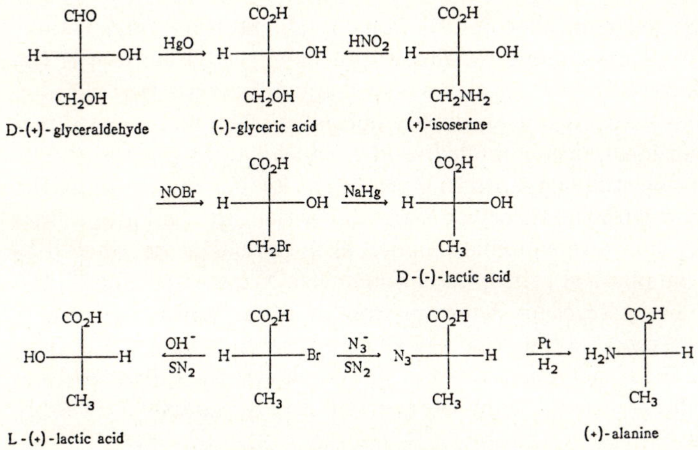

Fig. 5.2. The course of the chemical transformations showing that the configuration of natural (+)-alanine has been related to L-(−)-glyceraldehyde. All steps are stereospecific, those labeled Sn_2 invert the configuration at the asymmetrically substituted carbon atoms, the other steps proceed with configurational retention (adapted from Ref. [2]

made, there was no way of knowing whether the configuration of (+)-glyceraldehyde was indeed the one assumed (Fig. 5.1). The choice had a mere 50% chance of being correct.

– The approach is limited to molecules whose chirality is the result of asymmetric substitution of one or several carbon atoms. There are many chiral molecules which do not contain such an asymmetric carbon atom: molecules with asymmetrically substituted metal atoms, substituted biphenyls, plane-chiral compounds etc.

In 1951, J.M. Bijvoet and coworkers demonstrated [3], that a diffraction effect called *anomalous dispersion* can be used to determine the absolute configuration of molecules *without* recourse to a microscopic standard (such as (+)-glyceraldehyde). With this technique, he determined the absolute configuration of NaRb-(+)-tartrate (Fig. 5.1) and showed, that it was identical to the configuration deduced from chemical correlation with (+)-glyceraldehyde. In their communication to Nature the authors state "that

Emil Fischers convention ... appears to answer to reality. Obviously, the agreement between conventional and the real model then also embraces all compounds, the configurations of which – relative to tartaric acid - have been determined" [3]. This very fundamental result, (which among other things) avoided the need to rewrite many chemistry textbooks, has not fully received the recognition it deserves [4].

5.2 "Normal" X-Ray Diffraction

According to the Merriam–Webster Dictionary [5], the term "anomalous" implies a thing "not conforming to what might be expected because of the laws that govern its existence". In the case of X-ray diffraction, "the paradox is that 'anomalous scattering' is absolutely normal, while 'normal scattering' occurs only as an ideal, oversimplified model, which can be used as a first approximation when studying scattering problems" [6]. For the benefit of the reader not familiar with the theory of X-ray diffraction, we shall give a brief introduction into that 'oversimplified model' [8, 9, 11], before we proceed to a discussion of the physical principles of anomalous X-ray dispersion [7, 11]. On a heuristic level, structure determination by X-ray diffraction can be discussed with reference to the principles of an optical microscope (Fig. 5.3): the light from a light source is scattered by the object; the diffracted light impinges on the lense, which recombines the light rays to generate a (possibly enlarged) image.

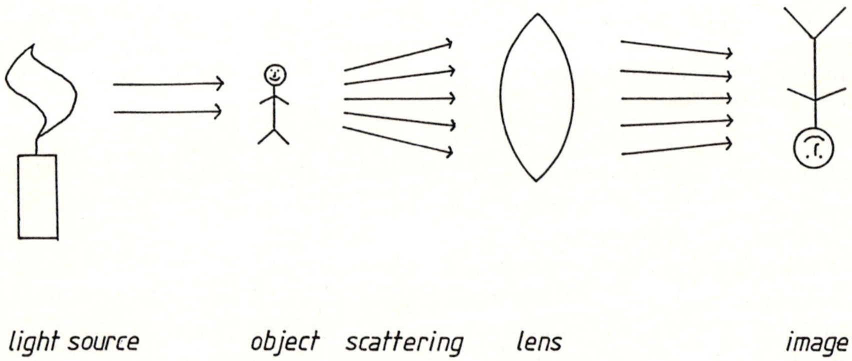

Fig. 5.3. Components of a conventional microscope

The resolving power of such an instrument is limited by the wavelength λ of the light, i.e. objects smaller than λ cannot be resolved. The wavelength of visible light ($\lambda \sim 10^{-6}$m ~ 10000 Å) far exceeds atomic or molecular dimensions. The use of light with $\lambda \sim 1$ Å$\sim 10^{-10}$m – which would be appropriate for atoms or molecules – is impeded by the unavailability of lenses for this kind of radiation (materials of sufficient refractive index show prohibitive absorbtion of X-rays). One therefore resorts to replacing the lense by an appropriate detector (e.g. a photographic film or a counter), i.e. one determines the *intensity* of the scattered radiation as a function of the direction and of the sample orientation. This information is called a *diffraction pattern.*

At this point it should be noted that the diffraction pattern does not contain all the information available to the lense to generate the image: naively speaking, knowledge of the intensity impinging at a particular point of the lense does not tell anything about the origin of the radiation, i.e. which fraction of the intensity was scattered by which part of the object. The latter information is called the *phase* of a scattered wave, and the problem of generating an image from the diffraction pattern is called the *phase problem.* In other words: recording the scattered intensity instead of directly recombining the scatterd light rays (by means of a lens) to form an image of the object constitutes a *loss of information.*

Figure 5.4 illustrates the physical reason for the scattering of X-rays and indicates the way to compute the scattered intensity, provided the three-dimensional arrangement of the atoms causing the scattering is known (which is of course the inverse of the usual situation, where – in the course of a structure determination – one starts with observing the scattered intensity and aims at elucidating the three-dimensional structure from these observations).

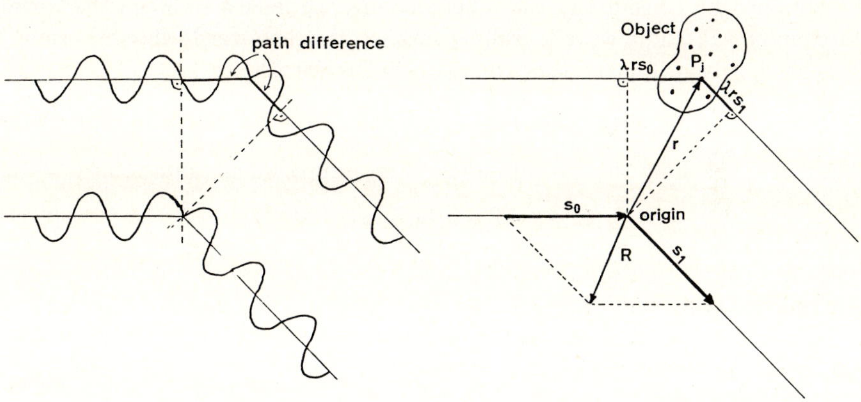

Fig. 5.4. *Left*: the superposition of the waves scattered by two atoms depends on the path difference. *Right*: Definition of vectors $\mathbf{r}_j$, $\mathbf{s}_0$, $\mathbf{s}_1$, and $\mathbf{R}$ and the computation of path differences

Let us assume an electromagnetic wave of wavelength λ travels parallel to the vector $\mathbf{s}_0$ (of length $1/\lambda$) and interacts with an object consisting of n scatterers (e.g. atoms) P_j at position vectors $\mathbf{r}_j$. As a result of this interaction, each of the scatterers will be excited to emit secondary radiation in every direction of space. The intensity scattered into the direction of, say, vector $\mathbf{s}_1$ (also of length $1/\lambda$) is the superposition of waves scattered by each of the scatterers.

At any moment, a harmonic wave can be represented by an amplitude f and a phase δ. It is convenient to represent such a wave by a complex number

$$f \exp(2\pi i\delta) = A + iB \tag{5.1}$$

which can be graphically represented by a two-dimensional vector (Argand diagram, Fig. 5.5). At any moment, the result of the superposition (interference) of waves of the same wavelength can be obtained by (vectorial) addition of the two complex numbers representing the two waves. How can we obtain f_j and δ_j for the waves scattered by P_j?

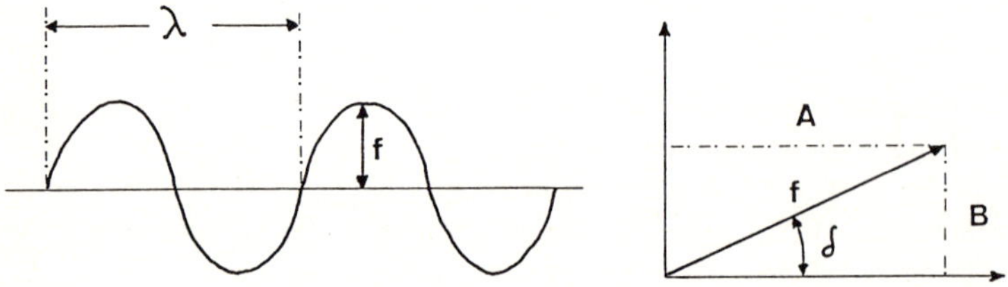

Fig. 5.5. Representation of an electromagnetic wave by an Argand diagram

f_j is proportional to the "scattering power" of the scatterer P_j[1], δ_j can be obtained by considering the difference in path length between the wave scattered by P_j and a wave travelling through the (arbitrarily chosen) origin. It is evident from Fig. 5.5 that this path difference Δx_j is

$$\Delta x_j = \lambda(\mathbf{s}_1\mathbf{r}_j - \mathbf{s}_0\mathbf{r}_j) = \lambda\mathbf{r}_j(\mathbf{s}_1 - \mathbf{s}_0) \ . \tag{5.2}$$

The phase is simply the path difference Δx_j expressed in units of λ:

$$\delta_j = \Delta x_j/\lambda = \mathbf{r}_j(\mathbf{s}_1 - \mathbf{s}_0) = \mathbf{r}_j\mathbf{R} \ . \tag{5.3}$$

The vector $\mathbf{R} = (\mathbf{s}_1 - \mathbf{s}_0)$, frequently called the *scattering vector*, points along the bisector of $\mathbf{s}_1$ and $-\mathbf{s}_0$.

Summing over all the complex waves scattered by each of the P_j yields the scattering of the whole object:

$$\mathbf{F}(\mathbf{R}) = \sum_{j=1}^{n} f_j \exp(2\pi i\mathbf{r}_j\mathbf{R}) \ . \tag{5.4}$$

$\mathbf{F}(\mathbf{R})$ is called the *structure factor*. The experimentally observable intensity is the square of the amplitude $\mathbf{F}(\mathbf{R})$ and hence

[1] If P_j are atoms, f_j is called *atomic scattering factor*. It is related to the number of electrons of atom j and to the angle between the incoming and the scattered wave

$$I(\mathbf{R}) = \mathbf{F}(\mathbf{R})\mathbf{F}^*(\mathbf{R}) \tag{5.5}$$

where $\mathbf{F}^*(\mathbf{R})$ denotes the conjugate complex of $\mathbf{F}(\mathbf{R})$. Since $\mathbf{F}(-\mathbf{R}) = \mathbf{F}^*(\mathbf{R})$, the diffraction pattern is always centrosymmetric, i.e.

$$I(\mathbf{R}) = I(-\mathbf{R}) \; . \tag{5.6}$$

This centrosymmetry of the diffraction pattern – irrespective of the symmetry of the object – is called *Friedel's Law* (Fig. 5.6).

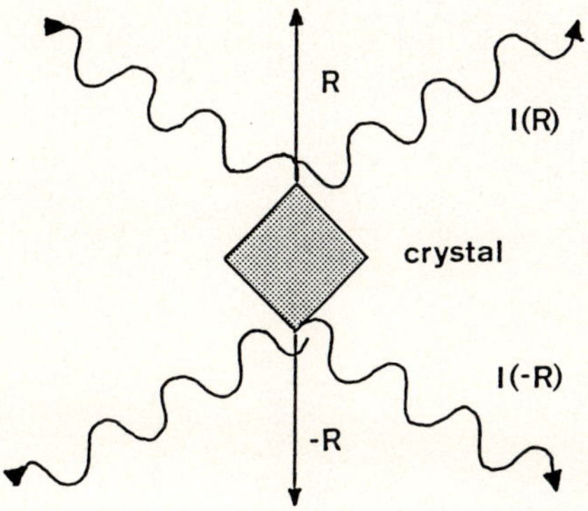

Fig. 5.6. Friedel's law $I(\mathbf{R}) = I(-\mathbf{R})$ and the physical setup to observe $I(\mathbf{R})$ and $I(-\mathbf{R})$

5.2.1 Scattering From a Crystal

Figure 5.7 shows a section[2] through a typical diffraction pattern $I(\mathbf{R})$ of an organic crystal. The diffracted intensity is concentrated at regularly arranged discrete points called reflections[3]. Each reflection can be assigned a triple of integers $[h, k, l]$ specifying its location $\mathbf{R}$ on the film such, that

$$\mathbf{R} = h\mathbf{a}^* + k\mathbf{b}^* + l\mathbf{c}^* \tag{5.7}$$

$\mathbf{a}^*$, $\mathbf{b}^*$ and $\mathbf{c}^*$ are called reciprocal lattice vectors; they are related to the lattice vectors of the crystal. Thus, Friedel's law for crystals reads

$$I(h, k, l) = I(-h, -k, -l) \; . \tag{5.8}$$

The validity of this law for the crystals whose diffraction pattern is shown in Fig. 5.6 is evident. For crystals, the calculation of the complex structure factor is somewhat simpler than for a generalized object:

$$\mathbf{F}(h, k, l) = \sum_{j=1}^{n} f_j \exp\{2\pi i(hX_j + kY_j + lZ_j)\} \tag{5.9}$$

[2] The way how such a diffraction pattern (called a precession photograph) is recorded can be found in any textbook of crystallography [2–4], but it is of no concern in the present context

[3] The occurrence of diffracted intensity only at discrete points is a consequence of the translation symmetry of crystals

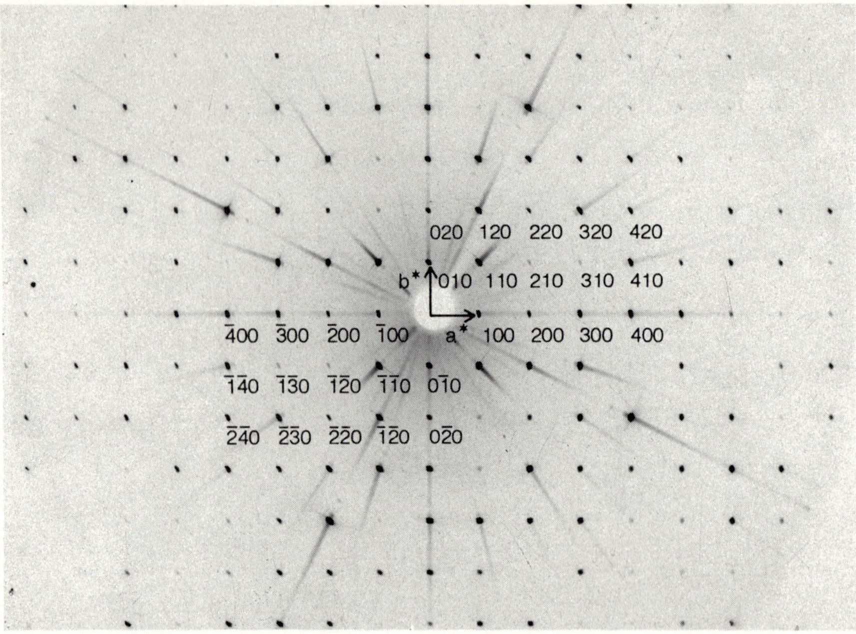

Fig. 5.7. Section $[h, k, 0]$ of the diffraction pattern of a typical organic crystal. The reciprocal lattice vectors $\mathbf{a}^*$ and $\mathbf{b}^*$ are indicated on the photograph, together with the indices for some reflections. The third reciprocal lattice vector, $\mathbf{c}^*$, runs out of the plane of the photograph

X_j, Y_j, and Z_j are the coordinates of atom j, and the summation only includes the atoms in one *unit cell*, i.e. atoms which are not related by translational symmetry. Between reflections, i.e. at locations $\mathbf{R}$ which are not integer multiples of the reciprocal lattice vectors, $\mathbf{F}$ becomes negibly small.

Crystal structure analysis consists of determining the atomic coordinates X_j, Y_j and Z_j of the atoms in the unit cell in such a way, that the calculated intensity values

$$I_{\text{calc}}(h,k,l) = \mathbf{F}(h,k,l)\mathbf{F}^*(h,k,l) \tag{5.10}$$

fit all the experimentally observed intensities $I_{\text{obs}}(h,k,l)$.

5.2.2 Friedel's Law and When It Breaks Down

Crystallographers like to divide their crystals into two classes: centrosymmetric and non-centrosymmetric crystals (Fig. 5.8). Centrosymmetric crystals contain a center of symmetry (usually at the origin of the unit cell), which means that for any atom at $\mathbf{r}$ there is an identical atom at $-\mathbf{r}$. As far as chiral objects are concerned, a center of symmetry has the same effect as a mirror plane: for any chiral object, the operation of the center of symmetry generates the opposite enantiomer. This means that only substances consisting of centrosymmetric molecules or racemates can form centrosymmetric crystals. Non-centrosymmetric crystals, on the other hand, may well accomodate optically active substances[4].

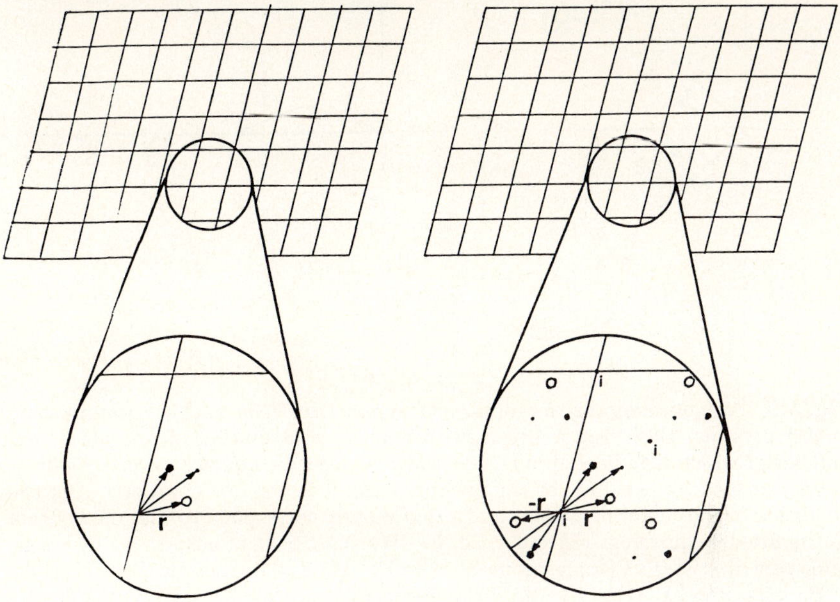

Fig. 5.8. Non-centrosymmetric (*left*) and centrosymmetric (*right*) crystals. Note that the locations of centers of symmetry are labeled i

Figure 5.9 shows Argand diagrams illustrating Friedel's law. For centrosymmetric crystals, the structure factor $\mathbf{F}(h,k,l)$ (Eq. 5.9) is always a real number, and hence $\mathbf{F}(h,k,l) = \mathbf{F}(-h,-k,-l)$.
The structure factor of non-centrosymmetric crystals is generally complex and $\mathbf{F}(h,k,l) = \mathbf{F}^*(-h,-k,-l)$. Since

$$I(h,k,l) = \mathbf{F}(h,k,l)\mathbf{F}^*(h,k,l) = \mathbf{F}(h,k,l)\mathbf{F}(-h,-k,-l)\,, \qquad (5.11)$$

Friedel's law should still be obeyed.

In 1930, Coster, Knol and Prins [10] showed in a classic experiment that there is a detectable difference in the reflection of opposite $[1,1,1]$ faces of Zincblende, i.e. that $I(1,1,1]$ differs from $I[-1,-1,-1]$[5].

In deriving Eqs. 5.4 and 5.9, we tacitly assumed that the scattering process *in itself* is not accompanied by any phase shifts, or that such phase shifts are the same for all scatterers. If this were not the case, i.e. if the individual

[4] There are non-centrosymmetric crystals which contain mirror planes or glide planes, and which therefore (as far as optically active substances are concerned) behave like centrosymmetric crystals

[5] Their experiment was slightly more involved, since they demonstrated that the ratio in the scattering from the two opposite faces changed when different diffraction orders (i.e. $[1,1,1]$, $[2,2,2]$ etc.) were observed at different wavelengths

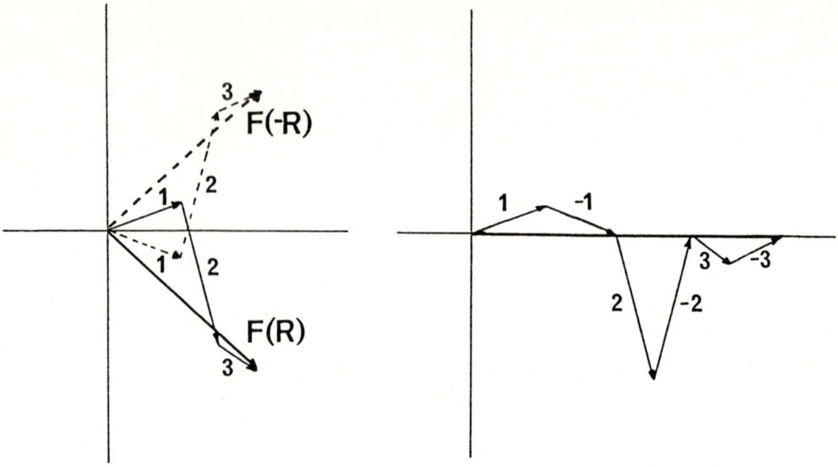

Fig. 5.9. Argand-diagrams for non-centrosymmetric (*left*) and centrosymmetric (*right*) crystals. Each vector originates from the contribution of one atom, the number of which is indicated in the drawing. A negative number j corresponds to atom j at location $-\mathbf{r}_j$, i.e. at the position related to the one of atom j after the operation of the center of symmetry. Dotted vectors correspond to the contribution to the Friedel-equivalent reflection, i.e. $\mathbf{F}(-\mathbf{R})$. Note that atoms at $-\mathbf{r}_j$ make the same contribution to $\mathbf{F}(\mathbf{R})$ as atoms at $\mathbf{r}_j$ to $\mathbf{F}(-\mathbf{R})$. For centrosymmetric crystals, $\mathbf{F}(\mathbf{R})$ is real and $\mathbf{F}(\mathbf{R}) = \mathbf{F}(-\mathbf{R})$, for non-centrosymmetric ones, $\mathbf{F}(\mathbf{R}) = \mathbf{F}^*(-\mathbf{R})$

phases δ were not completely determined by the path differences Δx_j alone, there would be no reason why Friedel's law should still hold.

Atoms which violate the above assumption are called *anomalous scatterers*. The intrinsic phase change of such atoms (relative to "normal" scatterers) can be taken into account by expressing their scattering factor as a complex number

$$f = |f| \exp(i\phi) = f' + if'' . \tag{5.12}$$

f' is real and positive (like the scattering factor of a "normal" atom). It can be shown (vide infra) that f'' is also positive, i.e. that the phase angle ϕ lies between 0 and $\pi/2$, which corresponds to an apparent retardation of the wave scattered by the anomalous scatterer relative to the waves scattered by normal atoms.

Figure 5.10 shows phase diagrams for a centrosymmetric and a non-centrosymmetric structure which contain one anomalous scatterer per unit cell: for the centrosymmetric structure, the structure factor $\mathbf{F}(h, k, l)$ becomes complex, but Friedel's law still holds; for non-centrosymmetric structures with a chiral arrangement of atoms in the unit cell, Friedel's law breaks down. The figure also demonstrates, that

$$|\mathbf{F}^+(h, k, l)| = |\mathbf{F}^-(-h, -k, -l)| \tag{5.13}$$

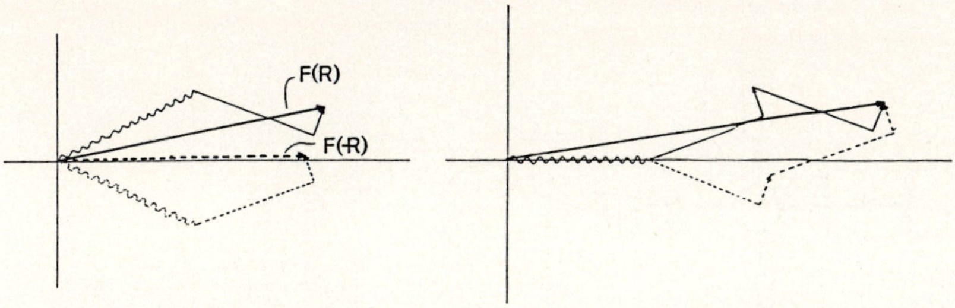

Fig. 5.10. Phase diagrams for structures with an anomalous scatterer. The *wavy line* represents the contribution of all the "normal" atoms, and it is assumed that there is one anomalous atom per unit cell for the non-centrosymmetric crystal *(left)* and two (centrosymmetrically related) anomalous scatterers for the centrosymmetric crystal *(right)*. In the centrosymmetric case, $\mathbf{F}(\mathbf{R}) = \mathbf{F}(-\mathbf{R})$ is still valid, although $\mathbf{F}(\mathbf{R})$ is now a complex number. In the non-centrosymmetric crystal, $|\mathbf{F}(\mathbf{R})|$ differs from $|\mathbf{F}(-\mathbf{R})|$, which amounts to a breakdown of Friedel's law. Note that the scattering of a crystal consisting of the opposite enantiomer is inverse, i.e. $\mathbf{F}^{+}(\mathbf{R}) = \mathbf{F}^{-}(-\mathbf{R})$, with $\mathbf{F}^{+}$ and $\mathbf{F}^{-}$ denoting the structure factors of the two enantiomers, respectively

i.e. that the intensity of reflection (h, k, l) for one enantiomer is the same as the intensity of the $(-h, -k, -l)$ reflection for the opposite enantiomer. Thus, if the (h, k, l) and the $(-h, -k, -l)$ reflections are correctly identified (i.e. if the reciprocal basis vectors $\mathbf{a}^*, \mathbf{b}^*$ and $\mathbf{c}^*$ form a right-handed coordinate system), it is possible to determine which of the two enantiomers is present.

5.2.3 Physical Origin of Anomalous Scattering

The scattering of X-rays by atoms is a quantum process, and a quantum-electrodynamic treatment of the scattering process is outside the scope of this article and beyond the competence of its author. However, *qualitatively*, the physical basis of the anomalous behaviour of certain atoms can be understood on the basis of a classical model.

In such a model, the incoming electromagnetic wave $E = E_0 \exp(i\omega t)$ interacts with electrons of mass m and charge e. The stimulated oscillation of the electrons will give rise to the secondary, scattered wave. Each electron is under the influence of a restoring force kx, exerted by the atomic nucleus and of a damping gx'. A mechanical analog of the situation is shown in Fig. 5.11.

The equation of motion for this electron is

$$mx'' + gx' + kx = eE_0 \exp(i\omega t) \tag{5.14}$$

Its steady-state solution, which can be found in any textbook on classical mechanics or in [11], is

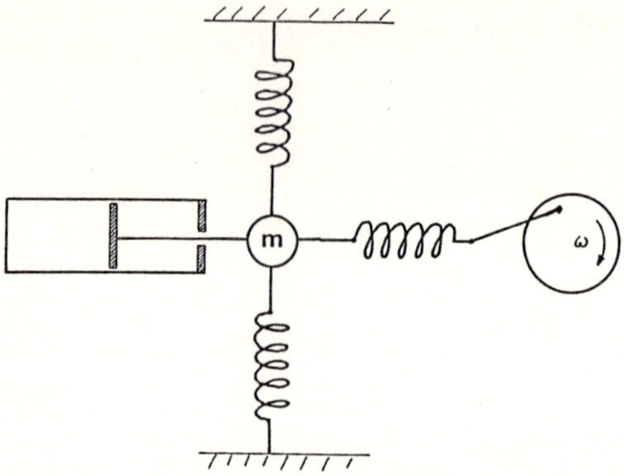

Fig. 5.11. Mechanical representation of the forces acting on a (classical) electron under the influence of an electromagnetic field

$$x = A \exp(i\omega t) \tag{5.15}$$

with

$$A = \frac{E_0 e}{k - m\omega^2 + ig\omega} \tag{5.16}$$

Usually, one substitutes $\omega_0 = (k/m)^{1/2}$, and calls ω_0 the *characteristic frequency* of the oscillator. Thus

$$A = \frac{E_0 e}{m(\omega_0^2 - \omega^2) + ig\omega} \tag{5.17}$$

Figure 5.12 shows the dependence of the modulus $|A|$ and the phase ϕ of $A = |A| \exp(i\phi)$ on ω. The curves have been drawn for arbitrary values of ω_0 and g. The width of the resonance peak of $|A|$ increases with increasing g, and so does the width of the transition range of the phase. $\phi = \pi$ for $\omega \gg \omega_0$, i.e. the scattering of "weakly" bonded electrons creates an intrinsic phase shift of π with respect to the incoming wave. In the other extreme, i.e. $\omega \ll \omega_0$, the intrinsic phase shift is zero. Since ω_0 is roughly equal to the ionization potential of the electron, it is typically much smaller than ω, with the exception of K and L electrons of heavy atoms. In other words: "normal" atoms contain only electrons with $\omega \gg \omega_0$, while "anomalous scatterers" contain some electrons for which this condition is no longer fulfilled.

The scattering factor f is usually defined as the amplitude relative to the one of the unbound electron, i.e. an electron with $k = 0$ and $g = 0$. Thus

$$f = \frac{m\omega^2}{m(\omega^2 - \omega_0^2) - ig\omega} \tag{5.18}$$

which can be rearranged to yield

$$f = m\omega^2 \left(\frac{m(\omega^2 - \omega_0^2) + ig\omega}{m^2(\omega^2 - \omega_0^2)^2 + g^2\omega^2} \right) = f' + if'' \tag{5.19}$$

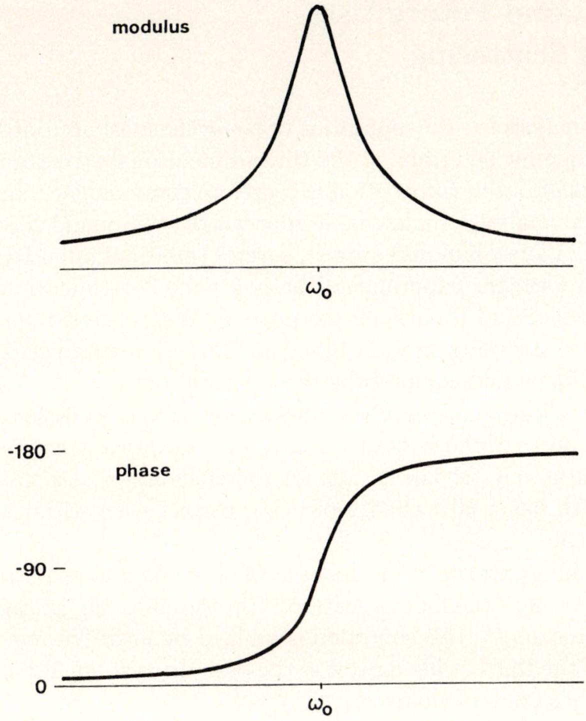

Fig. 5.12. Modulus $|A|$ and phase ϕ of the imaginary amplitude from Eq. 5.17, as a function of ω (calculated in arbitrary units with $E_0e = 1.0$, $m = 1.0$, $g = 1.0$ and $\omega_0 = 100.0$)

Since m, g and ω are positive, this demonstrates that *the coefficient f'' of the imaginary term is positive*, which was anticipated above.

The quantum-mechanical treatment yields qualitatively the same result, in particular the crucial positive sign of the imaginary term. Obviously, reversal of the sign reverses the result of the determination of the absolute structure. In the early 1970s, Tanaka claimed that the sign of f'' should indeed be reversed, and hence all absolute configurations determined until then should be changed into their antipodes. The claim was based on spectroscopic evidence as well as on theoretical considerations [11, 12]. Although both the spectroscopic and the theoretical arguments were subsequently disproved, the claim stimulated an independent experimental verification of the theoretical basis of the Bijvoet-technique [13]. Today, the technique to determine absolute configurations from anomalous diffraction data is beyond reasonable doubt, although there may be considerable experimental difficulties in its application (see below).

5.3 Past, Present, and Future Use of Anomalous Scattering

Single crystal structure analysis has the reputation of being the most accurate and most reliable technique for determining the three-dimensional structure of molecules. Its large-scale routine application is therefore commonplace. In a typical crystal-structure analysis of a low-molecular-weight compound (i.e. a compound with less than a few hundred atoms), several thousand intensity values are measured and form the experimental basis for the determination of several hundred positional and librational parameters. When carried out with sufficient care, crystal structure analysis based on diffraction data yields a vast amount of very reliable and comprehensive structural data.

The problem of determining the absolute configuration[6], which occasionally may crop up in the course of the structure analysis of an optically active compound, seems comparatively trivial: it calls for the elucidation of a single 1-bit parameter, which has a 50% chance of being correct even without performing any experiment.

A prerequisite for making this choice on the basis of crystallographic data is the existence of at least one "anomalous scatterer" in the unit cell, i.e. at least one atom with significant f''; this condition is fulfilled by most "heavy" elements, i.e. elements of the third or higher row of the periodic system. Table 5.1 lists f' and f'' for a selection of elements;

If a determination of absolute configuration is intended, the procedure is not much different from a "normal" structure analysis, with the exception that it is advisable to collect a more comprehensive set of intensity data. At the onset, the structure is determined "as usual", assuming the validity of Friedel's law and paying no particular attention to which of the possible two enantiomers one obtains. The last stage of structure analysis is the so-called refinement, which involves a "fine-tuning" of each structural parameter to produce an optimum fit between calculated and observed intensities. At this point, the observed Friedel differences are compared to what is calculated for the two enantiomers[7]. Table 5.2 shows this kind of analysis for the classical case of the NaRb-(+)-tartrate crystal structure [3].

Simple as it may seem, the technique of determining the absolute configuration of optically active compounds from anomalous diffraction data has not become as popular as conventional crystal-structure analysis: at the time of writing (spring 1990), the Cambridge Structural Data Base (CSD) [19], a computer-readable collection of all crystal-structures of organic or metal-organic compounds, contained 78641 entries; 2041 of them carried a "absolute configuration" flag, although about 25% of all the entries are structures in non-centrosymmetric space groups.

---

[6] "Absolute structure" is the more generally applicable though less widely used term [14]

[7] Several techniques are available for making this choice as sensitive and unbiased as possible [14, 16, 17, 18]

Table 5.1. Real and imaginary coefficient of the atomic scattering factor for a selection of elements (adapted from [15]). The data are for CuK_α and MoK_α-radiation and zero scattering angle. Note that f' decreases with increasing scattering angle, while f'' does not depend on the scattering angle. Therefore, the *relative* contribution of f'' to the total scattering increases with increasing scattering angle

	CuK_α		MoK_α	
Element	f'	f''	f'	f''
C	6.02	0.01	6.00	0.00
N	7.03	0.02	7.00	0.00
O	8.05	0.03	8.01	0.01
F	9.07	0.05	9.01	0.01
Ne	10.10	0.08	10.02	0.02
Na	11.13	0.12	11.03	0.03
Mg	12.16	0.18	12.04	0.04
Al	13.20	0.25	13.06	0.05
Si	14.24	0.33	14.07	0.07
P	15.28	0.43	15.09	0.10
S	16.32	0.56	16.11	0.12
Cl	17.35	0.70	17.13	0.16
Ar	18.37	0.87	18.16	0.20
K	19.37	1.07	19.18	0.25
Ca	20.34	1.29	20.20	0.31
Sc	21.29	1.53	21.34	0.37
Ti	22.19	1.81	22.25	0.45
V	23.04	2.11	23.27	0.53
Cr	23.80	2.44	24.28	0.62
Mn	24.43	2.81	25.30	0.73
Fe	24.82	3.20	26.30	0.85
Co	24.54	3.61	27.30	0.97
Ni	25.04	0.51	28.29	1.11
Cu	26.98	0.59	29.26	1.27
Zn	28.39	0.68	30.22	1.43
Ga	29.65	0.78	31.16	1.61
Ge	30.84	0.89	32.08	1.80
As	31.99	1.01	32.97	2.01
Se	33.12	1.14	33.82	2.22
Br	34.23	1.28	34.63	2.46

In addition to not being very popular, much criticism has been expressed about the way how most determinations of absolute structures have been carried out [14]. While, under favourable circumstances[8], differences in the crystallographic R-value[9] between the two enantiomers of 0.5% or more are not uncommon (in which case the determination of absolute structure is

[8] e.g. a Bromine atom as anomalous scatterer of MoK_α radiation

[9] The R-value is a conventional (though rather unfortunate) measure of the agreement between observed and calculated structure factors. For low-molecular-weight compounds, accurate structure determinations usually have $R < 0.05$, while structure analyses with $R > 0.1$ are often considered poor

Table 5.2. Comparison of selected Friedel-equivalent reflections for NaRb-(+)-tartrate (from [3]). The quotient $I(h,k,l)/I(-h,-k,-l)$ was calculated for the conventional tartrate molecule. "$>$" means that $I(h,k,l)$ was observed to be more intense than $I(-h,-k,-l)$

h	k	l	calculated $I(h,k,l)/I(-h,-k,-l)$	observed
1	6	1	1.30	$>$
1	7	1	0.83	$<$
1	8	1	1.25	$>$
1	9	1	1.41	$>$
1	11	1	0.66	$<$
2	7	1	3.00	$>$
2	10	1	0.84	$<$
2	13	1	0.51	$<$

commonplace), the choice of the correct enantiomeric structure can be experimentally very demanding if no strong anomalous scatterers are present [17].

Very few crystallographically assigned absolute configurations have had to be revised following reinvestigations[10], which is in fact surprising if one scrutinizes the relevant literature [14]: for 1988 (which at the time of writing, is the last year completely included), the CSD lists 6127 entries, 117 of which were labeled with the "absolute configuration" flag. 20 of these entries had fluorine or oxygen as the heaviest atom, among them 9 structures with $R > 0.05$ and two with R not specified[11]. Since detection of anomalous effects for structures consisting of only first-row elements is only feasible if extremely accurate data have been recorded [17], the experimental basis for the absolute configuration determination of some of the less accurate structures appears poor. The scarcity of absolute configuration revisions may thus be the result of few reinvestigations, or of the fact that the absolute configuration was known from chemical or spectroscopic data prior to the crystallographic investigation (which is the case for the majority of biologically occurring compounds).

There is yet another reason for the apparent paucity of absolute-configuration determinations: crystal structure analysis offers a simpler and more reliable path to the determination of the absolute configuration than the analysis of anomalous differences: if one succeeds in derivatizing the compound of unknown chirality with an optically active compound of known chirality, a "normal" structure determination of the complex will suffice to determine the

[10] Only one such rare example – mitomycin [20] – was quoted in a review by Jones [14]

[11] Refcodes DINMUI10 ($R = 0.066$), CODIUC10 (R not given), GASZIJ (0.059), GEBPUY (0.052), GEMZIH (0.066), GEZRAE (0.074), GEZREI (0.073), SABJOU (not given), SAFDOS (0.072), VAGCEL (0.054), VAGCIP (0.061)

absolute configuration of the unknown part relative to the part with known absolute configuration. Since such a structure analysis does not rely on small intensity differences between Friedel-equivalent reflections, this technique is more reliable and, in addition, it does away with the requirement for a heavy atom in the unit cell (which usually has an adverse effect on the accuracy of the structure analysis).

5.3.1 Outlook

The phenomenon of anomalous dispersion of X-rays has provided us with at least the one piece of information: we know the handedness of biological systems. Important as this information may be (it is one of the epistemological foundations of the present book), it has a somewhat philosophical touch and one can argue that it is of little utilitarian significance.

The effect of anomalous dispersion of X-rays has recently found an application in a different area of crystallography: it promises to revolutionize the experimental phase determination of X-ray reflections of macromolecular crystals.

For structures of this kind, which contain molecules with molecular weights of 10^4 to 10^6, the solution of the phase problem usually requires the application of a very tedious technique, called multiple isomorphous replacement (MIR): crystals of the macromolecule have to be soaked in solutions of heavy atoms in the hope to find conditions where heavy atoms diffuse into the crystal and specifically bind to the macromolecule without otherwise changing its crystal structure. Such a complex is called an isomorphous derivative, and searching for isomorphous derivatives may necessitate the screening of hundreds of different conditions.

At least two different isomorphous derivatives have to be at hand for an a-priori solution of the phase problem. Phases are obtained by comparing the observed structure factors of the derivatives with the corresponding quantities of the native structure; the differences between the two sets of data originate from the additional heavy atom, whose contribution to the protein structure factor yields an indication about the macromolecular phases. The details of this procedure can be found in any textbook on protein crystallography, e.g. [21].

It was noted above (Fig. 5.12) that the anomalous dispersion effect depends on the wavelength. If the native macromolecule already contains a heavy atom (e.g. a metal atom), the anomalous dispersion of this atom can be switched on and off, by a simple change in wavelength, yielding the same kind of phase information as an isomorphous derivative. In favourable circumstances, protein structures can be determined without any isomorphous derivatives; in most cases the use of anomalous diffraction data reduces the required number of isomorphous derivatives from two to one [22].

Anomalous dispersion effects are generally small, compared to the total scattering of a macromolecule even very small. The experimental requirements (tuneability, high intensity) for a successful application of anomalous dispersion techniques in macromolecular crystallography are therefore formidable. Synchrotron radiation sources, with intensities several orders of magnitude higher than conventional X-ray sources and with a continuous spectrum of wavelengths, are ideally suited for this kind of application, and they promise to make possible a renaissance in the use of anomalous dispersion.

5.4 References

1 Helferich B (1953) Angew Chem *65:* 45; Wichelhaus H Knorr L Duisberg C (1919) Ber dt Chem Ges *52A:* 129

2 Roberts JD Caserio MC (1965) Basic principles of organic chemistry. WA Benjamin New York

3 Bijvoet JM Peerdeman AF van Bommel AJ (1951) Nature (London) *168*: 271; Peerdeman AF van Bommel AJ Bijvoet JM (1951) Proc Roy Acad Amsterdam *B54:* 16

4 see footnote 21 on p. 130 in Dunitz JD (1979) X-Ray Analysis and the Structure of organic molecules. Cornell University Press

5 Merriam G & C (1972) The Merriam-Webster pocket dictionary of Synonyms

6 Caticha-Ellis S (1981) Anomalous Dispersion of X-Rays in Crystallography, published for the International Union of Crystallography by University College Cardiff Press Cardiff Wales

7 Ramaseshan S Abrahams SC (eds) (1975) Anomalous Scattering. Munksgaard Copenhagen

8 An introduction into crystal structure analysis can be found in any textbook on the subject, e.g. Luger P (1980) Modern X-Ray Analysis on Single Crystals. de Gruyter Berlin; Glusker JP Trueblood KN (1985) Crystal structure Analysis, 2nd ed. Oxford University Press New York; refs [9] and [11]; see also: Taylor CA (1980) A Non-Mathematical Introduction to X-ray Crystallography, published for the International Union of Crystallography by University College Cardiff Press Cardiff Wales

9 Stout GH Jensen LH (1989) X-ray Structure Determination, A Practical Guide, 2nd ed. John Wiley New York

10 Coster D Knol KS Prins JA (1930) Z Phys. *63:* 345

11 Dunitz JD (1979) X-Ray Analysis and the Structure of organic molecules. Cornell University Press

12 Tanaka J (1972) Acta Crystallogr *A28:* 229; Tanaka J Katayama C Ogura F Tatemitsu H Nakagawa M Chem Commun *1973:* 21; Tanaka J Ozeki-Minakata K Ogura F Nakagawa M (1973) Nature, Phys *241:* 22

13 Brongersma HH Mul PM (1973) Chem Phys Lett *19:* 217

14 Jones PG (1984) Acta Cryst *B30*: 660; Jones PG (1984) Acta Cryst *B40:* 662; Jones PG (1985) In: Sheldrick GM Krüger C Goddard R (eds) Crystallographic Computing.p 260. Clarendon Press Oxford

15 Ibers JA Hamilton WC (eds) (1974) International Tables for X-Ray Crystallography, Vol. IV. The Kynoch Press Birmingham England

16 Rodgers D (1981) Acta Cryst *A37:* 734

17 Rabinovich D Hope H (1980) Acta Cryst *A36:* 670

18 Flack HD (1983) Acta Crystl *A39:* 876

19 Allen FH Kennard O Taylor R (1983) Acc Chem Res *16:* 146

20 Tulinsky A van der Hende JH (1967) J Amer Chem Soc *89:* 2905; Shirahata K Hirayama N (1983) J Amer Chem Soc *105:* 7199

21 Blundell TL Johnson LN (1976) Protein Crystallography. Academic London

22 Hendrickson WA (1985) In: Sheldrick GM Krüger C Goddard R (eds) Crystallographic Computing, p 277. Clarendon Press Oxford

6 Chirality in Organic Synthesis – The Use of Biocatalysts

K. Faber and H. Griengl

6.1 Chirality in Organic Chemistry and Biochemistry

6.1.1 Explanation of Basic Terms

In this chapter not only the use of biocatalysts in reactions of chiral substrates will be overviewed but there will also be given some general information on the consequences of the phenomenon of chirality for chemistry, as this book is intended to have also non-chemists as potential readers.

The prerequisite for a compound to be termed chiral is that this molecule can exist in two different forms having exactly the same type of bonds, bond lengths, and bond angles with the only difference that one form of this molecule is the mirror image of the other, an observation that can also be made when looking at a pair of shoes or gloves (for a more detailed treatment of this topic see also the chapter by G. Derflinger). These two forms are called *enantiomers*, to be distinguished from *diastereomers* which are all the other stereoisomers, where again type and sequence of binding is the same but the arrangement of the atoms in space is different. It is important to recognize that diasteromers may be chiral or achiral.

The necessary and sufficient condition for chirality is a lack of reflectional symmetry which is discussed in more detail in the chapter by G. Derflinger in this book. This symmetry criterium can be fulfilled by reduction of the symmetry to a center, an axis or a plane in the molecule, either real or formal, leading to a *central*, *axial* or *planar* chirality, respectively. In addition to these and other special types of chirality – which are described in textbooks for stereochemistry [1, 2, 3, 4, 5] – *helical chirality* is important.

The enantiomers of a chiral compound have special arrangement of the atoms in space which is called *configuration*. In assigning a configuration rotations about bonds are normally not considered. The *absolute configuration* of organic compounds was determined for the first time by J.M. Bijvoet in 1949, using anomalous X-ray diffraction (see the chapter by C. Kratky). As configuration nomenclature for carbohydrates and aminoacids and in some special cases the D,L-System established by E. Fischer and M.A. Rosanoff is used [6, 7]. For more general application the R, S-System of Cahn–Ingold–Prelog was established, the principle for central chirality of which is outlined below [8].

Helical chirality is characterized by P and M for plus and minus sign of the helix. *Racemic forms* are composed of equal numbers of both enantiomers.

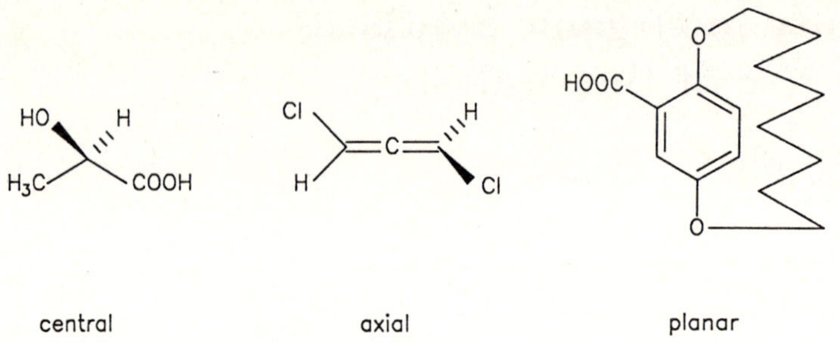

Scheme 6.1. The most important types of chirality

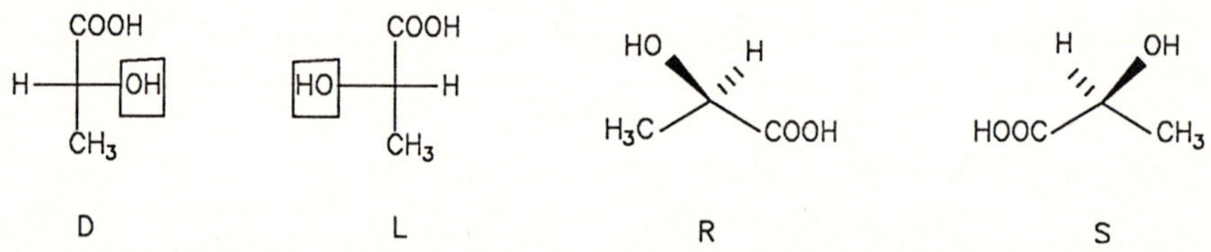

Scheme 6.2. Configurational nomenclature: D,L with reference to the side of the principal chain where the substituent is located (dexter or laevus); *R*, *S* with respect to clockwise (rectus) or counterclockwise (sinister) arrangement of substituents after application of priority rules

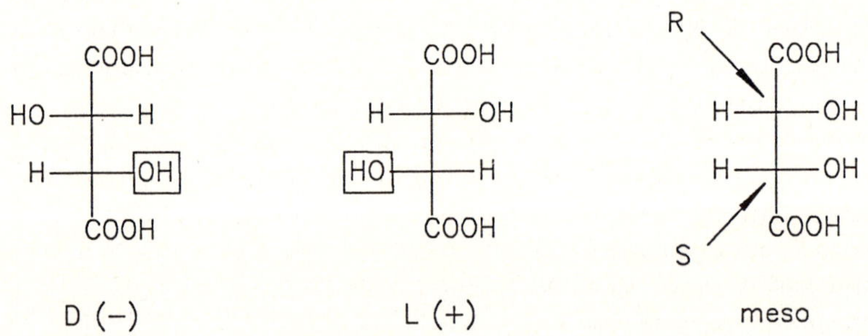

Scheme 6.3. Stereoisomeric forms of tartaric acid (framed: reference ligand)

Compounds whose individual molecules contain equal numbers of enantiomeric groups of opposite chirality, identically linked, but no other chiral groups, are termed *meso-compounds*. These molecules, as a whole, are achiral.

6.1.2 Comparison of Properties: Enantiomers and Diastereomers

The diastereomers of a given structure differ with respect to all physical and chemical properties. For the identification and separation of enantiomers (from racemates) it is of extreme importance to be aware of a basic principle of stereochemistry, that the interaction between two elements of chirality

Table 6.1. Diastereomeric recognition between two elements of chirality

Elements	Effect
glove + hand	fit or nonfit
screw + nut	fit or nonfit
chiral compound + polarized light	sign of optical rotation
chiral center + CIP-procedure	assignment of *R* or *S*

always results in diastereomeric recognition. Only by this correlation it is possible to assign the configuration at all. Some examples are given in Table 1.

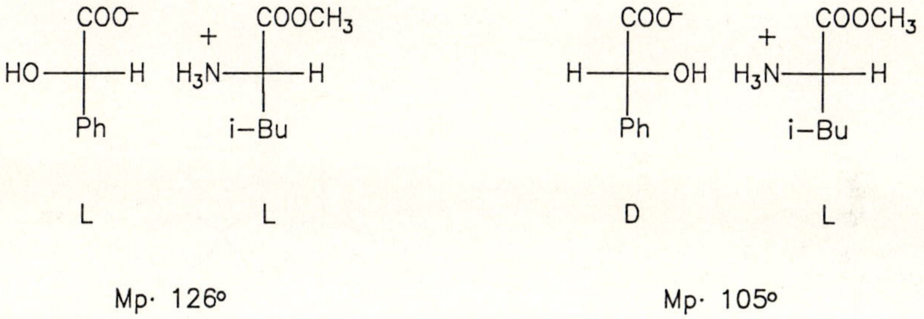

Scheme 6.4. Different properties of salts of L- and D-mandelic acid with L-leucin methyl ester [9].

With respect to chemical reactions enantiomers do not show any difference when interacting with achiral reactants, whereas with chiral reactants both reaction rate and product are different.

6.1.3 The Importance of Enantiomeric Purity

In nature the tremendous complexity and versatility of biochemical reactions occurring in plants, microorganisms and higher animals is governed by biocatalysts – the enzymes. These catalysts are highly selective with respect to substrate and reaction course. Mainly due to this fact, is life possible at all. Since all enzymes are chiral, enantiomers show different reactivity and reaction course. As a consequence, enantiomers are, with respect to biological systems, distinct species. Some examples are given in figure below.

As a consequence no racemates should be applied as drugs or agrochemicals, since in general only one enantiomer has the desired properties, whereas the other is responsible for side-effects or at least is an unnecessary burden for metabolism [10]. In some rare cases enantiomers can be interconverted within the biological system after administration. Then racemates can still be used. Although at present most active components are applied as racemates

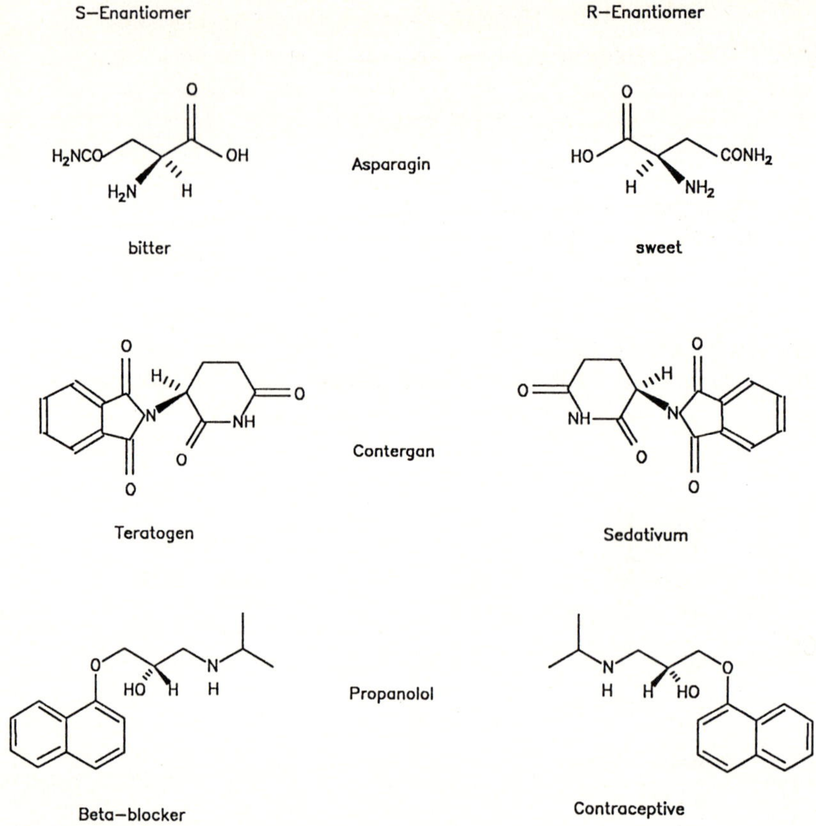

Scheme 6.5. Biochemical effects of enantiomers

this situation will change [11]. Therefore, an urgent and tremendous need for methods for obtaining chiral compounds enantiomerically pure exists.

6.1.4 Methods of Obtaining Enantiomerically Pure Chiral Compounds

In principle two approaches are possible: First, the racemate can be resolved using the principle of diastereomeric recognition outlined above. This can be performed by the classical method applying an enantiomerically pure chiral reagent as resolving agent [12], by crystallization in case of a racemate possessing a melting diagram of a racemic mixture [13], by chromatography using chiral colums (see the chapter by W. Lindner) or by enzymatic or microbial resolution (see Sect. 6.5.2).

As a second possibility, a prochiral compound can be transformed into a chiral compound more ore less enantioselectively, which means that one enantiomer is formed preferentially (see the chapter by E. Winterfeldt). A center of prochirality is characterized by a tetrahedral atom bearing two different and two identical (enantiotopic) ligands. Of these identical ligands, for prochirality nomenclature, that one which leads to an (R)-compound when considered to be preferred to the other by the sequence rule (without change in priority with respect to other ligands) is termed pro-R [14] or *Re* [15], and the other is termed pro-S or *Si*. A prochiral plane is characterized by a trigonal center bearing three different ligands where both faces of this plane are like mirror images. The side having the priority order of ligands in a clockwise fashion is termed *Re*, the other *Si*. Preferred reaction of one of the enantiotopic ligands or (for the second case) one enantiotopic face with a chiral reagent leads to enantioselection with respect to the chiral product.

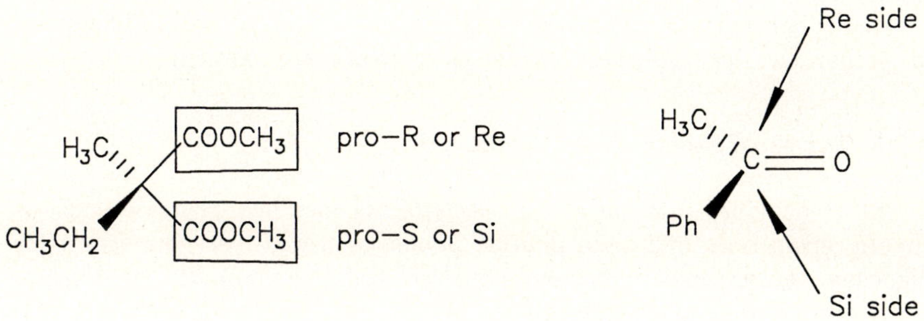

Scheme 6.6. Prochiral compounds, Re/Si-nomenclature

The same holds for *meso*-compounds where by chemical reaction the mirror symmetry within the molecule will be destroyed. As chiral reagents either any suited chiral compound or biocatalysts can be used.

6.2 Biocatalysts in Organic Chemistry General Remarks

6.2.1 Enzymes

Enzymes are proteins where only a small region of the whole molecule, the *active center*, is actually involved in the transformation. Enzymes are only catalysts and not reactants. Except for hydrolytic reactions, where the reagent is provided by the solvent water, for other transformations *coenzymes* are needed which are linked to the active site. Representative examples are given in Sect. 6.3.3.

In contrast to enzymes where catalytic amounts are sufficient and are not consumed during the reactions, coenzymes have to be applied in stoichiometric quantities. In living cells coenzymes are regenerated by biochemical metabolism being operative in the cell. When using isolated enzymes for chemical reactions the coenzyme has to be added in equimolar amounts. Taking into consideration the high price of most coenzymes this is only feasible for small scale experiments. For larger batches the coenzyme consumed has to be recycled (see Sect. 6.6.2).

6.2.2 Whole Cell Systems

The need for coenzyme recycling is avoided when whole cell systems are used for performing the biocatalytic transformation. Except for very few microorganisms which are easy to handle such as baker's yeast (*Saccharomyces cerevisiae*) [16], some experience in microbiology and special equipment is necessary here. There is a need for working under sterile conditions and some additional safety requirements have to be met. As a rule of thumb the substance concentration is in the range of 0.1–1.0 g/L of broth. Therefore, for preparative scale syntheses rather large vessels are required.

6.2.3 Types of Selectivities Achieved

In nature enzymes show *substrate selection.* For instance, proteases such as chymotrypsin only hydrolyse peptide bonds, a process very important for digestion. Interestingly, it is possible to use this enzyme in vitro to catalyse many other hydrolytic reactions, too [17].

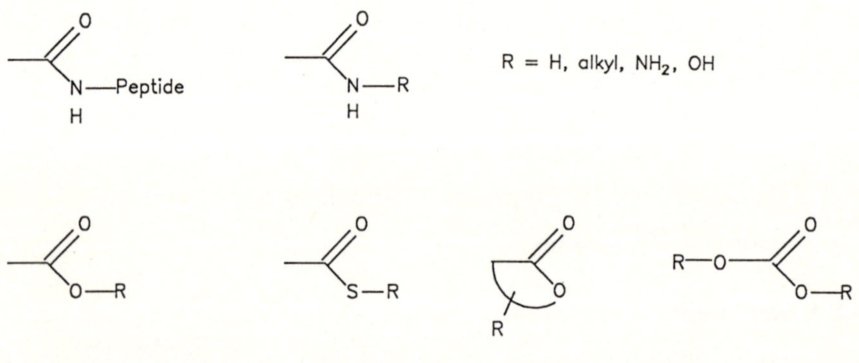

Scheme 6.7. Examples for functional groups susceptable to α-Chymotrypsin catalyzed hydrolysis [17]

Very often organic compounds carry two or more functional groups which all can a priori react under the conditions applied.

While by acid hydrolysis of L-*N*-acetylphenylalanine ethyl ester both the amide and the ester bonds are split, enzymatic reaction brings about *chemoselection*: [18, 19, 20]. Using α-chymotrypsin only the carboxylic ester bond is hydrolyzed. Another enzyme, hog kidney acylase, can catalyse the cleavage of the *N*-acyl group. In addition *enantioselection* is observed in

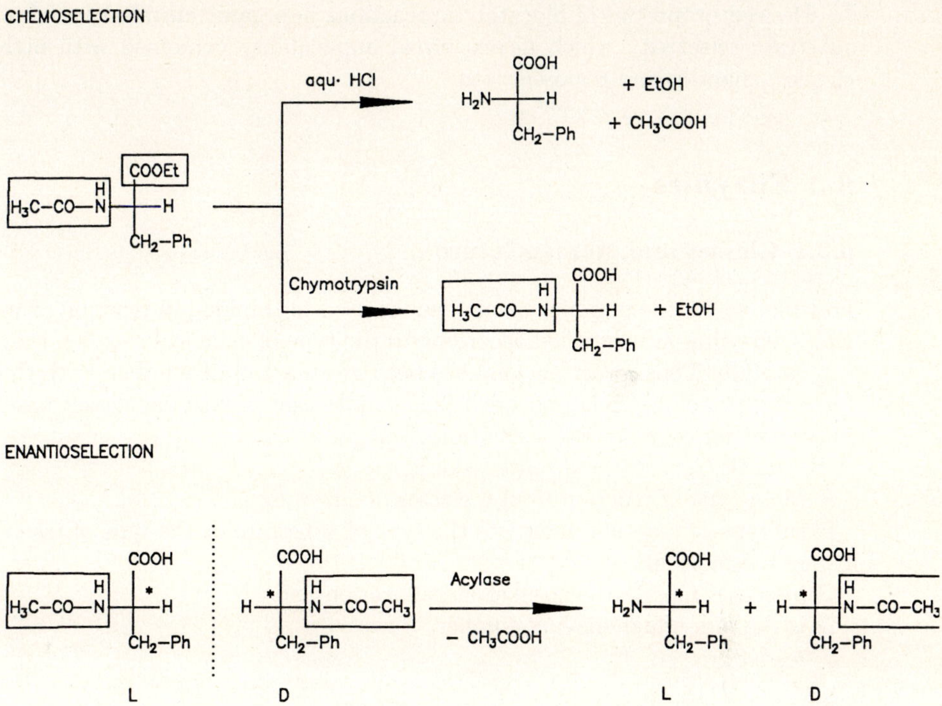

Scheme 6.8. Chemoselection and enantioselection as exemplified with derivatives of phenylalanine

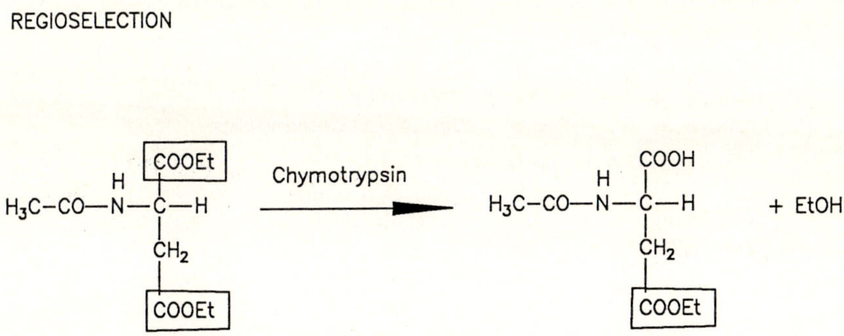

Scheme 6.9. Regioselective hydrolysis of diethyl *N*-acetylaspartate catalyzed by α-Chymotrypsin

both cases since only the L-form reacts. If two identical groups with different surrounding are present in the molecule, *regioselection* [21, 22, 23, 24, 25] can be achieved by enzymatic catalysis that only one group reacts preferentially, as exemplified for the α-chymotrypsin catalyzed hydrolysis of diethyl L-*N*-acetyl aspartate.

Desired properties of biocatalytic reactions in organic chemistry are low substrate selectivity which means broad applicability combined with high chemo-, enantio- and regioselection.

6.3 Enzymes

6.3.1 Classes and Nomenclature

Fortunately for the organic chemist who is used to thinking in reaction principles, enzymes are classified according to the type of chemical reaction they can catalyse: Thus, every enzyme has been given a 4-digit number, with the following principles being encoded [26]. A selection of enzyme classes most important for organic transformations is given below:
A.B.C.D (E.C. = enzyme commission)

- A Main type of reaction from 6 classes of enzymes,
- B subtype of reaction, indicates the type of substrate or the type of transferred molecule,
- C indicates mostly the cosubstrate allocation and
- D is the individual enzyme number.

Table 6.2. International classification of enzymes

1.	Oxidoreductases (redox reactions)		
1.1	acting on >CH-OH	1.4	acting on >CH-NH_2
1.2	acting on >C=O	1.5	acting on >CH-NH-
1.3	acting on >CH=CH<	1.6	acting on NADH, NADPH.
2.	Transferases (functional group transfer)		
2.1	C_1-units	2.4	Glycosyl units
2.2	Aldehydes or ketones	2.7	Phosphates
2.3	Acyl groups	2.8	Sulfur containing groups
3.	Hydrolases (Hydrolytic reactions)		
3.1	Ester bonds	3.5	other C-N bonds
3.2	Glycosidic bonds	3.6	Acid anhydrides
3.4	Peptide bonds		
4.	Lyases (Addition to double bonds)		
4.1	on >C=C<	4.3	on >C=N-
4.2	on >C=O		
5.	Isomerases		
5.1	Racemases		
6.	Ligases (σ-Bond formation)		
6.1	C-O	6.3	C-N
6.2	C-S	6.4	C-C

6.3.2 Properties and Stabilities

Enzymes are very efficient catalysts. Generally, chemical catalysts are employed in amounts of 0.1–1 mol%. In most enzymatic reactions about 10% of weight of enzyme versus substrate is used, which looks a lot at the beginning without further consideration: The enzymes used are with some exceptions only crude preparations containing about 1% of pure enzyme, which leads to an actual molar concentration of 10^{-4} of biocatalyst, assuming a general molecular weight of 100.000 for the enzyme and 100 for the substrate. Thus, the efficiency can be estimated as being up to 1000 times higher than chemical catalysis.

Enzymes can be generally cheap catalysts. Since most of enzymatic conversions can be performed with crude enzyme preparations, containing only a small fraction of pure protein, cheap animal or plant sources of biocatalysts can be applied. Due to this reason particularly hydrolytic enzymes are most widely used in organic chemistry [27].

Enzymes react under very mild conditions. The pH-range of enzyme catalyzed reactions is in the range of 5–8, the corresponding temperature is generally 20–40°C. Thus it is obvious, that under such mild conditions other functional groups present in the substrate can survive easily and fewer side-reactions, often hampering classic chemical transformations, are observed.

Enzymes can catalyse almost all reactions (see Sect. 6.3.1). With only a few exceptions, almost all types of chemical reactions known in organic chemistry can be catalyzed by enzymes, such as:

- *Hydrolysis* and *synthesis* of esters [28], lactones [29], amides [30], acid anhydrides [31], and nitriles [32],
- *oxidation* and *reduction* of alkanes [33], alkenes [34], aromates [33, 35], alcohols [36, 37] and ketones [38],
- *addition* and *elimination* of water [39], ammonia [40] and HCN to double bonds [41] and
- *alkylations* [42], *isomerisations* [43], *acyloin* [16] and *aldol reactions* [44]. Even *Michael-additions* are known [45].

6.3.3 Coenzymes

Coenzymes are compounds of relatively low molecular weight (compared to that of the enzyme) which are necessary for numerous types of reactions. They generally provide either redox equivalents (hydrogen, oxygen or electrons) or simply chemical energy stored as energy-rich functional groups, such as acid anhydrides, etc. [46, 47].

The most important are:

- *Nicotinamide adenine dinucleotide (phosphate)*, abbreviated as $NAD(P)^+$, an important hydrogen carrier for the redox reaction of polar or polarized C=X double bonds [48].

- *Flavines* are the corresponding counterpart for redox reactions of unpolar or unpolarized C=C double bonds.
- *Pyridoxal phosphate* is needed for transamination.
- *Adenosine triphospate* [ATP] serves as a general chemical energy storage, being available for the synthesis of energy requiring compounds [49].
- *Cobalamin* (vitamin B_{12}) can provide electrons and is necessary for biohydroxylation.
- *Thiamine pyrophosphate* represents a d_1-synthon and is therefore an example of a biological umpoled reagent.

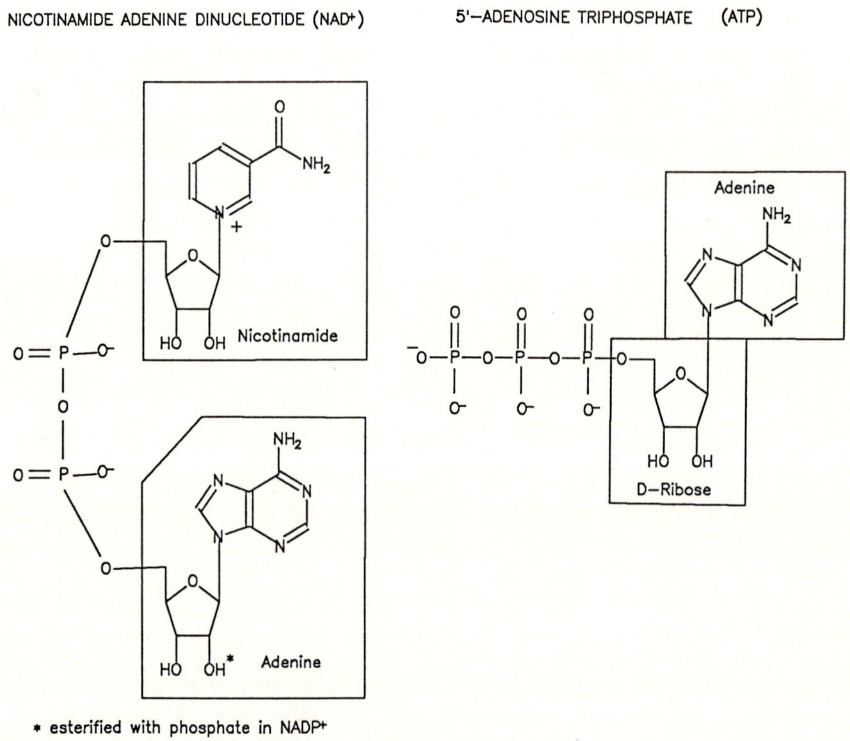

Scheme 6.10. Some important coenzymes

6.3.4 Enzyme Mechanisms

Unlike the majority of chemical catalysts in which only a single functional group is required, more groups (and sometimes also coordinated metal ions) have to work together at the active site of an enzyme to effect catalysis [50]. Although individual enzyme mechanisms have been elucidated in some cases, where the exact three-dimensional structure and the identity of the functional groups involved in catalysis are known, for most of the other enzymes assumptions are made about their molecular action. An illustrating example for the former case is the mechanism of serine hydrolases, such as trypsin or the lipase from *Mucor miehei* [51]

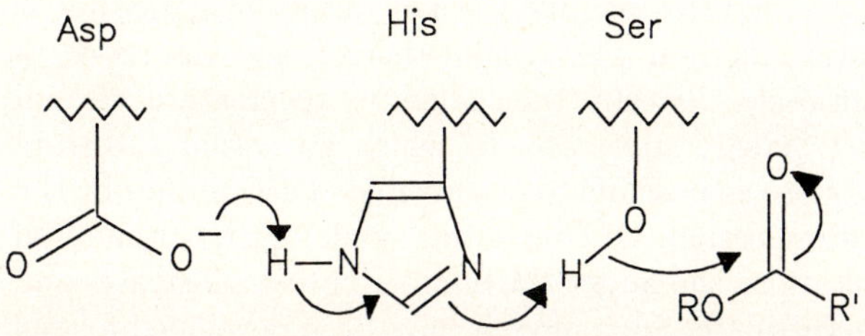

Substrate Ester

Scheme 6.11. The catalytic triade of serine hydrolases

Two additional groups (Asp and His) present in the active site, effect a decrease of the pK-value of the serine-OH (which is the actual reacting chemical operator) to enable it to perform a nucleophilic attack on the carbonyl group of the substrate. All three of the reacting groups working together are called the *catalytic triad* [51].

6.3.5 Active Site and Enzyme Models

Due to the ever increasing number of applications of esterases and lipases on non-natural organic substrates, a couple of "models" for individual enzymes have been developed in order to provide expectations of results when non-natural substrates are involved and to allow a chemist to redesign a substrate if the initial results were unsufficient. The most important principles underlaying these model conceptions are discussed here:

X-Ray Structure. A correct 3-dimensional "map" of the active site can be elucidated by X-ray crystallography of crystalline enzymes, or (in some cases) even of crystalline enzyme-substrate complexes [52]. Unfortunately, this can only be done with pure crystalline enzymes, which are clearly in the minority of those used for organic synthesis (e.g. α-chymotrypsin [53], and subtilisin [54]).

Molecular Modelling. If only the amino acid sequence of an enzyme is known either wholly or even in part, computer assisted calculations called *molecular modelling* can provide three-dimensional structures of enzymes [55]. This is accomplished by analogy calculations between known parts of the investigated enzyme with other enzymes, whose amino acid sequence and three-dimensional structure is already known. Of course the reliability of this method strongly depends on the amount of overlap (or similarity) in the amino acid sequence between both of the enzyme candidates. An identity of greater than 60% is considered to be quite reliable.

Substrate Model. If neither the amino acid sequence nor X-ray data are available for an enzyme, which is unfortunately the case in the majority of enzymes, one can proceed as follows:

A number of artificial substrates having a broad variety of structures is subjected to an enzymatic reaction. The results thereof, e.g. the speed of conversion and the enantioselection etc. then allow us to create a general structure of an imagined "ideal" substrate, to which an actual substrate structure should come as close as possible to ensure a good acceptance by the enzyme and a high enantioselection. Of course this crude but quick method gives more reliable expectations the larger the number of test-substrates and the more rigid their structures are. Such models have been developed for pig liver esterase (PLE) [56] and *Candida cylindracea* lipase [57].

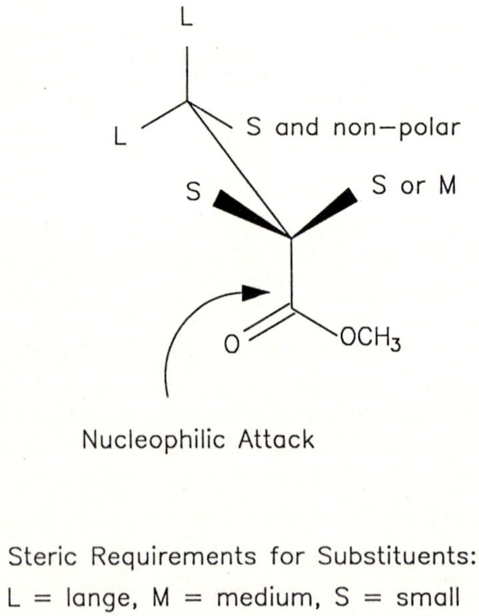

Scheme 6.12. Substrate model for pig liver esterase

Active Site Models. Instead of developing an ideal substrate structure one also has tried to assume the structure of the active site of the enzyme by the method described above. Of course this method has even more uncer-

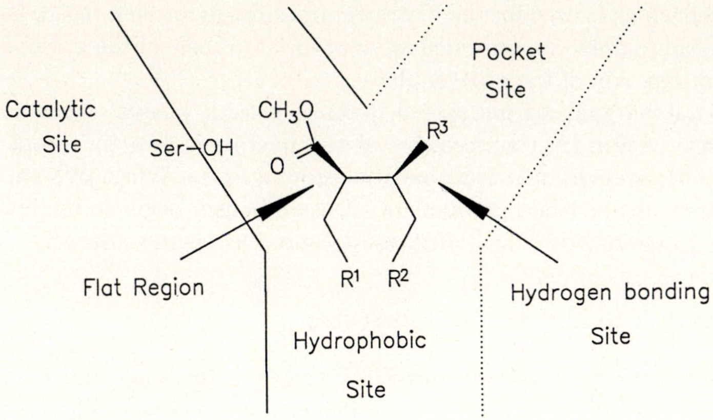

Scheme 6.13. Active site model for pig liver esterase

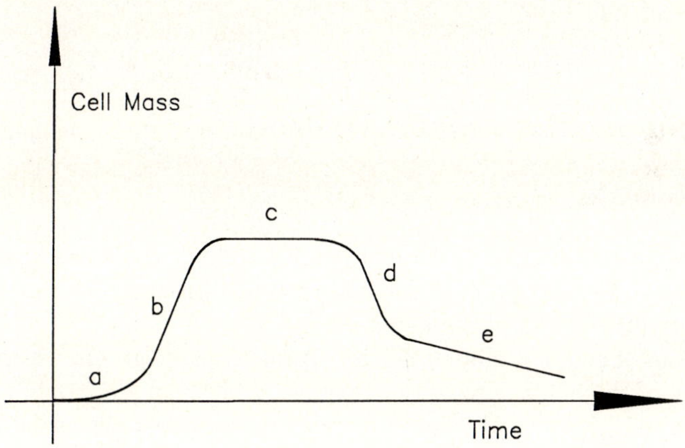

Scheme 6.14. Microbial growth phases [60]: *a* lag phase, *b* exponential growth phase, *c* stationary phase, *d* death phase, *e* survival phase

tainties than that described above. An illustrating example is the active site model for pig liver esterase developed by Ohno et al. [58].

6.4 Use of Whole Cell Systems

6.4.1 Principles

The classic method for performing biocatalytic reactions in organic chemistry is the use of whole cell systems [59, 27]. Here there is no need to isolate or to purify often unstable enzymes, and expensive coenzymes are provided by the metabolism of the cell in stoichiometric amounts. What one has to take into account is the fact that even the most simple microbial cell contains a multienzyme system. Therefore, to obtain the predominant action of one selected enzyme which is used for the biotransformation, as a rule it has to be induced. This can mainly be performed by appropriate choice of the growing conditions and selection of the growth phase.

In broth, microorganisms undergo a cycle of growth phases. After an initial lag phase caused by the necessity of the microorganism to become adapted to the novel environment exponential growth starts. When all nutrients begin to run out or when products of the metabolism begin to inhibit, the stationary phase begins. Then, after some time cells begin to die off.

6.4.2 Application to Unnatural Substrates

For biotransformation of unnatural substrates, either the growing culture or a suspension of "resting" cells from the stationary phase is used. Where no metabolism of the cells (e.g. for the preservation of coenzymes) is necessary, even dried cell preparations can be used. Some recent examples for the first case are given in Sect. 6.6.

6.5 Application of Biocatalytic Hydrolysis

6.5.1 General Remarks

According to the type of substrate which is hydrolysed, hydrolases are classified into subgroups. Most important for bioorganic transformations are: Proteases, esterases/lipases and phospholipases.

Their mechanism of action is similar [50, 61]: A nucleophilic group, which is an inherent part of the active site of the enzyme, attacks an electrophilic center (e.g. a carbonyl group) and thus forms an *acyl-enzyme* intermediate which can then be cleaved by any freely available nucleophile (usually water) to liberate the product from the enzyme and to regenerate the active site of the enzyme (see Sect. 6.3.4)

6.5.2 Resolution of Racemates

When a racemate is subjected to enzymatic hydrolysis, chiral recognition occurs: Due to the asymmetry of the active site of the enzyme, one enantiomer fits better than the other and it therefore reacts faster than the other. In the ideal case the rate difference is so extreme, that the well fitting enantiomer is quickly transformed and the other is not converted at all. Thus the reaction spontaneously stops at 50% conversion [62].

Theory of enzymatic resolution. In practice most cases of enzymatic resolution do not show the ideal situation described above, where one enantiomer is converted quickly and the other not at all. The difference in (or better: the ratio between) the reaction rates of the enantiomers is not indefinite, but is a finite one. What one observes in the rate of conversion of such

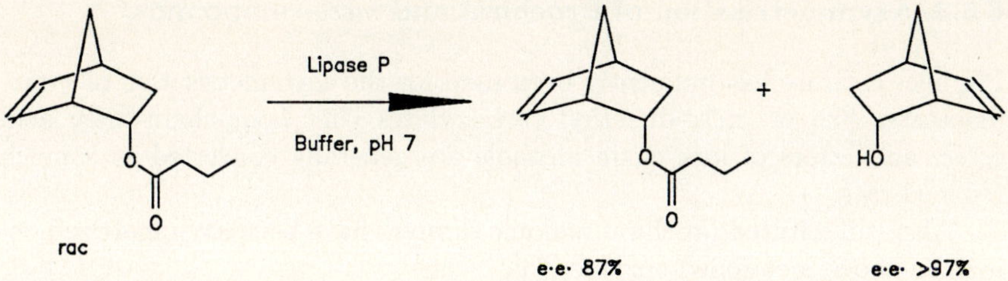

Scheme 6.15. Enzymatic resolution of a racemate by *Pseudomonas* lipase [63]

a case, is not a complete standstill at 50% but a clear decrease in the speed of reaction at this point. In these numerous cases, one encounters a number of dependencies [64]:

The reaction rate of both enantiomers varies with the degree of conversion, since the ratio of the two enantiomers does not remain constant during the reaction. Thus, the optical purity of both substrate and product does not remain constant either, but is a function of the conversion instead.

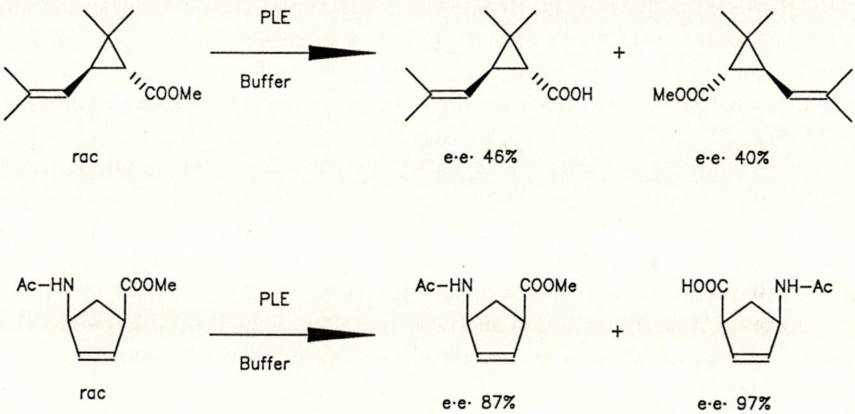

Scheme 6.16. Pig liver esterase catalyzed resolution of racemates [67, 68]

Generally, one always tries to make those reactions *irreversible* [65], if at all possible. The easiest way to do this is to add excess cosubstrate – about 20 equivalents are sufficient – to retain an irreversible reaction. Other techniques are more specialized and are discussed in the organic solvents section (Sect. 6.7.1). As an example to illustrate the technique described above, numerous racemic acetates have been resolved with pig liver esterase [28]. Even racemates of axial chirality are applicable [66].

6.5.3 Asymmetrisation of Prochiral and *meso*-compounds

Pig liver esterase has frequently been used for the asymmetrisation of *meso*-diacetates [69] or *meso*-dimethyl carboxylates [70]. Long-chain fatty acid esters and esters of long-chain alcohols are generally converted at a much reduced rate.

Also, substituted prochiral malonic diesters have been asymmetrized using the above mentioned enzyme [73].

Porcine pancreatic lipase (PPL) has been most widely used for the asymmetrisation of *meso*-diacetates of diols [75]: Is a very useful alternative for cyclopentane-systems, where PLE gives only low enantioselection [70].

6.5.4 Selective Protection and Deprotection

PLE effects mild hydrolysis at around pH 7 of acetates of primary and secondary alcohols [76] and of methyl and ethyl carboxylates [77]. This is particularly useful for the mild deprotection of acid or base sensitive compounds.

6.5.5 Mild Conditions

Nitrilases and Nitrile Hydratases. Nitrile hydratases catalyse the addition of water to a nitrile thus forming the corresponding amide, which is usually further converted by an amidase to yield the final carboxylic acid. Nitrilases are able to hydrolyse a nitrile directly to its corresponding carboxylic acid [32].

Most applications concerning these systems have been reported using whole cell systems, such as *Arthrobacter* and *Rhodococcus* species [79]. This reaction is particularly useful for the hydrolysis of nitriles bearing other acid or base sensitive functional groups, since the classical chemical hydrolysis of these compounds requires very harsh conditions.

An enantioselection found in these reactions, mainly attributed to the action of the amidase, was found only on α-aminonitriles leading to α-aminoacids [81].

6.6 Reduction and Oxidation Using Biocatalysts

6.6.1 Introduction

Oxidoreductases. They catalyse numerous redox reactions, among which the reduction of ketones leading to optically active secondary alcohols is the most important for preparative organic chemistry [82].

In contrast to hydrolases, where the water required for the reaction is always present in excess, with oxidoreductases a cofactor is required, which transfer the hydrogen redox equivalents to the substrate.

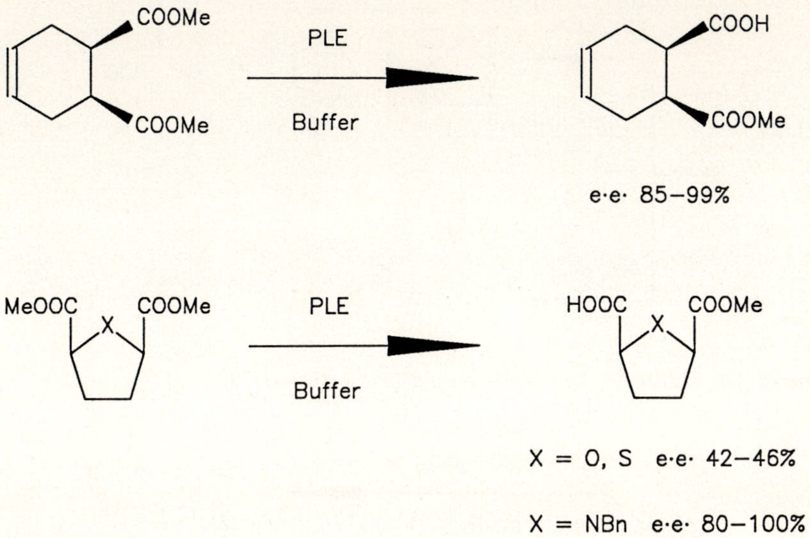

Scheme 6.17. Asymmetrisation of *meso*-diesters using pig liver esterase [70, 71, 72, 56]

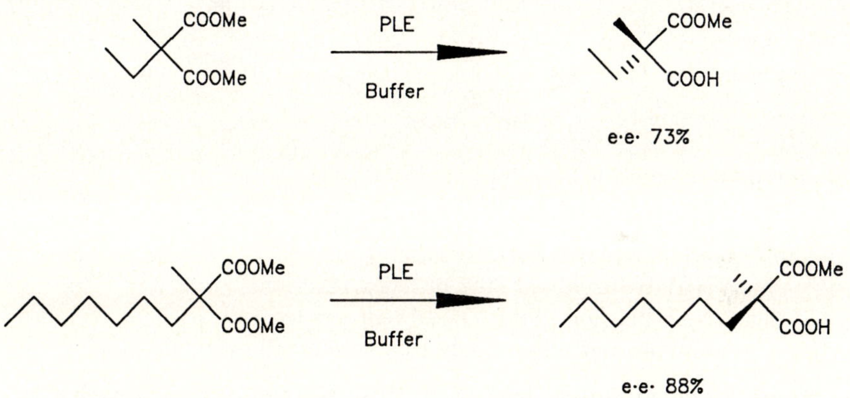

Scheme 6.18. Asymmetrisation of prochiral diesters by pig liver esterase [74]

Here two different types of reaction can be observed [83]:
a) A prochiral ketone is stereoselectively reduced i.e. the incoming hydride equivalent approaches from one side of the molecule (the *Re*- or the *Si*-side) thus forming an optically active secondary alcohol (see Sect. 6.6.3).
b) A racemic ketone is subjected to enzymatic resolution, leading to one enantiomer being reduced and the other remaining untouched.

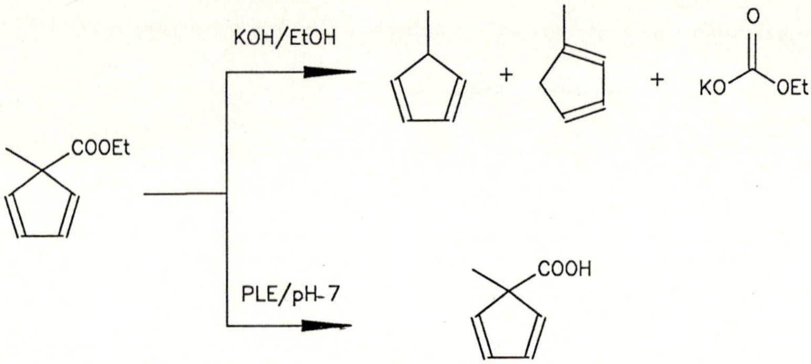

Scheme 6.19. Mild ester hydrolysis by pig liver esterase [78]

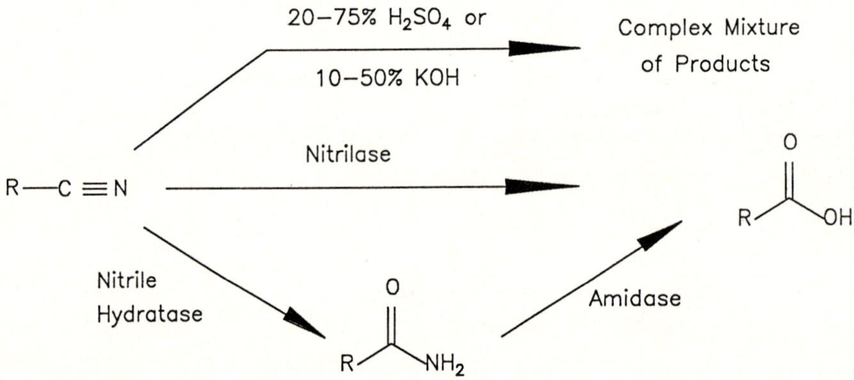

Scheme 6.20. Enzymatic hydrolysis of nitriles [80]

The most common required cofactors are:

NADH (for about 90% of redox reactions)
NADPH (for about 10% of redox reactions)
Flavine and others (only to a very small extent).

If one considers the net balance of these reactions, it is clear that cofactors are required in stoichometric amounts. Generally, these cofactors are chemically quite sensitive and expensive which makes their recycling necessary.

6.6.2 Enzymatic Cofactor Recycling

Although NAD^+ can be chemically reduced by sodium dithionite ($Na_2S_2O_4$) with a low number of cycles, it is insufficient for an effective reduction of cofactor-cost and modern cofactor recycling is always done enzymatically [84, 48].

In most cases two enzymes are employed: One for the reduction of the substrate and the other for the oxidation of the sacrificial cosubstrate or vice versa [85].

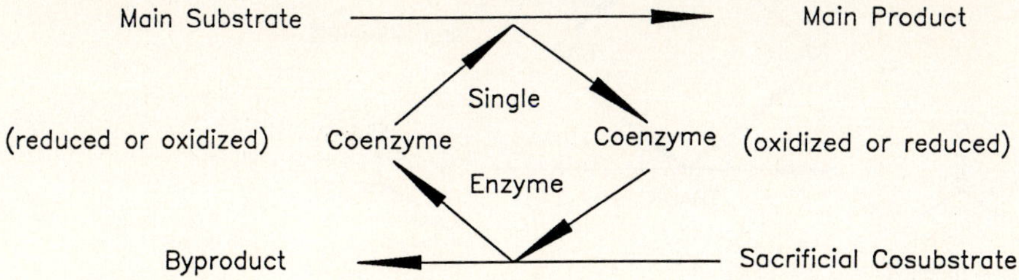

Scheme 6.21. Coenzyme recycling by the coupled substrate method

Only in selected cases is a single enzyme capable of catalyzing both reactions [86]. Using such a method, about 10^3 cycles can easily be accomplished, if both the cofactor and the enzyme(s) are free dissolved. Special biotechnological techniques such as co-immobilisation of enzyme(s) and coenzyme and the use of membrane reactors can lead up to 10^6 cycles of the coenzyme thus reducing the overall costs drastically [87].

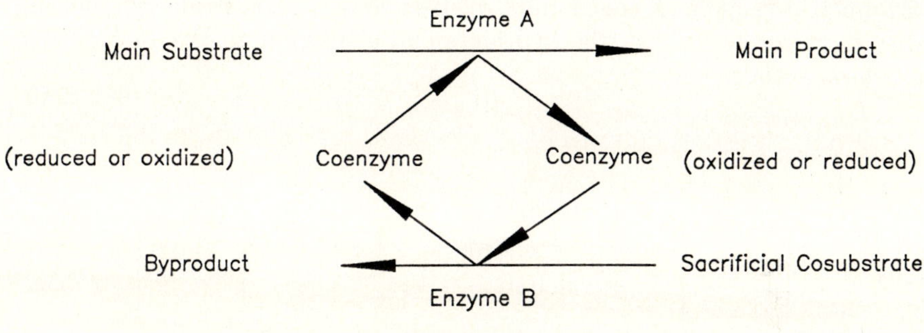

Scheme 6.22. Coenzyme recycling by the coupled enzyme method

6.6.3 Enantioface Differentiation in Reduction of Ketones

Oxidoreductions using enzymes. Besides horse liver alcohol dehydrogenase (HLADH) which has widely been used for the resolution of substituted cyclic ketones [88], an alcohol dehydrogenase obtained from *Thermoanaerobium brockii* (TBADH) has proved very useful for the asymmetrization of prochiral straight-chain ketones [86].

Oxidoreductions using whole microorganisms. Due to the limitations in cofactor recycling and the considerable costs involved, whole microorganisms (mainly yeasts) are frequently used for asymmetric reductions of ketones [16]. The advantage of this technique is:

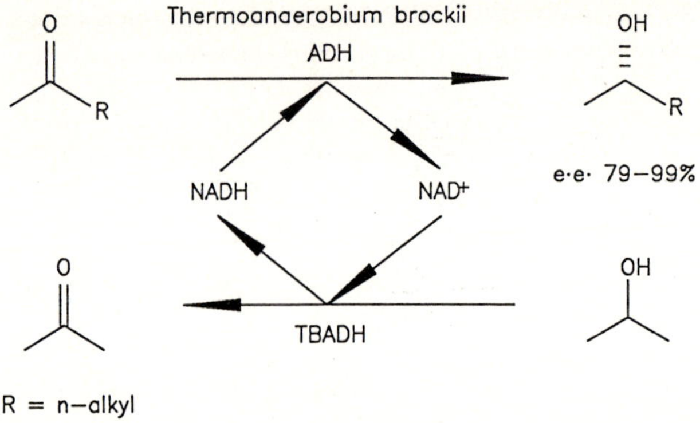

Scheme 6.23. *Thermoanaerobium brockii* catalyzed reduction of ketones [86]

yeast(s) are especially sturdy and easy to handle and are readily available. They contain numerous dehydrogenases and all the cofactors necessary for the conversion. Furthermore, cheap sugars such as glucose or saccharose can be used as the sacrificial cosubstrate.

Interestingly, a general trend in stereochemical preference for the produced optically active secondary alcohol can be found not only for yeasts but also for other microorganisms which are capable of reducing ketones in an asymmetric fashion. It is generally referred to as Prelog's rule [89], stating that the enzyme delivers the hydride anion equivalent to the *Re*-face of a prochiral ketone.

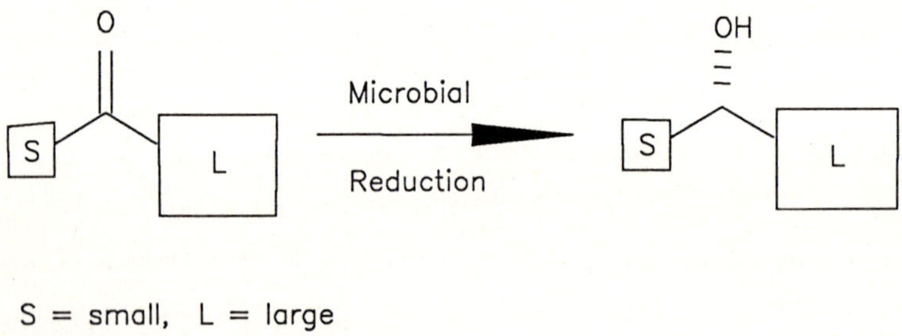

Scheme 6.24. Prelog's rule [90]

An interesting example of a divergent stereochemical outcome of an asymmetric reduction of a β-ketoester guided by a consequent substrate modification is given below: When the alcohol moiety of the substrate ester was gradually increased in size, the hydride equivalent was delivered from the opposite side of the substrate. By this means both enantiomers of the resulting β-hydroxyester were obtained with a single microorganism [91].

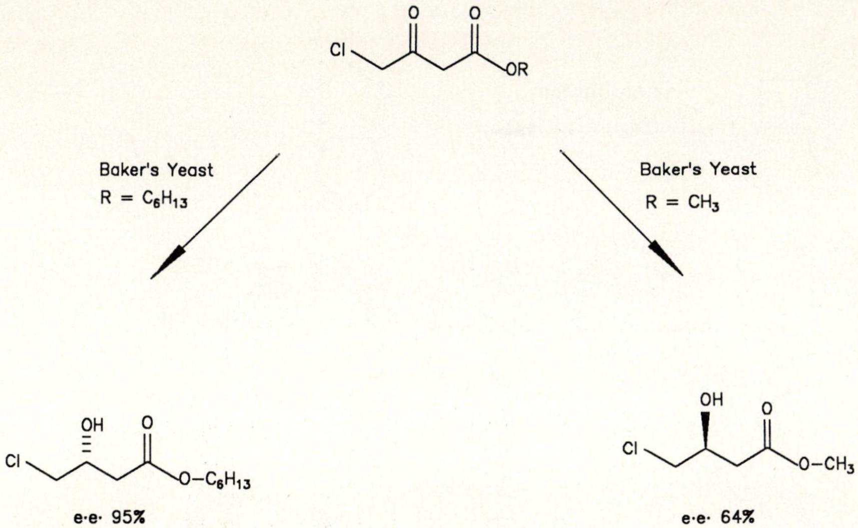

Scheme 6.25. Enantiodivergent stereoselective reduction of acetoacetates by baker's yeast [91]

6.6.4 Oxidation of Ketones

Mono-oxygenases, named after their characteristic of transferring a formal portion of a single O atom onto their substrates, catalyse the oxidation of ketones to yield lactones [38]. The chemical equivalent for this transformation is the Baeyer–Villiger reaction, which up to now cannot be performed in an asymmetric manner by purely chemical means. Therefore the enzymatic reaction is of high synthetic value. Since these sensitive enzymes need two different cofactors for the action which are not commercially available, the reaction is usually performed with whole microorganisms. The direction of oxygen insertion is generally the same as in the chemical Baeyer–Villiger reaction: In general the oxygen is inserted at the C-C bond towards the higher substituted side of the ketone.

6.6.5 Hydroxylation of Nonactivated Carbon Atoms

Mono-oxygenases also can convert aliphatic C-H bonds to C-OH functionalities, a reaction where no feasible chemical equivalent is available. Similar to above, these reactions are usually performed with whole microorganisms [93].

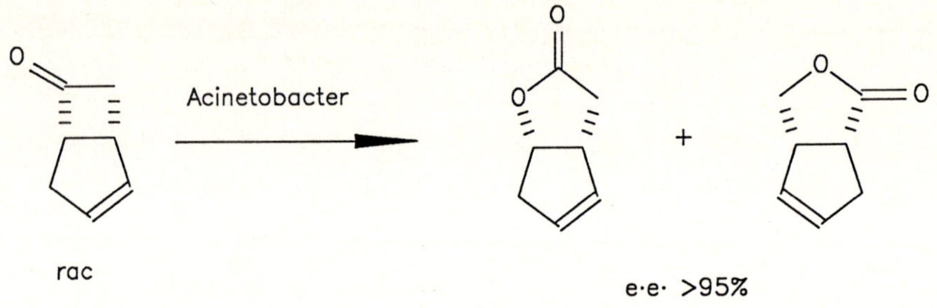

Scheme 6.26. Microbial Baeyer–Villiger reaction [92]

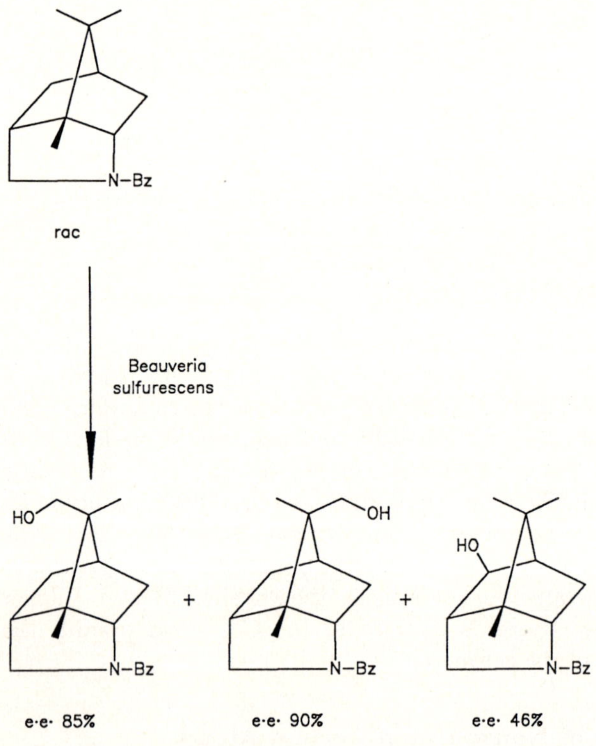

Scheme 6.27. Biohydroxylation using *Beauveria sulfurescens* [94]

6.6.6 Other Oxidations

Di-oxygenases transfer a formal portion of 2 [O] atoms onto the substrate producing two types of products:

a) Oxidation of C-H bonds lead either to a peroxide (which undergoes further reactions) or a 1,1-diol which generally collapses to a carbonyl group [95].

b) Oxidation of an aromatic C=C bond forms a *cis*-glycol, which is chiral, if the starting aromatic substrate carries substituents [96, 97].

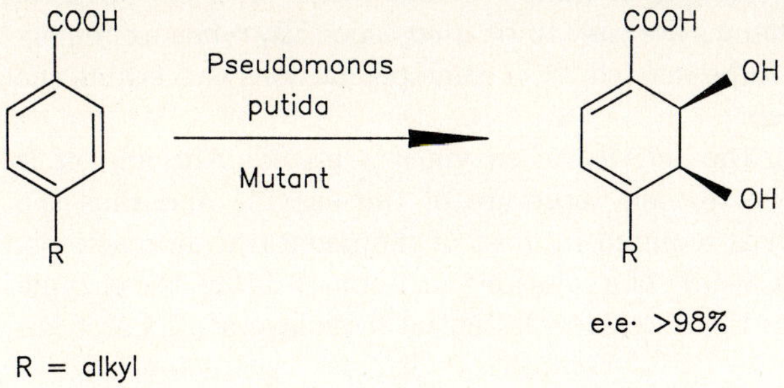

Scheme 6.28. Microbial oxidation of aromates [97]

6.7 Further Applications

6.7.1 Use of Organic Solvents, Transesterification

During the last few years, many different enzymatic reactions have been carried out in systems containing organic solvents in order to avoid some of the following disadvantages which are associated with the use of water [98]:

Most organic compounds are insoluble and on a technical scale removal of water is expensive due to its high boiling point and high heat of vaporisation. Furthermore, side reactions such as hydrolysis [99], polymerisation [100] or racemization [101] of sensitive compounds can be minimized in an organic environment.

The following types of organic solvent systems used for enzyme catalyzed transformations can be characterized:

a) Monophasic solution: The enzyme and an organic substrate are truly dissolved in a monophasic water/water-miscible organic solvent system (e.g. water-acetone). Generally, up to 10% of added organic cosolvent does not seem to impede the enzyme's activity too severely, however greater fractions of organic solvent should be avoided. In these systems, enzyme activity is often destroyed by the presence of the highly polar organic cosolvent and hence they are not often used [102, 103].

b) Biphasic solution: The enzyme is dissolved in the aqueous solution and the organic substrate and/or product being usually more soluble in the immiscible organic solvent, stays in the other phase (e.g. a water–hexane system). Of course, stirring in order to facilitate mass transfer is an essential factor here. Since enzymes are susceptible to shear forces, too much agitation leads to a loss in activity. Thus, such systems are not easy to be used from a physical standpoint [104, 105].

c) Solid lyophilized enzymes can be freely suspended in a lipophilic organic solvent, which contains substrate and/or product in dissolved state. Although freeze-dried enzymes are often referred to as "dry" they always contain a fraction of about 10% of tightly bound water on their surfaces. Quite amazingly, such systems are easy to be used especially when extremely lipophilic substrates are to be reacted [2], if some precautions concerning the choice of solvent are taken:

The role of water on the activity of enzymes is mixed: Although it is essential for retaining the tertiary structure of the enzyme and thus the activity of the enzyme, it is required in most of the denaturating reactions. Upon removing the "bulk water" of a solution (i.e. about $> 95\%$), the enzyme activity is retained unless the necessary "essential monolayers" of water are removed [106].

Organic solvents, added to a lyophilized enzyme, do not harm the enzyme activity, unless the very last necessary amount of water attached to the enzyme's surface is removed. As some measure of "compatibility" of organic solvents for enzyme reactions, the partition coefficient P is used [107]:

$$P = \frac{[c_{\text{octanol}}]}{[c_{\text{water}}]}$$

$\log P$	< 2	$2-4$	> 4
enzyme activity	low	moderate	high

Thus, one generally can conclude, that *lipophilic solvents* have *no detrimental effect* on an enzyme's activity, whereas *hydrophilic ones do.*

Potential advantages of employing enzymes in organic media. In hydrolytic reactions performed in an organic medium thermodynamic equilibria are shifted from hydrolysis to synthesis. This fact can be used to synthesize esters [108], lactones [109], amides and peptides [110].

Many side reactions involving water such as hydrolysis [99], polymerization [100] or racemization [101] are strongly suppressed, thus leading to better yields and/or purer products.

Immobilisation of enzymes is seldom necessary, since lyophilized enzymes are completely insoluble in organic solvents. They can easily be reused after recovery by simple filtration.

Most enzymes exhibit an enhanced thermal stability in organic solvents [111] due to the low or marginal concentration of water and last but not least microbial contamination, a problem in aqueous solution, can be ignored.

Interesterification. Historically, enzyme-catalyzed reactions performed in organic solvents are very old: Kastle and Loevenhart synthesized ethyl butyrate using a crude lipase preparation from porcine pancreas as early as 1900 [112].

Particularly in the case of diols, where asymmetric hydrolysis may be hampered by acyl migration of the obtained monoacetate in aqueous medium, thus leading to racemization of the product monoester [113, 114], interesterification performed in an organic solvent, where acyl migration is strongly suppressed, is a valuable technique for the preparation of optically active monoesters [113]. It is a general rule that it is always the same enantiomer (or the same enantiotopic group) which preferantially reacts. Thus, hydrolysis and interesterification – themselves being reactions of opposite directions of ester cleavage and formation – give rise to enantiomeric products as exemplified below.

Enol esters liberate free enols upon acyl transfer, which spontaneously rearrange to the more stable aldehydes or ketones, thus leading to a desirable irreversibility of the reaction. Vinyl acetate has shown to be most versatile here [116].

With the same results *acid anhydrides* can be employed as acyl donors as well [118], as long as a weak base is added to the system. The latter serves as an acid scavenger in order to protect the micro-environment of the enzyme [119].

6.7.2 Lyase-Catalyzed Additions to Double Bonds

Lyases catalyse the addition of small molecules such as water, ammonia or hydrogen cyanide onto C=X double bonds with X being C, O or N. Similarly to the asymmetric reduction of ketones, one or even two centers of asymmetry are created during this reaction.

Addition of NH_3. Aspartate ammonia-lyase catalyzes the asymmetric addition of NH_3 across the C=C double bond of fumaric acid yielding aspartic acid derivatives [40].

Addition of Hydrogen Cyanide. Mandelonitrile lyase (also called oxynitrilase) is used for the asymmetric addition of hydrogen cyanide onto the carbonyl group of aldehydes leading to chiral cyanohydrins. A wide variety of aromatic and even some aliphatic aldehydes is accepted. Whereas the enzyme isolated from almond bran always renders (*R*)-cyanohydrins [41], a different biocatalyst obtained from millet leads to the formation of their corresponding (*S*)-counterpart [120, 138].

6.7.3 C-C Bond Formation and Cleavage

Aldol-reactions. Aldol-type reactions have been accomplished with both fermenting microorganisms such as baker's yeast and with isolated enzymes. They resemble the asymmetric formation of a C-C bond combined with the creation of two new centers of chirality [44].

Aldolase-Reaction. Rabbit muscle aldolase, one of the key enzymes of glycolysis, has frequently been used for asymmetric aldol reactions on numerous non-natural aldehydes besides its natural substrate glyceraldehyde-3-phosphate [121]. It almost exclusively depends on dihydroxyacetone phosphate as cosubstrate, which is a quite sensitive and expensive compound. Hence, its preparation in situ by recycling techniques has served as a valuable tool [122].

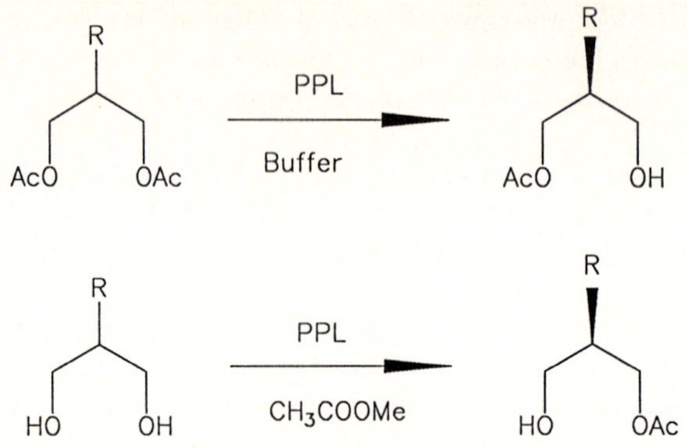

R = $(CH_2)_2CH{=}CH_2$, Ph e·e· 90–100%

Scheme 6.29. Enzymatic asymmetrisation of prochiral substrates by hydrolysis and interesterification using porcine pancreatic lipase [115]

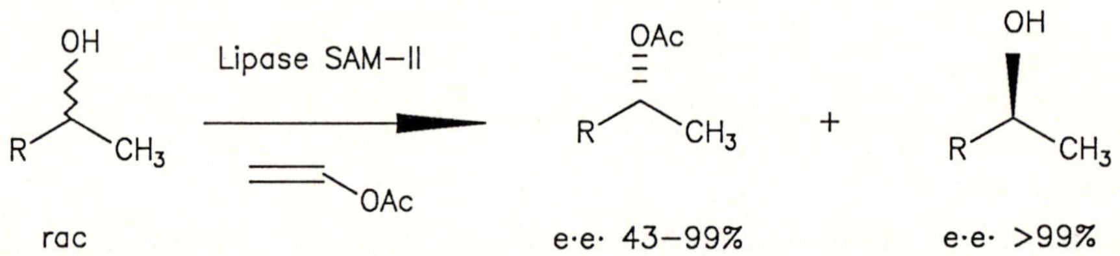

R = Ph, naphtyl, Bn

Scheme 6.30. Transesterification with a *Pseudomonas* lipase using enol esters [117]

6.7.4 Transferases

Transferases are a class of enzyme not yet widely used. An illustrative example is the exchange of pyrimidine bases on natural nucleosides by non-natural triazoles for the preparation of artificial nucleoside analogues which are used for the treatment of viral infections. If a natural nucleoside such as uridine is subjected to the action of fermenting *Enterobacter aerogenes* in presence of a triazole derivative, an enzyme-catalyzed exchange reaction between uracil and the non-natural base is obtained. The mechanism involves a phosphorolytic cleavage of uridine using ATP, and a subsequent S_N2 exchange of the phosphate moiety by the newly introduced heterocyclic base. By this means ribavirin, a potent antiviral agent, is produced on a technical scale [123].

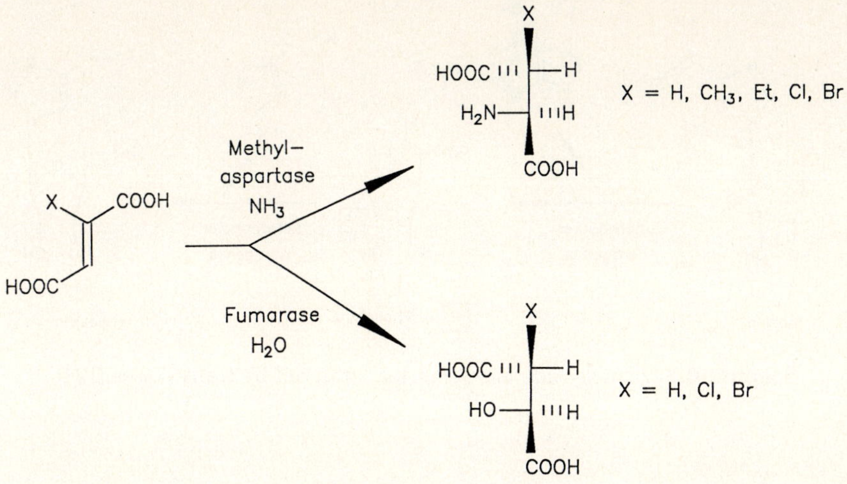

Scheme 6.31. Lyase catalyzed addition of NH_3 or H_2O onto fumarate [40, 39]

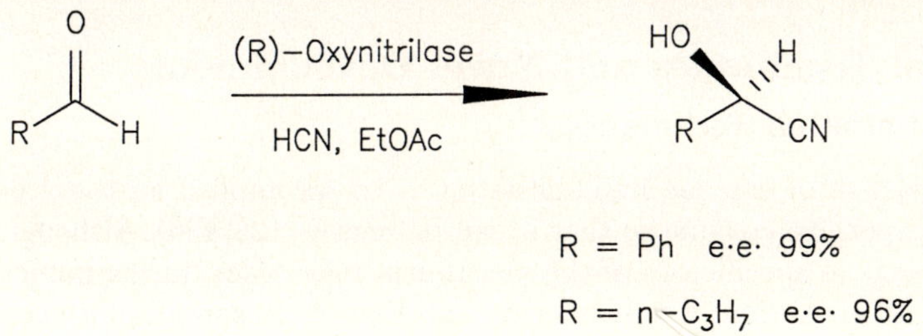

Scheme 6.32. Enzymatic addition of HCN onto aldehydes [41]

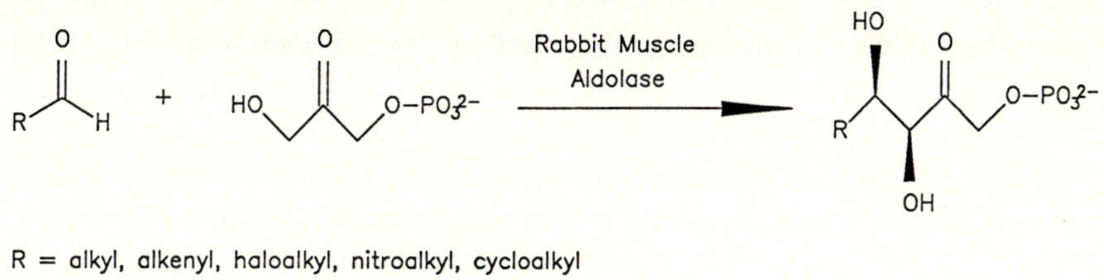

Scheme 6.33. Aldolase reaction [121]

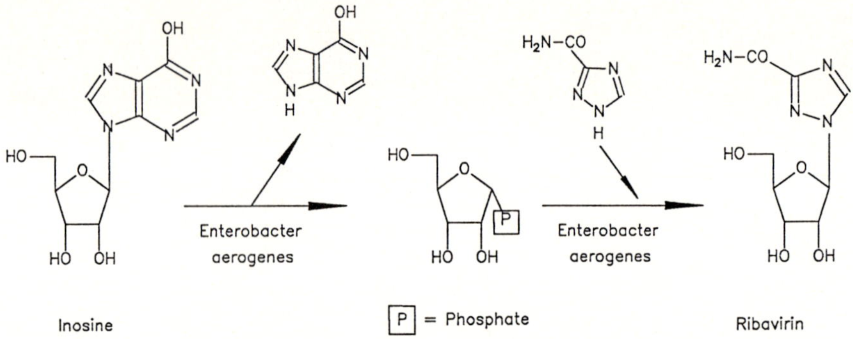

Scheme 6.34. Nucleoside base-transfer catalyzed by transferases [124]

Addition of Water. Likewise, fumarase can catalyse the addition of water onto fumaric acid derivatives to give optically active malic acid derivatives [39].

6.8 Special Techniques and Novel Developments

6.8.1 Immobilisation Techniques

The main objective of enzyme immobilisation is to accomplish an easy recovery of an expensive enzyme so that it can be reused [125, 126]. Although unlikely, a change in specificities of the biocatalyst may occur during immobilisation due to alteration of its conformation. However, examples for such a case are rare.

Adsorption. Simple physical adsorption of a biocatalyst onto a porous macroscopic carrier equipped with a highly polar surface such as diatomaceous earth, silica gel, glass beads, or even brick-dust can serve for an immobilisation technique. Such systems however, involving relatively weak binding forces are limited to non-aqueous media to prevent a leak of activity [127]. Whole microorganisms are often enclosed in gels such as Ca alginate or κ-carragenan for immobilisation.

Covalent Binding. Strong binding forces are obtained if a biocatalyst is covalently bound onto a macroscopic carrier. Here of course no leaking of activity is observed. The retained activities of the immobilized enzyme can be drastically reduced since the biocatalyst has to undergo a chemical reaction, which may change its conformation going in hand with a loss in activity [128].

One of the most modern methods for this purpose is the coupling of lysin residues of the enzyme surface onto polymers bearing reactive epoxy groups. Thus a stable C-N bond is formed which holds the biocatalyst firmly [128].

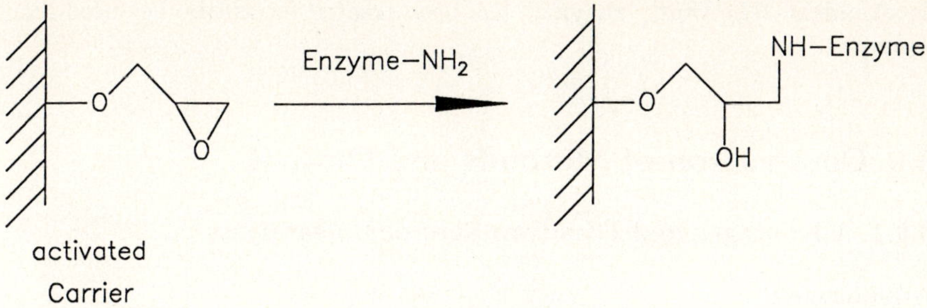

Scheme 6.35. Immobilisation by covalent attachment of an enzyme onto an epoxy resin

6.8.2 Artificial and Modified Enzymes, Enzyme Mimics

Up to now about 2500 enzymes have been found in nature. In order to understand the mechanism of action to obtain tailor-made enzymes for special applications and to have easier access to rare and expensive enzymes via analogs, novel approaches were developed.

One important aspect is the application of genetic engineering to obtain enzymes with ameliorated properties for biotechnology, such as improved thermostability, productivity and selectivity [129, 130]. Another target is to mimic the enzyme structure either by use of synthetic molecules such as cyclophanes or natural compounds such as cyclodextrins with the intention of either obtaining mechanistic information or having a synthetic or semi-synthetic enzyme as the ultimate goal [131, 132].

6.8.3 Catalytic Antibodies

Despite the large number of enzymes available, naturally reactions may still exist which cannot be catalyzed at all or only with difficulty with enzymes. A recent development may open novel dimensions here [133, 134]. Antibodies are formed in organisms when they are attacked by substances (called antigens) which might cause diseases.

Proteins such as enzymes are then bound to the antigen as markers for defense of the organism by the immune system. The versatility for this formation of antibodies is almost unlimited. Every catalytic action of enzymes is caused by a decrease of transition state energy. If it were possible to generate antibodies which can stabilize the transition state of a reaction, a catalytic acceleration of the corresponding chemical reaction by the antibody might be achieved. The reason is that by binding the antibody to the transition state the energy would be expected to be lowered. Of course, antibodies can only be formed against transition state analogs, e.g. by mimicking the tetrahedral transition state of ester hydrolysis by a phosphonate moiety. By this technique not only rate accelaration but also some regio- and enantioselectivity has been achieved. Further developments are antibody catalyzed Diels–Alder reactions, cyclization or redox reactions, an area which is in rapid development. The word "abzyme" has been created for antibodies used like enzymes.

6.9 Comparison of Methods and Outlook

6.9.1 Advantages and Disadvantages of Biocatalysts

Advantages

Enzymes are chemoselective. Most enzymes are very selective towards a single functional group and other functionalities, which might be damaged during side-reactions by chemical catalysts due to a lack of selectivity. Thus other functional groups can survive better [18] Therefore, less protective group chemistry is required if one uses biocatalysts leading to better overall yields.

Enzymes are regioselective. Enzymes can often distinguish between two identical functional groups which are only different by their chemical environment in the molecule. For example, hydroxy groups in sugars or steroids can selectively be protected by enzymatic acylation [135] or *exo–endo* positioned groups in bicyclic systems can selectively be transformed [136], a goal which is difficult to reach by pure chemical means.

Enzymes are enantioselective. Since all enzymes are made from L-amino acids, they represent chiral catalysts. Upon catalysis one observes chiral recognition of the chirality of the substrate by that of the enzyme thus leading to asymmetric transformations [64]. This advantage has probably brought one of the largest impact on modern asymmetric synthesis.

Disadvantages

Enzymes are temperature sensitive. High temperatures generally above 45°C cause denaturation, low temperatures – below 20°C – lead to a rapid decline in activity. Thus there is only a narrow temperature range open for performing reactions which severely may limit the applicability of biocatalysts.

Enzymes are pH-sensitive. Both a low (< 5) or a high (> 8) pH causes denaturation of an enzyme which adds to another limitation.

Substrate and product inhibition. Most enzymes are subject to either substrate or product inhibition leading to a low reaction rate at elevated concentrations. Thus the overall productivity of a process may be limited by a low concentration tolerance. To circumvent these obstacles, one adds the substrate gradually and removes the formed product continuously during the reaction.

Allergies. Some enzymes can cause allergies, they should be regarded as chemicals and therefore, handled with care.

6.9.2 Future Developments and Trends

In particular hydrolases – esterases, lipases and proteases – are simple-to-use catalysts for the preparation of optically active alcohols, amines and acids. The area is sufficiently well researched to be of potential use to a wide range of synthetic problems [27].

Dehydrogenases and microorganisms such as yeast(s) can readily be used for stereo- or enantioselective reduction of ketones to furnish the corresponding optically active secondary alcohols. Although much has been accomplished on a laboratory scale, further research, particularly on coenzyme recycling, has to be done before these methods can be used for processes on a technical scale [90].

The synthesis of optically active phosphate esters is now possible and this strategy should be seriously considered by chemists entering this area of work [49].

A wide variety of transformations is possible by means of enzymes or whole microorganisms, where the analogous reaction using pure classical chemical methods is not applicable or leads to low yields and selectivities: Baeyer–Villiger reactions cannot be performed in an asymmetric manner using pure chemical methods, but mono-oxygenases can [38]. The same is true for biohydroxylations [33]. Nitrile converting enzymes are seldom enantioselective, but they constitute a valuable alternative to the harsh chemical conditions required for the hydrolysis of nitriles [32].

Finally, microorganisms can synthesize extremely complicated optically active molecules from cheap sources [137].

6.10 References

1 Testa B (1979) Principles of organic stereochemistry. Marcel Dekker New York

2 Dale J (1978) Stereochemistry and conformational analysis. Verlag Chemie New York

3 Kagan H (1975) La Stereochimie Organique. Presses Universitaires de France Paris

4 Rademacher P (1987) Strukturen Organischer Moleküle, Physikalische Organische Chemie, vol 2. Verlag Chemie Weinheim

5 Kagan H (1977) Stereochemie, Grundlagen und Methoden Thieme Stuttgart

6 Fischer E (1891) Ber dtsch chem Ges *24:* 2683

7 Rosanoff MA (1906) J Am Chem Soc *28:* 114

8 Cahn R Ingold CK Prelog V (1966) Angew Chem Int Ed Engl *5:* 385

9 Weil K Kuhn W (1946) Untersuchungen über stereoisomere Salze des Leucinmethylesters. Helv Chim Acta *29:* 784

10 Ariens E J (1986) Stereochemistry: A Source of Problems in Medicinal Chemistry. Med Res Rev *6:* 451

11 Ariens EJ Van Rensen JJS Welling W (1988) Stereospecificity of Bioactive Agents: General Aspects as Exemplified by Pesticides and Drugs, in Stereoselectivity of Pesticides. Elsevier Amsterdam 1988 p 46

12 Newman P (1978, 1981, 1984) Optical Resolution Procedures for Chemical Compounds, vols 1–4. New York

13 Jaques J Collet A Wilen S (1981) Enantiomers, Racemates and Resolutions. Wiley, New York

14 Hanson KR (1966) Applications of the Sequence Rule. I. Naming the Paired Ligands g,g at a Tetrahedral Atom X_{ggij}; II. Naming the two Faces of a Trigonal Atom Y_{ghi}. J Am Chem Soc *88:* 2731

15 Prelog V Helmchen G (1972) Pseudoassymmetrie in der organischen Chemie. Helv Chim Acta *55:* 2581

16 Servi S (1990) Baker's Yeast as a Reagent in Organic Synthesis. Synthesis, 1

17 Jones JB Beck JF (1976) Asymmetric Synthesis and Resolutions Using Enzymes. In: Jones JB Sih CJ Perlman D (eds) Applications of Biochemical Systems in Organic Chemistry (Techniques of Chemistry) vol 10. Wiley New York, p 107

18 Chinsky N Margolin AL Klibanov AM (1989) Chemoselective Enzymic Monoacylation of Bifunctional Compounds. J Am Chem Soc *111:* 386

19 Sime JT Pool CR Tyler JW (1987) Regioselective Enzymic Hydrolysis in the Isolation of Isomers and Mupirocin. Tetrahedron Lett *28:* 5169

20 Waldmann H (1988) The Use of Penicillin Acylase for Selective N-Terminal Deprotection in Peptide Synthesis Tetrahedron Lett *29:* 1131

21 Chen S-T Wang K-T (1987) The Synthesis of β-Benzyl L-Aspartate and γ-Benzyl L-Glutamate by Enzyme-Catalyzed Hydrolysis. Synthesis 581

22 Guibe-Jampel E Rousseau G Salaun J (1987) Enantioselective Hydrolysis of Racemic Diesters by Porcine Pancreatic Lipase J Chem Soc, Chem Commun 1080

23 Kitaguchi H Tai D-F Klibanov A M (1988) Enzymatic Formation of an Isopeptide Bond Involving the ε-Amino Group of Lysine. Tetrahedron Lett *29:* 5487

24 Papageorgiou C Benezra C (1985) Use of Enzymatic Hydrolysis of Dimethyl Malonates for a Short Synthesis of Tulipalin B and of Its Enantiomer. J Org Chem *50:* 1144

25 Sweers H M Wong C-H (1986) Enzyme-Catalyzed Regioselective Deacylation of Protected Sugars in Carbohydrate Synthesis J Am Chem Soc *108:* 6421

26 International Union of Biochemistry, Enzyme Nomenclature. Academic Press Orlando 1984

27 Davies H G Green RH Kelly DR Roberts SM (1989) Biotransformations in Preparative Organic Chemistry. Academic Press London

28 Ohno M Otsuka M (l989) Chiral Synthons by Ester Hydrolysis Catalyzed by Pig Liver Esterase. Org Reactions *37:* 1

29 Gutman AL Zuobi K Guibe-Jampel E. (1990) Lipase-Catalyzed Hydrolysis of "γ-Substituted α-Aminobutyrolactones Tetrahedron Lett *30:* 2037

30 Hartsuck JA Lipscomb WN (1971) In: Boyer P D (ed) The Enzymes. vol 3. Acad Press New York p 1

31 Yamamoto Y Yamamoto K Nishioka T Oda J (1988) Asymmetric Synthesis of Optically Active Lactones from Cyclic Acid Anhydrides Using Lipase in Organic Solvents. Agric Biol Chem *52:* 3087

32 Nagasawa T Yamada H (1989) Microbial Transformations of Nitriles. Trends in Biotechnol *7:* 153

33 van den Tweel WJJ (1988) Strategies in the Selection of Novel Biocatalysts for the Conversion of Aromatic Compounds. In: Ratledge C Szentirmai A Barabas G Kevei F (eds) Proceedings of the 4th Int Workshop on Microbial Physiology & Manufactoring Industry, p 49

34 Dawson JH Sono M (1987) Cytochrome P-450 and Chloroperoxidase: Thiolate-ligated Heme Enzymes. Spectroscopy Determination of their Active Site Structures and Mechanistic Implications of Thiolate Ligation. Chem Rev *87:* 1255

35 Vigne B Archelas A Fourneron JD Furstoss R (1986) Microbial Transformations. Part 4. Regioselective Para Hydroxylation of Aromatic Rings by the Fungus Beauveria sulfurescens. The Metabolism of Isopropyl N-Phenyl Carbamate (Propham). Tetrahedron *42:* 2451

36 Lemiere GL Lepoivre JA Alderweireldt FC (1985) HLAD-Catalyzed Oxidations of Alcohol with Acetaldehyde as a Coenzyme Recycling Substrate. Tetrahedron Lett *26:* 4527

37 Irwin AJ Lok KP Huang KW-C Jones JB (1978) Enzymes in Organic Synthesis. Influence of Substrate Structure on Rates of Horse Liver Alcohol Dehydrogenase-Catalysed Oxidoreductions. J Chem Soc, Perkin Trans *I:* 1636

38 Walsh CT Chen Y-C J (1988) Enzymatic Baeyer-Villiger Oxidation by Flavine-dependent Monooxygenase. Angew Chem *100:* 342; Angew Chem Int Ed Engl *27:* 333

39 Findeis MA Whitesides GM (1987) Fumarase Catalysed Synthesis of L-*threo*-Chloromalic Acid and its Conversion to 2-Deoxy-D-ribose and D-*erythro*-Sphingosine. J Org Chem *52:* 2838

40 Akhtar M Botting NB Cohen MA Gani D (1987) Enantiospecific Synthesis of 3-substituted Aspartic Acids via Enzymic Amination of Substituted Fumaric Acids. Tetrahedron *43:* 5899

41 Effenberger F Ziegler Th Förster S (1987) Enzymkatalysierte Cyanhydrin-Synthese in organischen Lösungsmitteln. Angew Chem *99:* 491; Angew Chem Int Ed Engl *26:* 458

42 Buist PH Dimnik GP (1986) Use of Sulfur as a Chemical Connector. Tetrahedron Lett *27:* 1457

43 Xiao X Sen SE Prestwich GD (1990) Vinyl Oxirane Analog of (3*S*)-2,3-Epoxysqualene: A Substrate for Oxidosqualene Cyclases from Yeast and from Hog Liver. Tetrahedron Lett *30:* 2097

44 Toone EJ Simon ES Bednarski MD Whitesides GM (1989) Enzyme-Catalyzed Synthesis of Carbohydrates. Tetrahedron *45:* 5365

45 Kitazume T Ikeya T Murata K (1986) Synthesis of Optically Active Trifluorinated Compounds: Asymmetric Michael Addition with Hydrolytic Enzymes. J Chem Soc, Chem Commun 1331

46 Stryer L (1985) Biochemie. Vieweg Braunschweig

47 Lehninger AL (1987) Prinzipien der Biochemie. de Gruyter Berlin

48 Lee LG Whitesides G M (1985) Enzyme-Catalyzed Organic Synthesis: A Comparison of Strategies for in Situ Regeneration of NAD from NADH. J Am Chem Soc *107:* 6999

49 Langer RS Hamilton BK Gardner CR Archer MC Colton CK (1976) Enzymatic Regeneration of ATP. AIChE Journal *22:* 1079

50 Fersht A (1985) Enzyme Structure and Mechanism. Freeman New York

51 Brady L Brzozowski AM Derewenda ZS Dodson E Dodson G Tolley S Turkenburg JP Christiansen L Huge-Jensen B Norskov L Thim L Menge U (1990) A Serine Protease Triad Forms the Catalytic Center of a Triacylglycerol Lipase. Nature *343:* 767

52 Jansonius JN (1987) Crystallography in Molecular Biology. In: Moras D Drenth J Strandlberg B Suck D Wilson K (eds) Plenum Press New York, p 229

53 Cohen SG (1969) On the Active Site and Specificity of α-Chymotrypsin. Trans N Y Acad Sci *31:* 705

54 Schubert Wright C (1972) Comparison of the Active Site Stereochemistry and Substrate Conformation in α-Chymotrypsin and Subtilisin BPN′. J Mol Biol *67:* 151

55 Burkert U Allinger NL (1982) In: Caserio MC (ed) Molecular Mechanics. ACS Monograph vol 177 Am Chem Soc Washington

56 Mohr P Waespe-Sarcevic N Tamm C Gawronska K Gawronski JK (1983) A Study of Stereoselective Hydrolysis of Symmetrical Diesters with Pig Liver Esterase. Helv Chim Acta *66:* 2501

57 Oberhauser Th Faber K Griengl H (1989) A Substrate Model for the Enzymatic Resolution of Esters of Bicyclic Alcohols by *Candida cylindracea* Lipase. Tetrahedron *45:* 1679

58 Ohno M Kobayashi S Adachi K (1986) Creation of Novel Chiral Synthons with Pig Liver Esterase: Application to Natural Product Synthesis and the Substrate Recognition. In: Schneider MP (ed.) Enzymes as Catalysts in Organic Synthesis. Reidel Dordrecht p 123

59 Kieslich K (1984) Biotransformations. In: Reed HJ Rehm G (eds) Biotechnology, vol 6. Verlag Chemie Weinheim

60 Goodhue CT (1982) The Methodology of Microbial Transformation of Organic Compounds. Microb Transform Bioact Compd *1:* 9

61 Jones JB (1986) In: Frey PA (ed) Mechanisms of Enzymatic Reactions: Stereochemistry. Elsevier Amsterdam p 8

62 Klempier N Faber K Griengl H (1989) Biocatalytic Preparation of Enantiomerically Pure *endo*-Bicyclo[3.3.0]oct-7-en-2-ol. Synthesis 933

63 Oberhauser Th Bodenteich M Faber K Penn G Griengl H (1987) Enzymatic Resolution of Norbornane-Type Esters. Tetrahedron *43:* 3931

64 Chen C-S Sih CJ (1989) General Aspects and Optimization of Enantioselective Biocatalysts in Organic Solvents: The Use of Lipases. Angew Chem Int Ed Engl *28:* 695

65 Chen C-S Fujimoto Y Girdaukas G Sih CJ (1982) Quantitive Analysis of Biochemical Kinetic Resolutions of Enantiomers. J Am Chem Soc *104:* 7294

66 Ramaswamy S Hui RAHF Jones JB (1986) Enantiomerically Selective Pig Liver Esterase-catalyzed Hydrolysis of Racemic Allenic Esters. J Chem Soc, Chem Commun 1545

67 Schneider M Engel N Boensmann H (1984) Enzymatische Syntheses chiraler Bausteine aus Racematen: Herstellung von $(1R,3R)$-Chrysanthemum-, Permethrin- und Caronsäure aus racemischen Diastereomerengemischen. Angew Chem *96:* 52; Angew Chem Int Ed Engl *23:* 64

68 Sicsic S Ikbal M Le Goffic F (1987) Chemoenzymatic Approach to Carbocyclic Analogues of Ribonucleosides and Nicotinamide Ribose. Tetrahedron Lett *28:* 1887

69 Wang Y-F Sih CJ (1984) Bifunctional Chiral Synthons via Biochemical Methods. 4. Chiral Precursors to (+)-Biotin and (−)-A-Factor. Tetrahedron Lett *25:* 4999

70 Jones JB Hinks RS Hultin PG (1985) Enzymes in Organic Synthesis. 33. Stereoselective Pig Liver Esterase-Catalysed Hydrolysis of *meso*-Cyclopentyl-, Tetrahydrofuranyl- and Tetrahydrothiophenyl-1,3-diesters. Can J Chem *63:* 452

71 Kurihara M Kamiyama K Kobayashi S Ohno M (1985) Diversified Synthetic Approaches to the Carbapenem Antibiotics Based on Symmetrization-Asymmetrization Concept. Tetrahedron Lett *26:* 5831

72 Björkling F Boutelje J Hjalmarsson H Hult K Norin T (1987) Highly Enantioselective Route to (*R*)-Proline Derivatives via Enzyme-Catalysed Hydrolysis of *cis*-N-Benzyl-2,5-bismethoxycarbonylpyrrolidine in an Aqueous Dimethyl Sulfoxide Medium. J Chem Soc Chem Commun 1041

73 Luyten M Müller S Herzog B Keese R (1987) Enzyme-Catalyzed Hydrolysis of some Functionalized Dimethyl Malonates. Helv Chim Acta *70:* 1250

74 Björkling F Boutelje J Gatenbeck S Hult K Norin T Szmulik P (1985) Enzyme Catalysed Hydrolysis of Dialkylated Propanedioic Diesters, Chain Length Dependent Reversal of Enantioselectivities. Tetrahedron *41:* 1347

75 Hemmerle H Gais H-J (1987) Asymmetric Hydrolysis and Esterification Catalyzed by Esterases from Porcine Pancreas in the Synthesis of Both Enantiomers of Cyclopentanoid Building Blocks. Tetrahedron Lett *28:* 3471
76 Jongejan JA Duine JA (1987) Enzymatic Hydrolysis of Cyclopropyl Acetate, a Facile Method for Medium- and Large-Scale Preparations of Cyclopropanol. Tetrahedron Lett *28:* 2767
77 Lin CH Alexander DL Chidester CG Gorman RR Johnson RA (1982) 10-Nor-9,11-secoprostaglandins Synthesis Structure and Biology of Endorphine Analogues. J Am Chem Soc. *104:* 1621
78 Burger U Erne-Zellweger D Mayerl C (1987) The Hydrolysis of Ethyl 1-Methyl-2,4-cyclopentadiene-1-carboxylate by Nonenzymatic and Enzymatic Methods. Carbon-Carbon vs Carbon-Oxygen Bond Cleavage. Helv Chim Acta *70:* 587
79 Bengis-Garber C Gutman AL (1988) Bacteria in Organic Synthesis: Selective Conversion of 1,3-Dicyanobenzene into 3-Cyanobenzoic Acid. Tetrahedron Lett *29:* 2589
80 Jallageas J-C Arnaud A Galzy P (1980) Bioconversions of Nitriles and their Applications. Adv Biochem Engineering/Biotechnology vol 14 Springer Berlin Heidelberg p 1
81 Vo-Quang Y Marais D Vo-Quang L Le Goffic F Thiery A Maestracci M Arnaud A Galzy P (1987) Bacteria in Organic Synthesis: γ-Alkoxy-α-amino Acids from related α-Aminonitriles. Tetrahedron Lett *28:* 4057
82 Lemiere GL Van Osselaer TA Lepoivre JA Alderweireldt FC (1982) Enzymatic in vitro Reduction of Ketones. Part 8. A New Model for the Reduction of Cyclic Ketones by Horse Liver Alcohol Dehydrogenase (HLAD). J Chem Soc, Perkin Trans *II:* 1123
83 Lemiere GL (1986) Alcohol Dehydrogenase Catalysed Oxidoreduction Reactions in Organic Chemistry. In: Schneider MP (ed) Enzymes as Catalysts in Organic Synthesis. Reidel Dordrecht p 19
84 Bowen R Pugh SYR (1985) Redox Enzymes in Industrial Fine Chemical Synthesis. Chem Ind *10:* 323
85 Drueckhammer DG Sadozai SK Wong CH Roberts SM (1987) Biphasic One-pot Synthesis of two Useful and Separable Compounds Using Nicotinamide Cofactor-requiring Enzymes: Synthesis of (*S*)-4-Hydroxyhexanoate and its Lactone. Enzyme Microb Technol *9:* 564
86 Keinan E Hafeli EK Seth KK Lamed R (1986) Thermostable Enzymes in Organic Synthesis. 2. Asymmetric Reduction of Ketones with Alcohol Dehydrogenase from *Thermoanaerobium brockii.* J Am Chem Soc *108:* 162
87 Wichmann R Wandrey C Buckmann AF Kula M-R (1981) Continuous Enzymatic Transformation in an Enzyme Membrane Reactor with Simultaneous NAD(H) Regeneration. Biotechnol Bioeng *23:* 2789
88 Lepoivre JA (1984) Stereospecific Enzymatic Reduction of Carbonyl Functions with Horse Liver Alcohol Dehydrogenase. Janssen Chim Acta *2:* 20
89 Prelog V (1964) Specification of the Stereospecificity of some Oxido-Reductases by Diamond Lattice Sections. Pure Appl Chem *9:* 119
90 Crout DHG Christen M (1989) Biotransformations in Organic Synthesis. Modern Synthetic Methods *5:* 1
91 Zhou B Gopalan AS VanMiddlesworth F Shieh W-R Sih CJ (1983) Stereochemical Control of Yeast Reductions 1. Asymmetric Synthesis of L-Carnitine. J Am Chem Soc *105:* 5925
92 Alphand V Archelas A Furstoss R (1989) Microbial Transformations. 16. One-step Synthesis of a Pivotal Prostaglandin Chiral Synthon via Highly Enantioselective Microbiological Baeyer-Villiger Type Reaction. Tetrahedron Lett *30:* 3663
93 Fonken GS Herr ME Murray HC Reineke LM (1967) Microbiological Hydroxylation of Monocyclic Alcohols. J Am Chem Soc *89:* 672

94 Archelas A Fourneron J-D Furstoss R Cesario M Pascard C (1988) Microbial Transformations. 8. First Example of a Highly Enantioselective Microbiological Hydroxylation Process. J Org Chem *53:* 1797

95 Corey EJ Nagata R (1987) Synthesis of Three New Dehydroarachidonic Acid Derivatives and their Oxidation by Soybean Lipoxygenase. Tetrahedron Lett *28:* 5391

96 Rossiter JT Williams SR Cass AEG Ribbons DW (1987) Aromatic Biotransformations 2. Production of Novel Chiral Fluorinated 3,5-Cyclohexadiene-*cis*-1,2-diol-1-carboxylates. Tetrahedron Lett *28:* 5173

97 Taylor SJC Ribbons DW Slawin AMZ Widdowson DA Williams DJ (1987) Biochemically Generated Chiral Intermediates for Organic Synthesis: The Absolute Stereochemistry of 4-Bromo-*cis*-2,3-dihydroxycyclohexa-4,6-diene-1-caboxylic Acid from 4-Bromobenzoic Acid by a Mutant *Pseudomonas putida.* Tetrahedron Lett *28:* 6391

98 Klibanov AM (1990) Asymmetric Transformations Catalyzed by Enzymes in Organic Solvents. Acc Chem Res *23:* 114

99 Feichter C Faber K Griengl H (1990) Chemoenzymatic Preparation of Optically Active Long-Chain 3-Hydroxyalkanoates. Biocatalysis *3:* 145

100 Hammond DA Karel M Klibanov AM Krukonis VJ (1985) Enzymic Reactions in Supercritical Gases. Appl Biochem Biotechnol *11:* 393

101 Wang Y-F Chen ST Liu KKC Wong C-H (1989) Lipase-catalyzed Irreversible Transesterification Using Enol Esters: Resolution of Cyanohydrins and Syntheses of Ethyl (*R*)-2-Hydroxy-4-phenylbutyrate and (*S*)-Propanolol. Tetrahedron Lett *30:* 1917

102 Guanti G Banfi L Narisano E (1989) Enzymes as Selective Reagents in Organic Synthesis: Enantioselective Preparation of "Asymmetrized Tris(hydroxyme thyl)-methane". Tetrahedron Lett *30:* 2697

103 Björkling F Boutelje J Gatenbeck S Hult K Norin T Szmulik P (1986) The Effect of Dimethyl Sulfoxide on the Enantioselectivity in the Pig Liver Esterase Catalyzed Hydrolysis of Dialkylated Propanedioic Acid Dimethyl Esters. Bioorg Chem *14:* 176

104 Lilly MD (1982) Two-liquid-phase Biocatalytic Reactions. J Chem Technol Biotechnol *32:* 162

105 Carrea G (1984) Biocatalysis in Water-Organic Two-Phase Systems. Trends in Biotechnology *2:* 102

106 Zaks A Klibanov AM (1988) The Effect of Water on Enzyme Action in Organic Media. J Biol Chem *263:* 8017

107 Laane C Boeren S Vos K Veeger C (1987) Rules for Optimization of Biocatalysts in Organic Solvents. Biotechnol Bioeng *30:* 81

108 Adelhorst K Björkling F Godtfredsen SE Kirk O (1990) Enzyme Catalyzed Preparation of 6-0-Acylglucopyranosides. Synthesis 112

109 Zhi-Wei G Sih CJ (1988) Enzymatic Synthesis of Macrocyclic Lactones. J Am Chem Soc *110:* 1999

110 Glass JD (1981) Enzymes as Reagents in the Synthesis of Peptides. Enzyme Microb Technol *3:* 2

111 Zaks A Klibanov AM (1984) Enzymatic Catalysis in Organic Media at 100°C. Science *224:* 1249

112 Kastle JH Loevenhart AS (1900) Lipase, the Fat-Splitting Enzyme, and the Reversibility of its Action. Am Chem Soc *24:* 491

113 Ader U Breitgoff D Klein P Laumen KE Schneider MP (1989) Enzymatic Ester Hydrolysis and Synthesis. Two Approaches to Cycloalkane Derivatives of High Enantiomeric Purities. Tetrahedron Lett *30:* 1793

114 Liu KK-C Nozaki K Wong C-H (1990) Problems of Acyl Migration in Lipase-Catalysed Enantioselective Transformation of *meso*-1,3-Diol Systems. Biocatalysis *3:* 169

115 Ramos-Tombo GM Schär H-P Fernandez i Busquets X Ghisalba O (1986) Synthesis of Both Enantiomeric Forms of 2-Substituted 1,3-Propanediol Monoacetates Starting from a Common Prochiral Precursor, Using Enzymatic Transformations in Aqueous and in Organic Media. Tetrahedron Lett *27:* 5707
116 Degueil-Castaing M De Jeso B Drouillard S Maillard B (1987) Enzymatic Reaction in Organic Synthesis: 2 – Ester Interchange of Vinyl Esters. Tetrahedron Lett *28:* 953
117 Laumen K Breitgoff D Schneider MP (1988) Enzymic Preparation of Enantiomerically Pure Secondary Alcohols. Ester Synthesis by Irreversible Acyl Transfer Using a Highly Selective Ester Hydrolase from *Pseudomonas* sp. An Attractive Alternative to Ester Hydrolysis. J Chem Soc, Chem Commun 1459
118 Bianchi D Cesti P Battistel E (1988) Anhydrides as Acylating Agents in Lipase-Catalyzed Stereoselective Esterification of Racemic Alcohols. J Org Chem *53:* 5531
119 Berger B Rabiller CG Konigsberger K Faber K Griengl H (1990) Enzymatic Acylation using Acid Anhydrides: Crucial Removal of Acid. Tetrahedron Asymmetry *1:* 541
120 Niedermeyer W Kula M-R (1990) Enzyme catalyzed Synthesis of (*S*)-Cyanohydrins. Angew Chem Int Ed Engl *29:* 386
121 Bednarski MD Simon ES Bischofberger N Fessner W-D Kim M-J Lees W Saito T Waldmann H Whitesides GM (1989) Rabbit Muscle Aldolase as Catalyst in Organic Synthesis. J Am Chem Soc *111:* 627
122 Wong C-H Whitesides GM (1983) Synthesis of Sugars by Aldolase-Catalyzed Condensation Reactions. J Org Chem *48:* 3199
123 Shirae H Yokozeki K Kubota K (1988) Enzymatic Production of Ribavirin from Pyrimidine Nucleosides by *Enterobacter aerogenes* AJ 11125. Agric Biol Chem *52:* 1233
124 Utagawa T Morisawa H Yamanaka S Yamazaki A Hirose Y (1986) Enzymatic Synthesis of Virazole by Purine Nucleoside Phosphorylase of *Enterobacter aerogenes.* Agric Biol Chem *50:* 121
125 Rosevear A Kennedy JF Cabral JMS (1987) Immobilised Enzymes and Cells. Hilger Bristol
126 Suckling CJ (1977) Immobilized Enzymes. Chem Soc Rev *6:* 215
127 Isobe M Sugiura M (1977) Purification of Microbial Lipases by Glass Beads Coated with Hydrophobic Materials. Chem Pharm Bull *25:* 1987
128 Burg K Mauz O Noetzel S Sauber K (1988) Neue synthetische Träger zur Fixierung von Enzymen. Angew Makromol Chemie *157:* 105
129 Peberdy JF (1987) Genetic Engeneering in Relation to Enzymes. In: Rehm HJ Reed G (eds) Biotechnology vol 7a. Verlag Chemie Weinheim p 33
130 Kaiser ET (1988) Catalytic Activity of Enzymes with Modified Active Center. Angew Chem Int Ed Engl *27:* 902
131 Page UI Williams A (eds) (1987) Enzyme Mechanisms. Royal Soc of Chemistry London
132 Pike VW (1987) Synthetic Enzymes. In: Rehm HJ Reed G (eds) Biotechnology, vol 7a. Verlag Chemie Weinheim p 28
133 Schultz PG (1989) Antibodies as Catalysts. Angew Chem Int Ed Engl *28:* 1283
134 Schultz PG Lerner RA Benkovic SJ (1990) Catalytic Antibodies. Chem Eng News May *68:* Nr. 22 S. 26
135 Hennen WJ Sweers HM Wang Y-F Wong C-H (1988) Enzymes in Carbohydrate Synthesis: Lipase Catalyzed Selective Acylation and Deacylation of Furanose and Pyranose Derivatives. J Org Chem *53:* 4939
136 Sicsic S Leroy J Wakselman C (1987) Geometric Selectivity of Pig-Liver Esterase and its Application to the Separation of Fluorinated Bicyclic Esters. Synthesis 155

137 Kieslich K (1976) Microbial Transformations of Non-Steroid Cyclic Compounds. Thieme Stuttgart

138 Effenberger F Horsch B Förster S Ziegler Th (1990) Enzyme Catalysed Synthesis of (S)-Cyanohydrins and Subsequent Hydrolysis to (S)-α-Hydroxycarboxylic Acids. Tetrahedron Lett *31:* 1249

7 Preparation of Homochiral Organic Compounds

E. Winterfeldt

7.1 Introduction

Due to the extent to which the overlap of organic chemistry and particularly organic synthesis with biology, biochemistry, and medecine is increasing, the necessity of preparing homochiral compounds of a given and predictable absolute configuration is becoming more and more important. Not only because the biological activity of chemical compounds is linked to their absolute configuration in a well-defined way, thus rendering the preparation of homochiral products a conditio sine qua non for medicinal chemistry and for plant protection chemistry, but also since the investigation of compound-enzyme interaction, of receptor chemistry and of all types of chiral recognition above all need the availability of pure enantiomers as do all the current efforts to probe reaction mechanisms – particularly of biogenetic key steps – and the attempts to determine a scientific relationship between optical rotation and absolute configuration. The challenge to develop reliable methods for the preparation of homochiral compounds has been met by synthetic organic chemistry and particularly in the last twenty years we have seen remarkable progress in the efficiency of enantioselective transformations and an unusual increase in the efforts to prepare homochiral compounds.

This goal is, at the moment, generally reached in three different ways which may be characterized by the following.

7.2 Separation Techniques

First, one has to mention the various separation techniques [1], simply because they were the first to provide pure enantiomers from racemates. In the general procedure the racemic compound is treated with a homochiral base or acid to form diastereomeric salts in the hope that one of them will crystallize with preference. The obvious disadvantage is that this technique is highly empiric, a thorough investigation of various acids and bases as well as of a series of solvent systems generally being necessary, additionally if there is no possibility of racemizing the unwanted enantiomer, one loses 50% of the material in the process. On the other hand the operation is extremely simple and can be

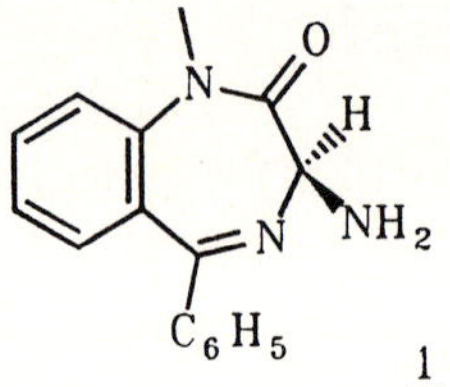

1

Scheme 7.1.

done even with comparatively large amounts of material. It becomes of particular interest if the separation can be done under equilibrating conditions. Very low solubility of one of the salts given, this then can lead to a nearly complete transformation of the racemic mixture into one enantiomer. Quite efficient examples of this type have been reported from the area of heterocyclic amines [2, 3].

Amine **1** for instance was obtained as a salt with camphorsulfonic acid in the presence of 3,5-dichlorobenzaldehyde, which obviously forms the Schiff base for fast equilibration. This way 91% of the material is transformed into one pure enantiomer.

Besides this, quite a number of chromatographic separation techniques [4] are of course available, which either make use of diastereoisomeric derivatives deliberately prepared for this purpose, or which operate on a chiral base thus separating racemic mixtures directly. Unfortunately, no universal column which could be used for any type of racemate seems to be available at the moment. For preparative use, one again runs into difficulties if no reracemisation of unwanted enantiomers can be achieved.

7.3 Homochiral Building Blocks from Natural Products

Second there is the rather old technique of transforming easily and cheaply available homochiral natural products such as sugars, amino acids, alkaloids, and terpenes in a sequence of diastereoselective reactions into a useful and hopefully highly flexible configurationally well-defined intermediate for the preparation of pure enantiomers. This means that in principle here is no enantioselectivity whatsoever involved as one is starting from a homochiral material right at the beginning. The crucial and decisive aspect with the use of these compounds is the so called "chemical distance", which is the number of steps needed to transform the natural product into the intermediate wanted. If this chemical distance is comparatively long the advantage of the low price for the starting material is lost very quickly. This means that one has to look closely at those compounds that are hopefully structurally very close to the material wanted. Another very often quite annoying drawback with these compounds is their lack of constitutional flexibility – one can for

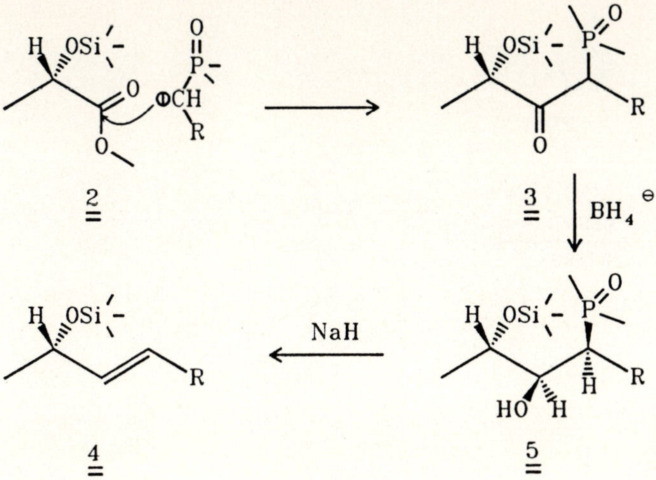

Scheme 7.2.

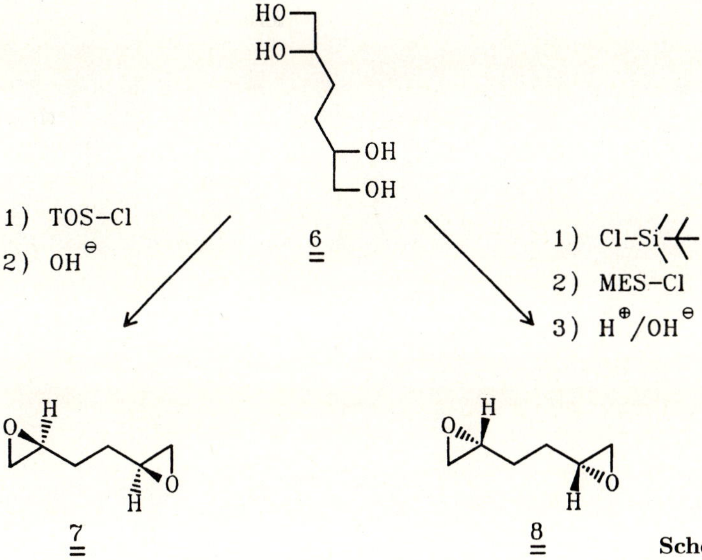

Scheme 7.3.

instance prepare quite a number of homochiral allylic alcohols of the general type **4** by using Warren's phosphine oxide technique [5], starting from silyl protected lactic acid ester **2** [6] and although the method turns out to be simple and reliable, it is of course restricted to those alcohols that are methyl substituted.

Another important restriction is the fact that in general only one absolute configuration will be available which means of course a lack of configurational flexibility and normally you can only prepare one enantiomer. In this case a very useful solution to this dilemma would be an operation which may be called enantiodivergent synthesis. In this case one makes use of a se-

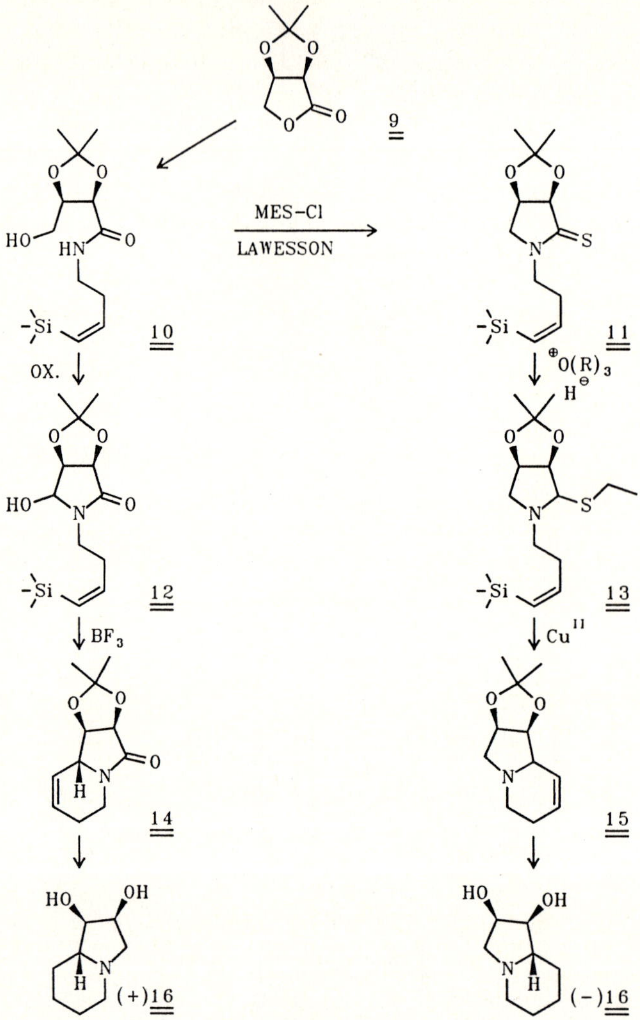

Scheme 7.4.

lected differentiation of functional groups in the molecule thus using them to make different types of bonds fo finally arrive at different overall configurations starting from a common homochiral intermediate. Typical examples are the recently communicated formation of the bis-epoxides **7** and **8** from the mannitol derived alcohol **6** [7].

By appropriate manipulations of the primary versus the secondary hydroxy groups one can prepare **7** as well as its enantiomer **8**. In the second example L. Overmann [8] managed to prepare both enantiomers of *cis*-indolizidine diol **16** starting from readily available D-isoascorbic acid which can be transformed in quantity into lactone **9**. Subsequent directed selectivity in the crucial cyclization step generating the indolizidine system opens the road to both mirror images. Another effort aiming at higher configurational flexibility makes use of the possibilities for self-reproduction of chirality [9]. Starting from optically active compounds of type **17** one can easily prepare and separate the cyclic acetal derivatives **18** and **19** after the chiral information has been exported in the acetal centre one may safely destroy the original configuration by enolization and still get back to homochiral reaction products of type **21**. With diastereoisomer **19** at hand there is, of course, reliable access to thc opposite configuration of the α-hydroxy acid derivative. One may easily imagine various other starting materials for exercises of this type and corresponding examples have been published [10].

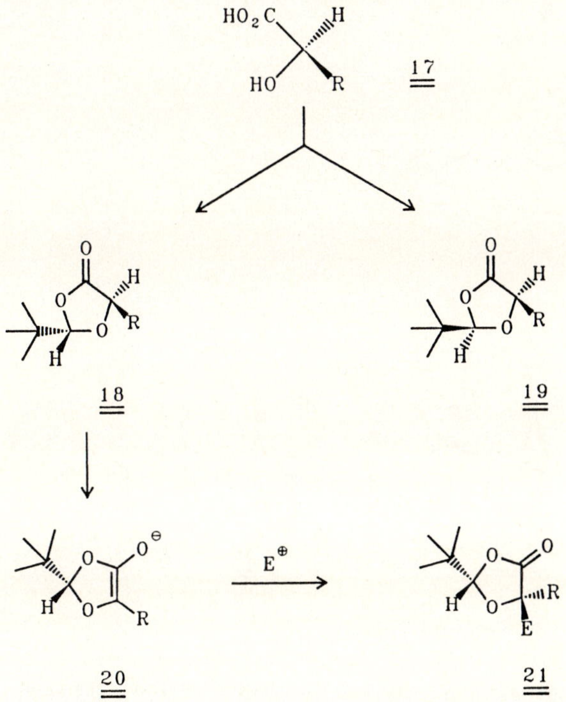

Scheme 7.5.

Finally one has to stress the point that a great number of biological active natural products are of course quite often biogenetically derived from simple easily available very fundamental naturally occurring compounds. In this case there will be a great desire to also use these compounds for a synthetic venture. Either with the aim of biomimetic synthesis, which hopefully operates along the lines of the biogenetic scheme, or with the intention of at least using the absolute configuration of the natural starting material. A lot of work in this direction has been done with sugars [11] and it is very tempting to do very similar things with amino acids. With these compounds, however, a quite serious problem has to be solved first. Although all important alkaloids are biosynthetically linked to amino acids [12, 13] their carboxy group is in most cases lost in the process, which means that in order to use amino

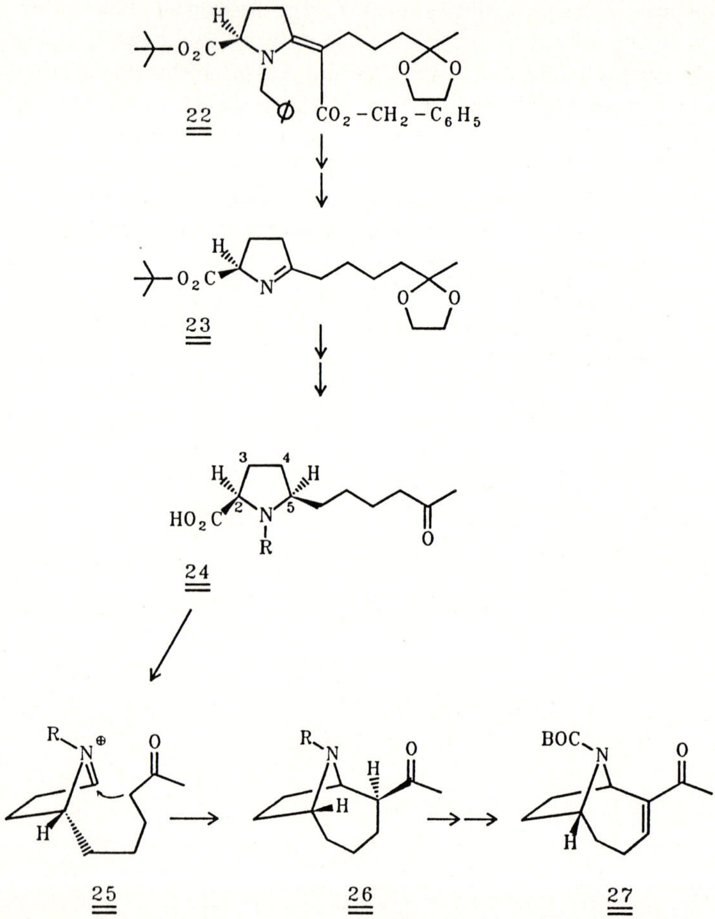

Scheme 7.6.

acids one has to have a convenient and hopefully comparatively mild technique to get rid of this functional group after the chiral information gained from the crucial carbon atom carrying the carboxy group has been safely deposited somewhere in the carbon framework of the potential intermediate. Classic purely thermal or copper catalyzed decarboxylation processes do not look very promising for this task and so the more modern decarbonylation reactions using phosphorous oxychloride or oxalylchloride as well as electrochemical decarboxylations [15] have become quite popular. A very elegant application of this concept can be found in H. Rapoport's synthesis of anatoxine [16] which started from glutamic acid. Intermediate **22** – accessible from pyroglutamate in a few steps – does decarboxylate after reductive debenzylation to form imine **23**.

The hydrogenation of this double bond is then very efficiently directed by the bulky *tert*-butylester to yield, after deprotection, **24** with remarkable stereoselectivity. Having introduced this well defined sp^3-centre at carbon atom 5 the time has come to convert the original centre of chirality (C_2) into an iminium salt, ready for Mannich cyclization (see **25**). This operation then leads to the correct carbon framework as seen in **26** which by exchange of the protecting group and dehydrogenation can be converted into boc-protected anatoxine.

Next to these quite fundamental achievements there is of course a very general worldwide effort to use the carboxy group of amino acids for the preparation of useful intermediates like aldehydes [17], ketones [18], alcohols [19] and amines [20]. There are also a number of easily available and cheap interesting starting materials in the terpene series but again the problem of chemical distance plays a crucial role and this may be one explanation that mainly comparatively simple cyclic and bicyclic compounds like menthol, carvon, pulegon, carene, and camphor, have been used as precursors for chiral building blocks. Some of these compounds and this is particularly true for camphor have also been used with great success for the generation of auxiliaries, which we have to deal with in the next section.

7.4 Auxiliary Modified Substrates

In the previous sections, compounds with a given absolute configuration were available right from the beginning and the homochiral compound was reached by separation and chemical transformation. When for the third important technique it comes to the chemical generation of pure enantiomers from prochiral starting materials as for instance homochiral alcohols (**29**) from ketones (**28**) or substituted ketones (**31**) from enolates (**30**) the interaction between substrate and reagent has to be directed by either substrate or reagent including catalysis or any other entity – solvents included – which will influence the formation of the transition state, in order to make sure that the generation of sp^3-centre takes place from one space sector exclusively or at least predominantly.

An inspection of the enolates of type **30** additionally gives an idea of the importance of configurational details of the substrate, as *E*- and *Z*-stereoisomers give rise to opposite enantiomers even if the attack of the electrophile takes place with face selectivity. To direct the attacking agent efficiently into one space sector preferentially, one has to place some chiral information very close to the prochiral centre with the intention of influencing the activation barriers for the formation of both in principle possible diastereomeric transition states so strongly, that the reaction path will only or preferentially run through the low-barrier transition state. The directing group can, in principle, be placed into the substrate or into the reagent or catalyst or both but even a chiral solvent cage could have a directing effect. This has been shown very generally to be the case but for preparative use the enantiomeric excess obtained this way is unsatisfactory. At this stage a very general remark on the state of the art of this field has to be made.

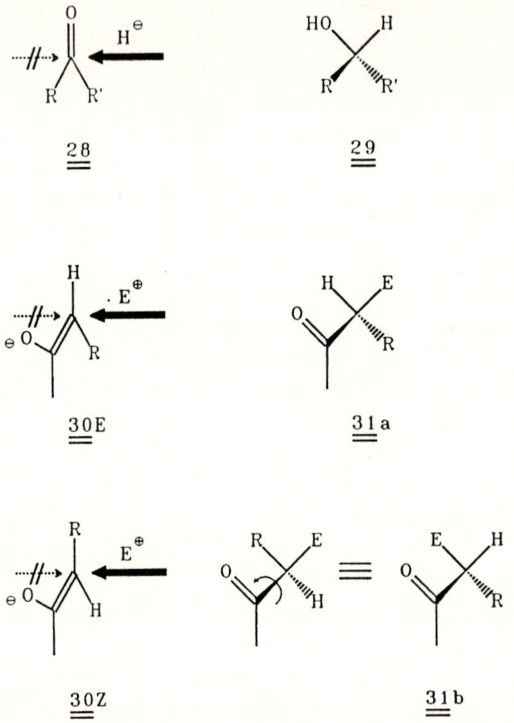

Scheme 7.7.

In the early years of enantioselective synthesis it was very common to consider selective transformations a success even if the selectivity was quite low, as one was mainly trying to prove that induction can in principle be achieved. This has of course changed completely in the last twenty years. Nowadays these transformations are being looked at as a preparative tool and this demands enantiomeric excesses in the order of 90% or more to make sure that after one isolation and purification procedure one will have individual enantiomers available with a degree of purity that at least corresponds to most natural products. Another very important difference to the early days is that nowadays one is aiming at predictable routes to homochiral compounds. As compared to the old days much more is known about intermediates and transition states of the reactions involved and one tries to make sure to influence the steric outcome in a well-defined way and reach individual enantiomers at will. To this end a number of facts have to be considered that will be of high importance for highly efficient transformations.

First of all, one is well advised to pick organic reactions that are well-known for passing through highly organized transition states which have strict sterical demands to all centres involved. This explains why processes with cyclic and quite rigid transition states like cycloadditions, sigmatropic rearrangements, aldoladditions, and metalloorganic processes in general are extremely popular in this field. Additionally one may fix the decisive centres involved rigidly by either operating at rather low temperatures or lowering the degrees of rotational freedom by chelate formation with appropriate counter kations or by additional σ-bonds that have to be broken again at a later stage when the auxiliary is regained. Examples will be given later. In any case these general demands are certainly responsible for the fact that mainly cyclic and bicyclic auxiliaries like those derived from proline (**32**) and camphor (**33**) or those with bulky substitients (from valine **34**) have been used.

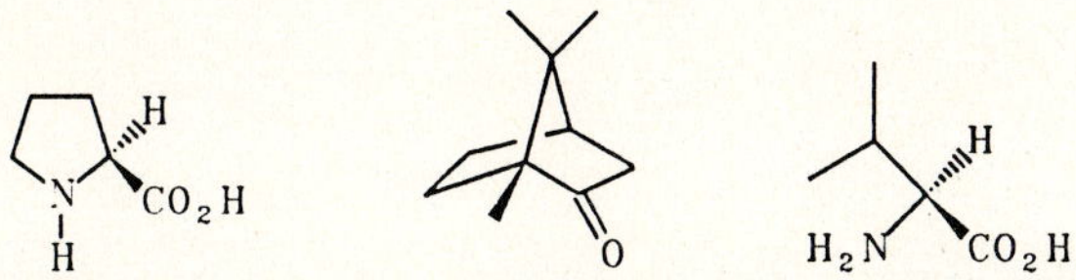

29 33 34

Scheme 7.8.

There are quite a number of reports on the general techniques in this field [21–24] and as there is certainly no room in this chapter for a complete treatment of all the very modern and elegant work in this area, a few selected examples, which are meant to illustrate the importance of the above mentioned directing effects, will suffice.

Purely restricted rotation in cooperation with efficient shielding of one side of the substrate is at work with derivatives like **35** [25] and **36** [26].

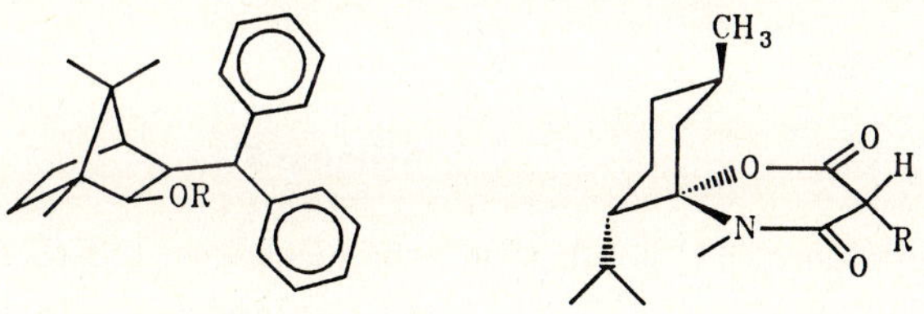

35 36

Scheme 7.9.

Although this is not assisted by any chelate formation, **35**(R^*) worked extremely well in our enantioselective formation of spiropiperidines of type **38** from diester **37** [27].

As this cyclization process was shown to proceed through the cyclohexenone **39** there is obviously a high preference for β-attack on this electron poor double which as models show is due to the camphor residue. In a very similar way the rigid substutited ring system of **36** favours α-attack on anions of type **40** derived from this species.

A quite efficient cooperation of chelation and restricted rotation can be inspected at a number of now quite popular synthetic methods like the already widely used D. Evans protocol, which starting from a cyclic system

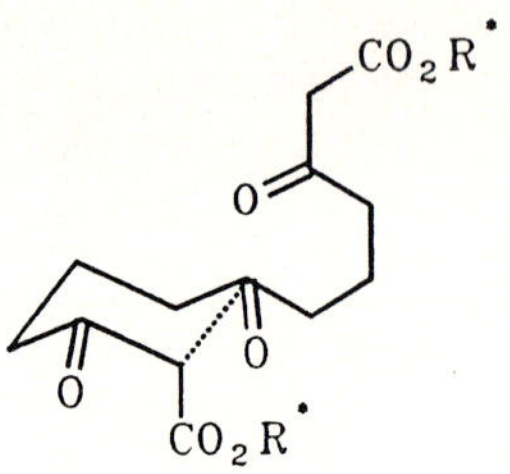

37

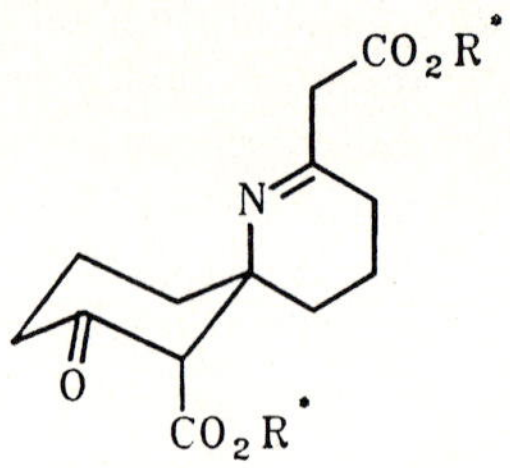

38

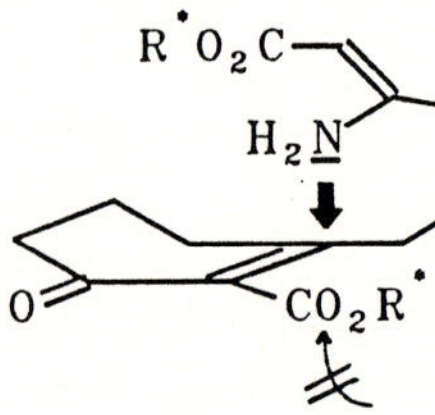

39

Scheme 7.10.

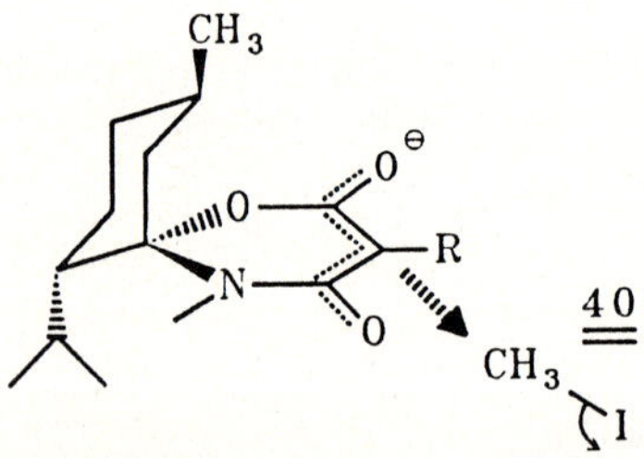

40

Scheme 7.11.

with a bulky directing substituent (**41**) is taken into the nicely tied together boronenolate **42** [28].

Interestingly the reaction with the synthetically highly flexible aldehyde **43** did not only proceed with a remarkable *syn*-selectivity (98:2) to generate **44**, but may also be manipulated mechanistically (increasing amount of Lewis acid) to switch over to a very efficient production of the corresponding *anti*-isomer. Processes of this type which lend themselves to manipulated selectivity are extremely popular of course as they guarantee configurational flexibility.

Further examples in this area are the amides investigated by Helmchen and Oppolzer [29] (see **45**), the glycin equivalent **46** [30], the chiral carbonyl derivative **47** [31] which in its Lewis acid activated form **48** [32] may also be attacked by nucleophiles as well as the A.L. Meyers version of tetrahydroisoquinoline (**49**) and tetrahydrocarboline **50** [33].

With these last two examples the directing power of a rigid chelate can nicely be demonstrated. If one omits the ether-handle from the chelating side

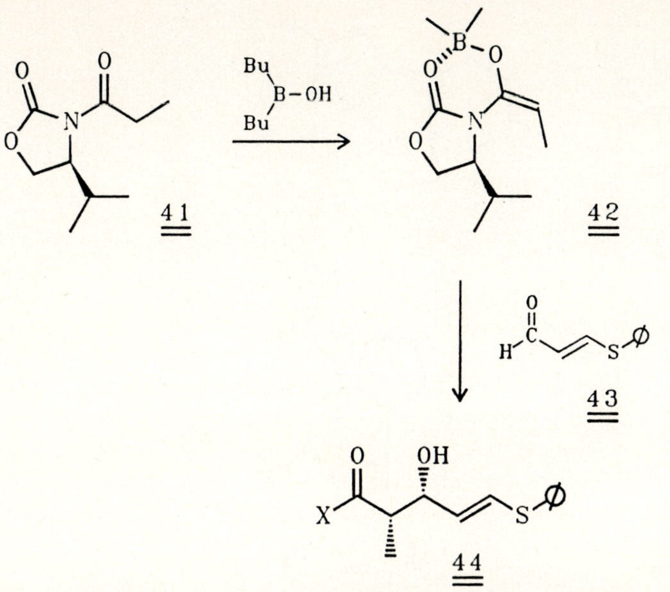

Scheme 7.12.

chain, the enantiomeric excesses in simple alkylation reactions drop sharply. The camphor sulfonic acid derivative **45** does on the one hand offer excellent opportunities for highly selective nucleophilic additions but chelates of this type also represent the electron-poor 2π-systems for cycloaddition processes. That this is a very general phenomenon with concerted reactions in general may be judged from the Diels-Alder additions with **51** and **52** [34] and the ene-reactions with **53** [35].

With reactions like these there is no strict differentiation into substrate and reagent anymore and so we shall have to return to reactions of this type when we are going to address chiral reagents. For the time being we have to focus briefly on those compounds where additional rigidisation and firm organization of the substrate is not achieved by chelation but by an additional σ-bond, to form polycyclic intermediates, which may be particularly useful if their overall bent conformation allows for application of the concave-convex principle. This principle simply implies that with molecules of the general structure **54** reagents will always preferentially attack from the upper side of the molecule.

This principle operates in a great number of cases and so we shall just pick the example of the directed Birch-reduction to demonstrate this effect. The proline modified salicyclic acid derivative **55** gives rise, after Birch-reduction and interception of the enolate with allylic bromide, to the enolether of a β-ketoester **56** with remarkable diastereoselectivity (98:2) [36] and it should be mentioned at this stage that if a ring open chiral amide is used the direction of the electrophilic attack may be reversed.

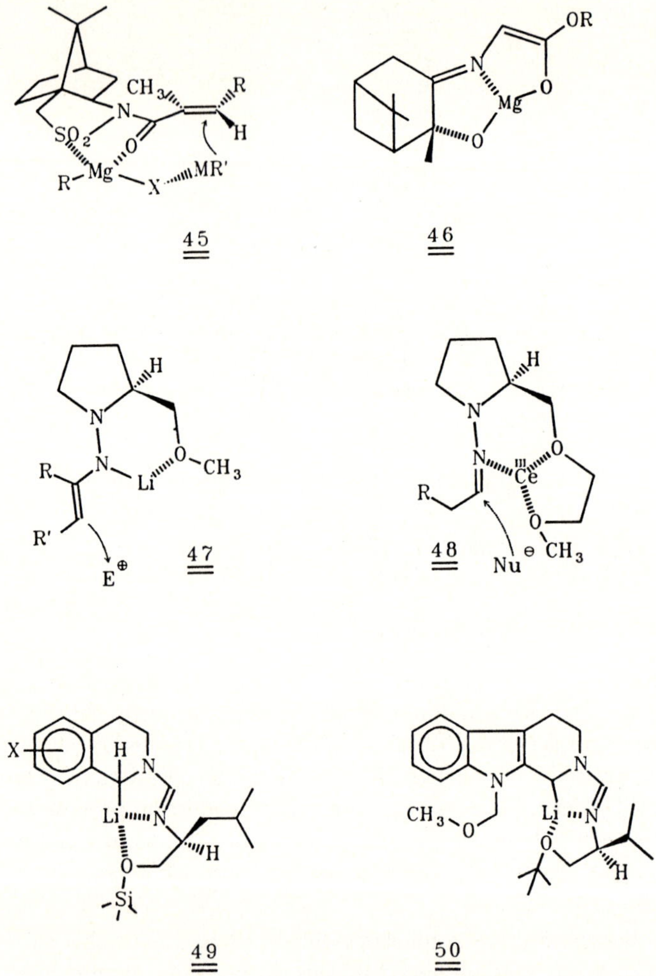

Scheme 7.13.

Another very useful example of a polycyclic rigid framework, which operates at the same time as a protecting group for a ketone, was devised by A.J. Meyers [37] and for γ- and δ-keto acids. In the case of γ-keto acids treatment with valinole gives rise to the bicyclic lactam **57**.

In subsequent alkylation reactions the electrophile attacks the more or less planar enolate from the α-side, off the two directing β-substituents. This gives a very convenient opportunity to change the absolute configuration at the carbon atom next to the carbonyl group by simply changing the sequence of electrophiles thus generating either **59** or its diastereomer **60**. Both, after hydrolysis and functional group manipulation, can be converted into the two enantiomers of a homochiral cyclopentenone (**61**).

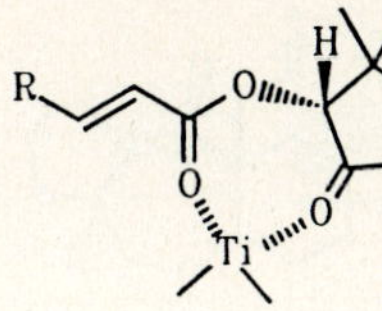

51

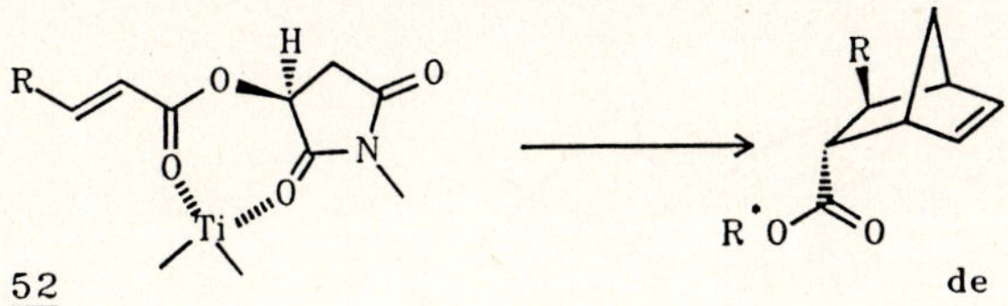

52

de = 99: 1

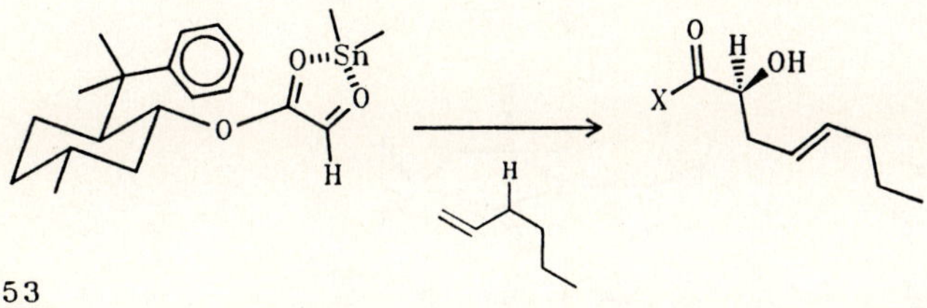

53

Scheme 7.14.

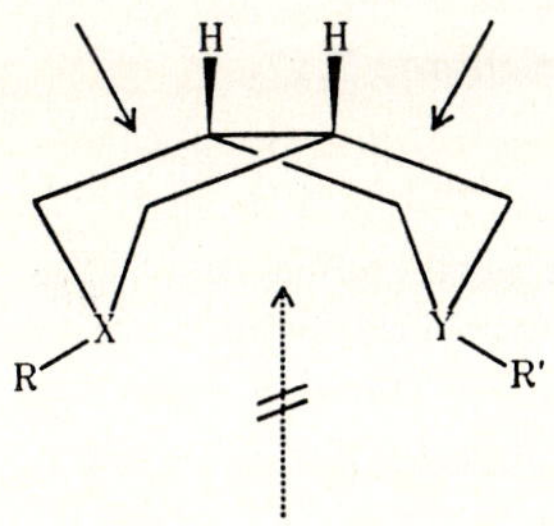

54

Scheme 7.15.

In all these cases there is the directed construction of one special configuration which is considered the stepping stone for a sequence of diastereoselective transformations. One could expect even more from a system that would keep a prochiral starting material safely locked in a chiral surrounding of a given absolute configuration for quite a while, thus allowing for a sequence of quite different operations to be done with the substrate, which

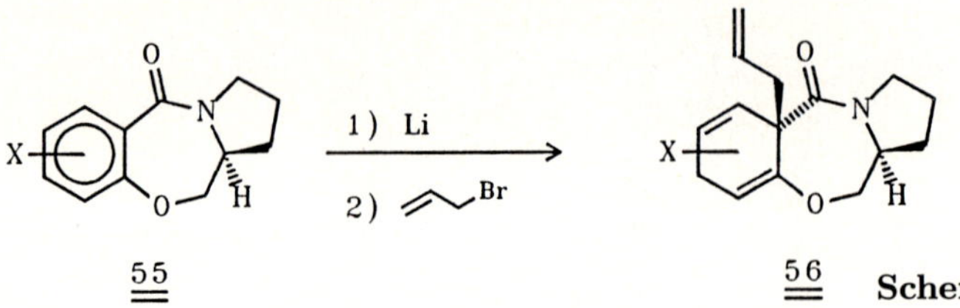

Scheme 7.16.

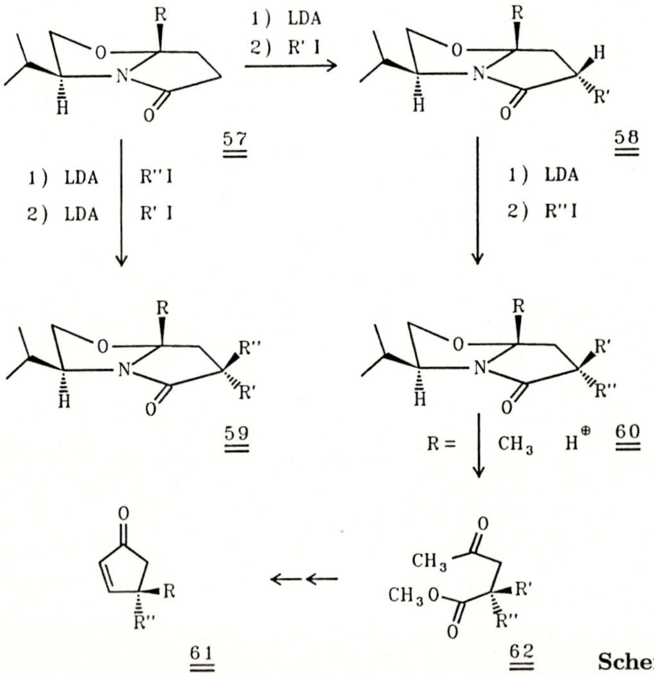

Scheme 7.17.

then would be released from its homochiral cage as a pure enantiomer. This procedure would have resemblance to an enzyme-complex catalyzing a series of reactions. The problem is how to fix and how to untie the substrate and the chiral template. A very simple and convenient solution of this problem is certainly to use a homochiral 4π-system as the template which then could be added to the substrate 2π-system. After a sequence of reactions one could then separate both parts again in a thermal or catalyzed retro-Diels-Alder reaction, to thus regenerate the chiral diene and set free a homochiral reaction product. But to use this concept properly a number of demands have to be met.

First of all one needs a proper 4π-system which by virtue of its substituents "L" and "S" should direct the stereoselectivity to get face-selective attack from the side of the smaller residue "S". Next there should be a donor

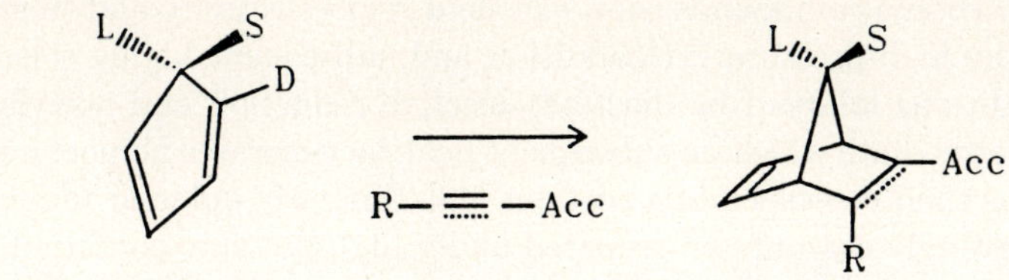

63 64

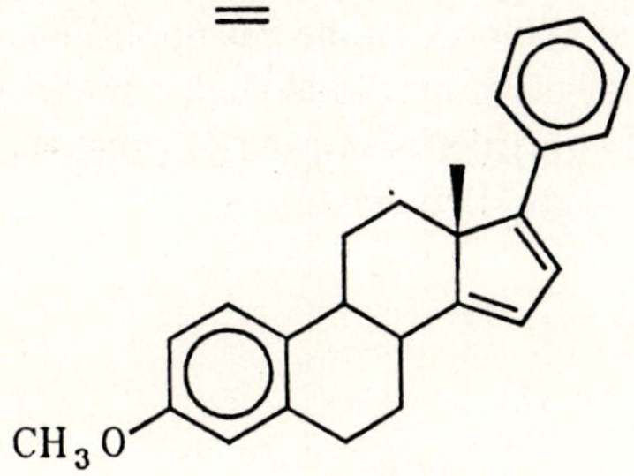

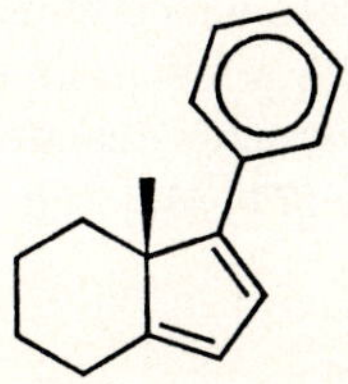

65 66

Scheme 7.18.

substituent to according to frontier orbital theory take care of the regioselectivity (see **64**) and in case of an olefinic 2π-substrate there should also be high *endo*-selectivity, as with *endo*-adducts particularly high stereoselectivity may be expected for the following steps. As Solo [38] noticed high stereoselectivity and also excellent *endo*-selectivity for ring-D cyclopentadienes in the steroid series, the cyclopentadienes **65** and **66** were prepared [39] and were readily shown to meet all these demands.

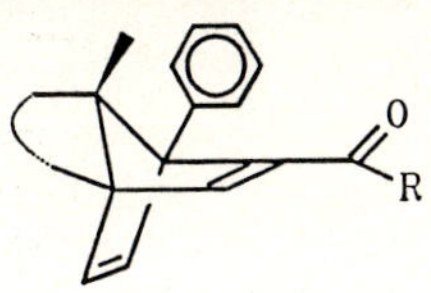

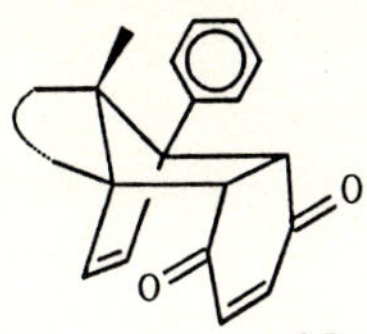

67 68

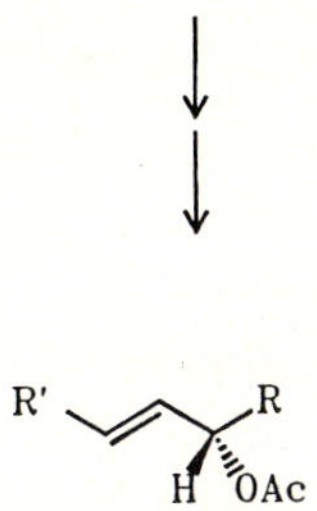

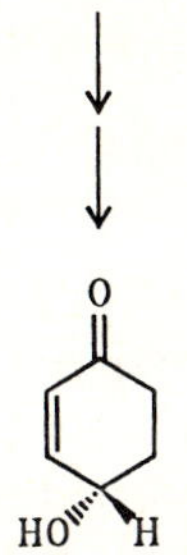

69 70

Scheme 7.19.

Acetylenic carbonyl compounds show excellent regioselectivity and stereoselectivity in the high-pressure cycloaddition and subsequent highly selective cuprate additions followed by diastereoselective reduction and acetylation yield an intermediate which on subsequent heating generates homochiral allylic alcohols as their corresponding acetates [40]. Similarly quinone adduct **68**, which interestingly can only be prepared under high-pressure conditions, is formed with extremely high stereoselectivity as the endo-adduct exclusively [41]. Hydrogenation followed by selectride reduction then gives rise to the intermediate which eliminates the homochiral cyclohexenone **70** upon heating. To cite a few more examples of enantioselective methods that somehow mimic enzymatic reactions the two chemical equivalents of NADH prepared by Kellog (**71**) [42] and Davies (**72**) [43] are a good choice.

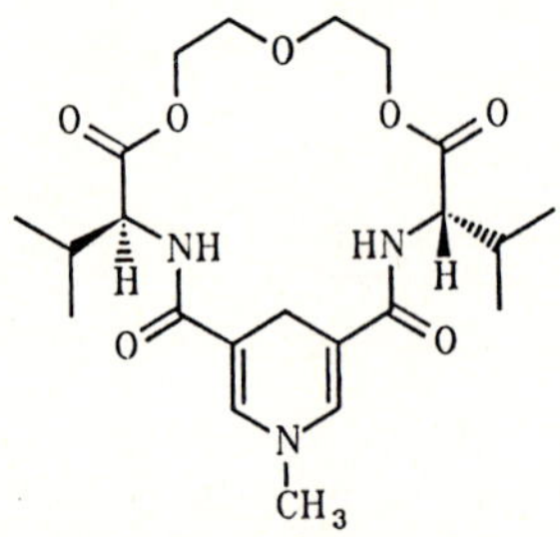

71

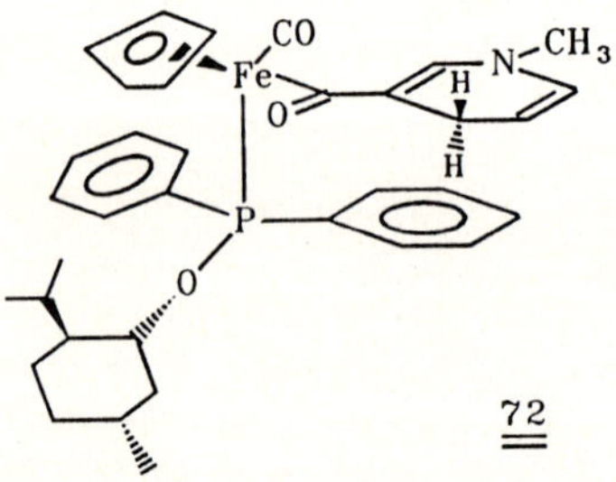

72

Scheme 7.20.

7.5 Homochiral Reagents

In connection with concerted reactions and cycloadditions there arose the problem already to strictly differentiate between substrate and reagent. Nevertheless there is quite a number of compounds in the literature by now which one may safely call a homochiral reagent. A very typical and highly flexible example is the titanium complex of tartrate modified tert.butylhydroperoxide which was introduced by Sharpless [44]. By now this is not only a very convenient tool for the preparation of chiral epoxides from allylic and homoallylic

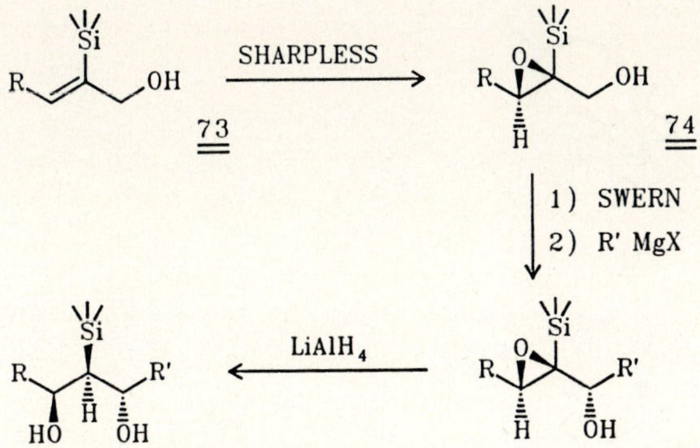

Scheme 7.21.

alcohols, but may – as diol **75** proves –, also be used with different substituted olefins, with the aim to further elaborate the corresponding epoxide **74**, which is the product of an enantioselective epoxidation of olefins of type **73** [45].

In the next example the homochiral epoxide **77**, again prepared with the Sharpless reagent, undergoes a regioselective intramolecular ring opening process via carbamate **78** to afford after oxidation hydroxyamino acids of type **79** [46].

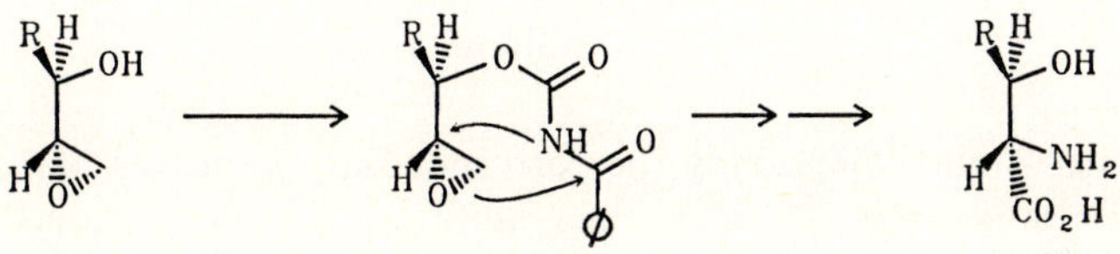

Scheme 7.22.

There are also very efficient reducing reagents which have been reviewed very thoroughly [47] but a few examples may be mentioned. First of all the quite popular and broadly investigated hydroboranes **80** [47] and **81** [47] have to be mentioned and additionally the two aluminumhydride complexes with bis-naphthol (**82**) [48] and with the so called Darvon alcohol (**83**) [49].

This very general principle, according to which a metalloorganic reagent is buttoned into a tailor made chiral coat is seen at work also with other metalloorganic reagents that supply nucleophilic carbon atoms and with deprotonating reagents as well. In all these cases a quite rigid, chiral set of ligands will enforce face-selective attack on a prochiral carbonyl group or CH_2-group. Quite remarkable results have been obtained for instance with the various chiral versions of zincdiethyl [50]. Sugar modified titaniumallyl compounds (e.g. **84**) [51, 52] delivered their C_3-unit with excellent enantioselectivity too and as structural data have been made available by an X-ray investigation [52] the results can be discussed on a firm configurational basis.

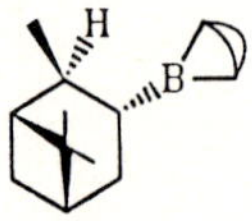

80

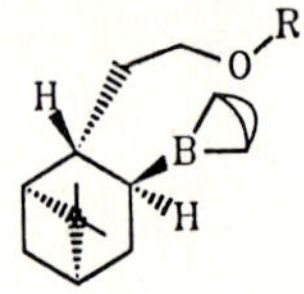

81

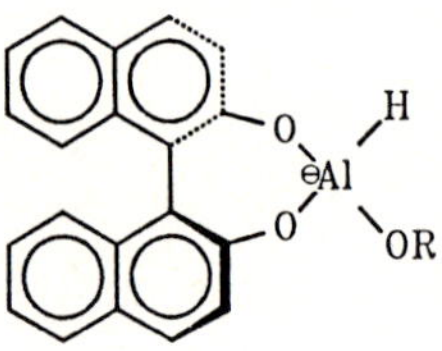

82

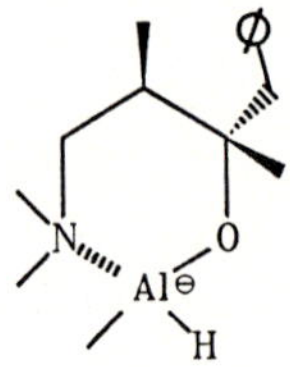

83

Scheme 7.23.

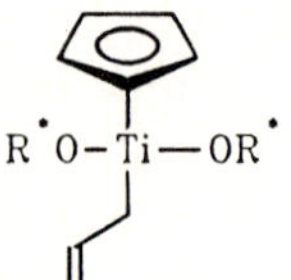

R* = SUGAR DERIVATIVE

84

Scheme 7.24.

In connection with allylic compounds the corresponding boronates naturally have to be mentioned too [53] since their addition to carbonyl compounds gives rise to configurational well-defined homoallylic alcohols which are synthetic equivalents of the corresponding aldols. Typical examples are **85** [54] and **86** [55].

In both cases the chiral information is incorporated into the diol residue but it may also be located in the allylic residue as for instance in **87** and this reagent has been used for an investigation probing the so called matched-mismatched combinations [56].

As the well-defined configuration of **87** has to fit the crucial three allylic carbon atoms into a chair like rigid transition state one can easily envisage some differences in space demand if either **87** or its corresponding enantiomer have to react with the chiral aldehyde **88**. If **87** fits nicely (matched pair) then its enantiomer is bound to have problems (mismatched pair). It has to be stressed at this stage that this is certainly a very general phenomenon which one will encounter whenever homochiral substrates are treated with homochiral reagents and a few very impressive examples have been presented in S. Masamunes brilliant article [56]. As formula **89** shows **87/88** do represent the matched pair indeed, as **89** is the only reaction product in 66% yield [57]. As may be expected treatment of **88** with the antipode of **87** gives rise to two reaction products in a 10:1 ratio (mismatched!). Just to prove the

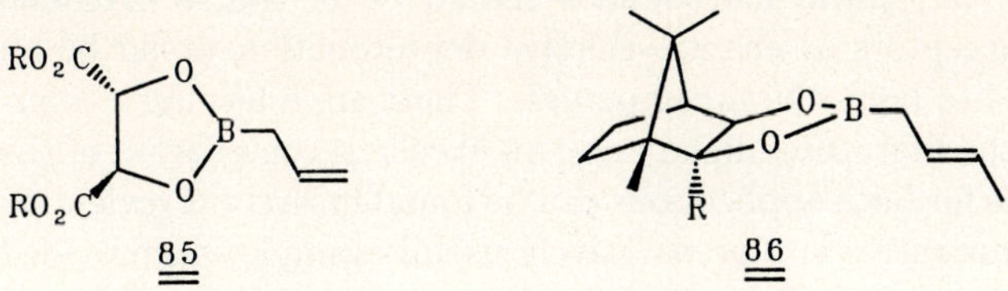

Scheme 7.25.

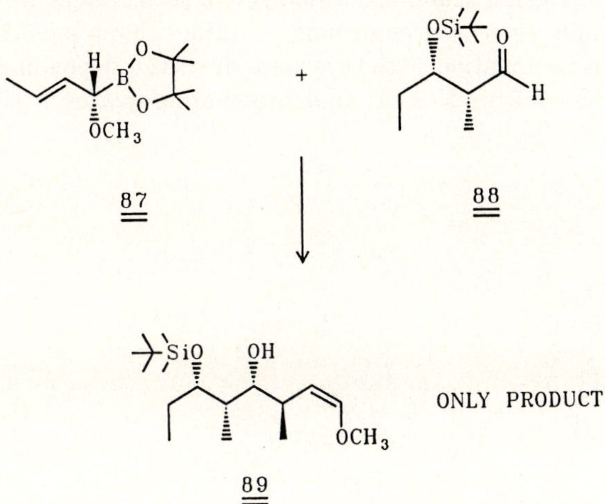

Scheme 7.26.

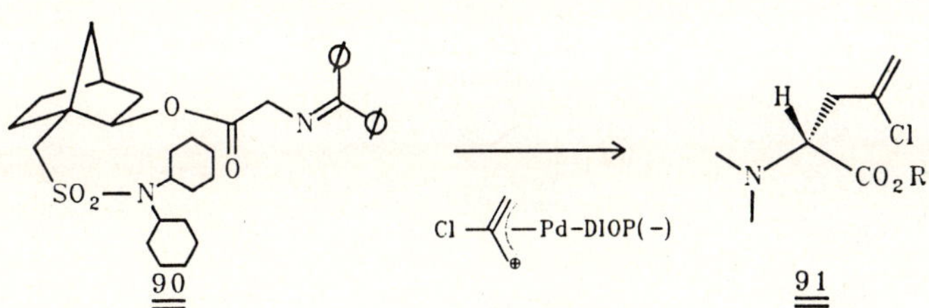

Scheme 7.27.

generality of this concept another completely different example should be looked at. The Pd^0-catalized alkylation of glycine derivative **90** shows high selectivity in the formation of **91**, only if the (-)-diop complex is used as the reagent [58].

This proves that chiral metal complexes in general can be quite useful for various transformations and S.L. Blystones [59] broad and informative review article provides quick entry into the field.

For preparative use, chiral nucleophiles should be of course extended into chiral proton acceptors as enantioselective deprotonation would be an extremely easy route to homochiral compounds. There are a few quite interesting examples in the literature already and an excellent compilation of the various chiral ligands for both applications can be found in a recent review article [60]. A quite illustrative and preparatively useful example was provided by Koga in the bicyclooctane field [61]. As these compounds (see **92**) are characterized by a very rigid and conformational quite stable bicyclic framework the approach to the two CH_2-groups flanking the carbonyl group will be very much influenced by the particular space demand of the chiral deprotonating agent and in a quite systematic screening of chiral lithium amides two completely different deprotonation pathways were disclosed, depending on the constitution and the configuration of the deprotonating species.

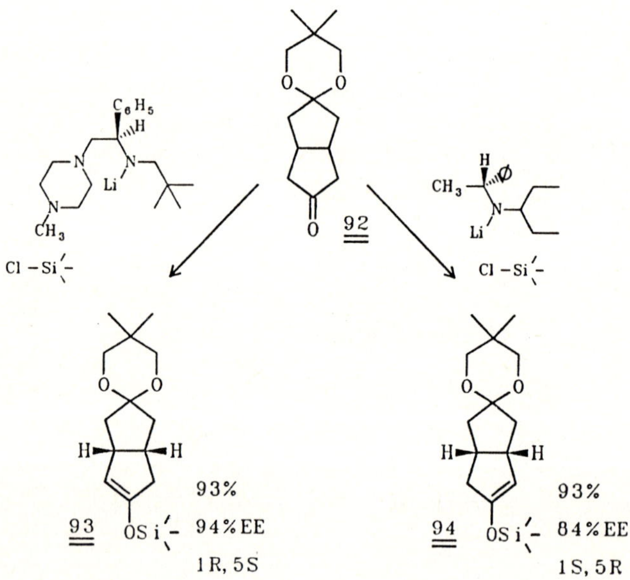

Scheme 7.28.

The corresponding enolates, which are intercepted as the silylethers **93** or **94**, are formed with a remarkable selectivity. Even more exciting than an enantioselective deprotonation would be a face-selective enolate protonation to generate homochiral compounds with a chiral proton source. If one would start from a racemic mixture this deprotonation-reprotonation sequence would correspond to a deracemization procedure. Most of the pioneering work in this field was done by L. and P. Duhamel [62] and by protonating enolates of amino acid derivatives with chiral amines enantiomeric excesses as high as 70% were achieved [62c, 62d].

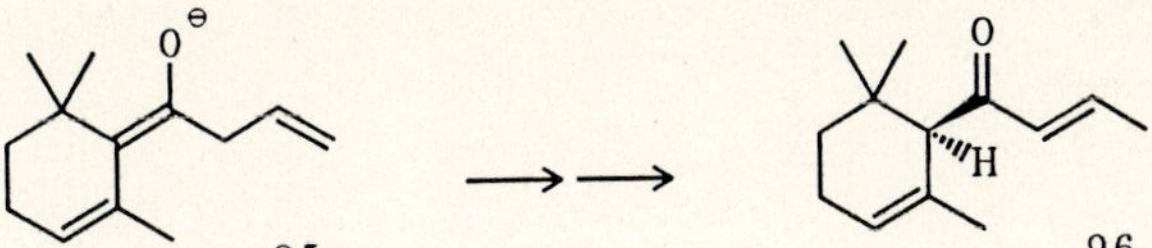

95 96 Scheme 7.29.

Another very interesting and challenging example was provided by Fehr [63], who on protonation of a mixed magnesium/lithium salt of enolate **95** with a chiral amine was able to prepare $R(+)$-a-damascone (**96**) with more than 80% EE.

7.6 Homochiral Catalysts

The preparation of homochiral catalysts – one of the most exciting areas in this field – is developing rapidly at the moment [64]. A very early and quite stimulating result was the enantioselective synthesis of the Hajos-Wiechert ketone (**97**) nearly twenty years ago in a proline catalyzed Robinson annelation [65]. This compound was originally prepared as a chiral building block for steroid synthesis but has in the meantime developed into a very general chiral intermediate [66].

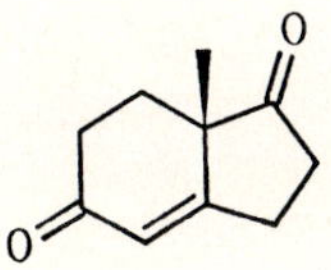

97 **Scheme 7.30.**

At more or less the same time another important breakthrough was achieved with Monsanto's L-dopa synthesis [67]. Initially developed for the synthesis of a-amino acids from the corresponding dehydro precursor this process is developing into a very general route from unsaturated acids of type **98** to their corresponding homochiral dihydro derivatives **99** [67, 68]. The hydrogenation is catalyzed by a rhodium complex formed with chiral phosphine ligands like **100**–**104** and catalysts of this type also operate very efficiently in the hydrogenation of β-ketoesters and in enantioselective isomerization of double bonds [69].

As in a number of cases spectacular selectivities were observed and as directed hydrogenations or reductions are of course highly important for enantioselective synthesis, one is not surprised to notice many efforts in this field [64]. There is for instance the remarkably efficient hydrosilation process developed by Brunner who synthesized a great number of nonphosphane ligands as for instance the thiazoline derivative **104**.

Corey [71] and Pfaltz [72] developed catalysts for borohydride reductions (**105** and **106** respectively), which are at the moment restricted, however, to selected sets of substrates.

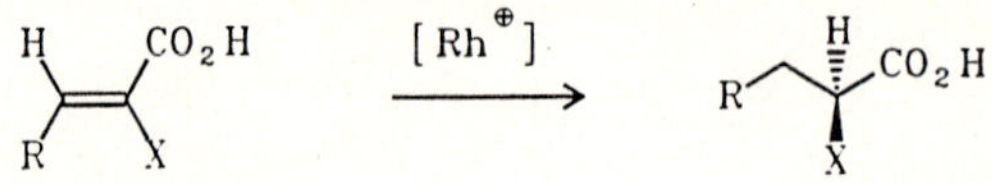

98 99

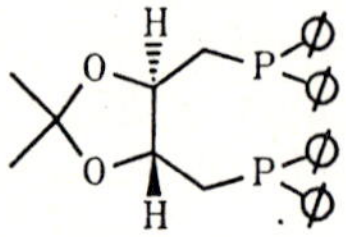

100 (–) DIOP

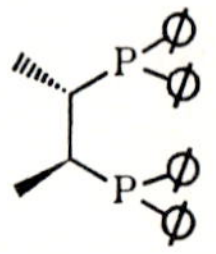

101 CHIRAPHOS

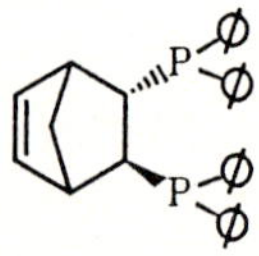

102 NORPHOS

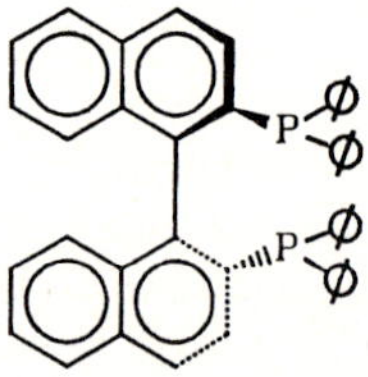

103 (+) BINAP

Scheme 7.31.

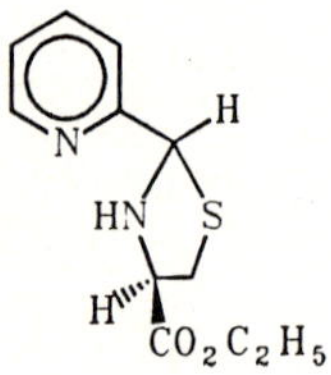

104

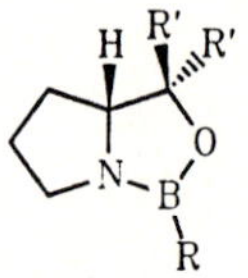

105

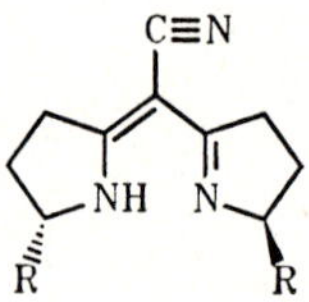

106

Scheme 7.32.

Of very high preparative importance in the future are catalysts that can be used in carbon-carbon bond forming reactions and so one can see a bright future for Lewis acids catalysts like **107** [73] and **108** [74] which may be used for Diels-Alder and aldol reactions as well as Hayashi's gold catalyst [75] which was also developed for this process.

In principle of course, all the well-known enzymatic conversions fall into this section of the chapter but as there is a special chapter exclusively devoted to this chemistry there is no need to give examples here. It should be mentioned here, however, that enantiotopic group selectivity, which was for quite a while believed to be a privilege of enzymes, may also be exercised with the techniques mentioned above and the scope of this is nicely reviewed in a recent paper by Ward [76]. This definitely proves that synthetic chemistry is well prepared to meet the challenges of enantioselective synthesis and has in less than twenty years provided powerfull tools in this field.

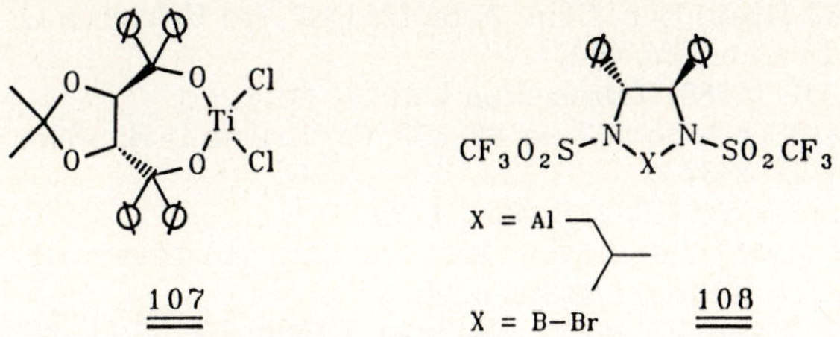

Scheme 7.33.

7.7 References

1 Jaques J (1981) Enantiomers, racemates, resolutions. J. Wiley New York
2 Boyer SK Pfund RA Portmann RE Sedelmeier GH Wetter HF (1988) Helv Chim Acta *71:* 337
3 Reider PJ Davies P Hughes DL Grabowski EJJ (1987) J Org Chem *52:* 955
4 Morrison JD (1983) Asymmetric synthesis. Academic New York, vol 1, p 59
5 Buss AD Warren S (1981) J Chem Soc, Chem Commun *100*
6 Beckmann M Hildebrandt H Winterfeldt E (1990) Tetrahedron Asymmetry *1:* 335
7 Machinaga N Kibayashi Ch (1990) Tetrahedron Lett *31:* 3637
8 Heitz MP Overman LE (1989) J Org Chem *54:* 2591. Further examples of enantiodivergent syntheses (see in Ref. [3])
9 Seebach E Naef R Calderari G (1984) Tetrahedron *40:* 1313
10 Seebach M Misslitz U Uhlmann P (1989) Angew Chem *101:* 484
11 Hanessian S (1983) Total Synthesis of Natural Products – the "Chiron" Approach. Pergamon Press Oxford
12 Mothes K Schütte HR Luckner M (1985) Biochemistry of Alkaloids. VEB Deutscher Verlag der Wissenschaften Berlin
13 Cordell GA (1981) Introduction to Alkaloids – A Biogenetic Approach. J. Wiley New York
14 Christie BD Rapoport H (1985) J Org Chem *50:* 1239 and further work cited
15 Renaud R Seebach D (1986) Synthesis 424
16 Sardina FJ Howard MH Koskinen AMP Rapoport H (1989) J Org Chem *54:* 4654
17 Reetz MT Binder J (1989) Tetrahedron Lett *30:* 5425
18 (a) Effenberger F Steegmüller D (1988) Chem Ber *121:* 117. (b) Radunz HE Reißig H-U Schneider G Riethmüller A (1990) Liebigs Ann 705
19 (a) Mikami K Kaneko M Loh TP Terada M Nakai T (1990) Tetrahedron Lett *31:* 3909. (b) Reetz MT Drewes MW Lennick K Schmitz A Holdgrün X (1990) Tetrahedron Asymmetry *1:* 375
20 (a) Corey EJ Ohtani M (1989) Tetrahedron Lett *30:* 5227. (b) Saari WS Fisher Th E (1990) Synthesis 453. (c) Chung JYL Wasicak JT (1990) Tetrahedron Lett *31:* 3957
21 ApSimon JW Seguin RP (1979) Tetrahedron (Report) *35:* 2797
22 ApSimon JW Collier TL (1986) Tetrahedron (Report) *42:* 5157
23 Mori K (1989) Tetrahedron (Report) *45:* 3233
24 Davies FA Sheppard AC (1989) Tetrahedron (Report) *45:* 5703
25 Hakam Kh Thielmann M Thielmann Th Winterfeldt E (1987) Tetrahedron *43:* 2035
26 Sato M Hisamichi H Kitazawa N Kaneko Ch Furuya T Suzaki N Inukai N (1990) Tetrahedron Lett *31:* 3605
27 Winterfeldt E (1988) Bull Soc Chim Belg *97:* 705
28 Nagao Y Dai WM Ochiai M Tsukagoshi S Fujita E (1990) J Org Chem *55:* 1148

29 (a) Oppolzer W Kingma AJ (1989) Helv Chim Acta *72:* 1337. (b) Helmchen G Wegner G (1985) Tetrahedron Lett *26:* 6047

30 Solladié-Cavallo A Simon MC (1989) Tetrahedron Lett *30:* 6011

31 (a) Enders D Lohray BB (1987) Angew Chem *99:* 359. (b) Enders D Bhushan V (1988) Tetrahedron Lett *29:* 2437

32 Weder T Edwards JP Denmark SE (1989) Synlett *1:* 20

33 (a) Loewe MF Meyers AI (1985) Tetrahedron Lett *26:* 3291. (b) Loewe MF Boes M Meyers AI (1985) Tetrahedron Lett *26:* 3295

34 (a) Poll T Abdel Hady AF Karge R Linz G Weetman J Helmchen G (1989) Tetrahedron Lett *30:* 5595. (b) Linz G Weetman J Abdeal Hady AF Helmchen G (1989) Tetrahedron Lett *30:* 5599

35 Whitesell JK Lawrence RM Chen HH (1986) J Org Chem *51:* 4779

36 (a) Schultz AG Sundaraman P (1984) Tetrahedron Lett *25:* 4591. (b) Schultz AG Puig S (1985) J Org Chem *50:* 915

37 (a) Meyers AI Lefker BA Sowin TJ Westrum LJ (1989) J Org Chem *54:* 4243. (b) Meyers AI Romine JL Fleming SA (1988) J Am Chem Soc *110:* 7245. (c) Meyers AI Busaca CA (1989) Tetrahedron Lett *30:* 6973. (d) Meyers AI Busaca CA (1989) Tetrahedron Lett *30:* 6977

38 (a) Solo AJ Singh B Kapoor JN (1969) Tetrahedron *25:* 4579. (b) Solo AJ Eng S Singh B (1972) J Org Chem *37:* 3542

39 Matcheva K Beckmann M Schomburg D Winterfeldt E (1989) Synthesis 814

40 Beckmann M Winterfeldt E reported at the 2nd Irsee Conference, March 1990, see also references in Ref [6]

41 Unpublished results from the authors laboratory

42 Kellogg RM (1984) Topics Curr Chem *101:* 3

43 Davies SG Skerlj RT Whittaker M (1990) Tetrahedron Lett *31:* 3213

44 Schinzer D (1989) Nachr aus Chem u Techn *37:* 1294

45 Rama Rao AV Bose DS Gurjar MK Ravindranathan T (1989) Tetrahedron *45:* 7031

46 Jung ME Jung YH (1989) Tetrahedron Lett *30:* 6637

47 (a) Brown HC Singaram B (1988) Accounts of Chem Res *21:* 287. (b) Midland MM (1989) Chem Rev *89:* 1553

48 Noyori R Tomino I Tanimoto Y Nishizawa M (1984) J Am Chem Soc *106:* 6709

49 (a) Yamaguchi Y Mosher HS (1973) J Org Chem *38:* 1870. (b) Brinkmeyer RS Kapoor V (1977) J Am Chem Soc *99:* 8339

50 (a) Joshi NN Srebnik M Brown HC (1989) Tetrahedron Lett *30:* 5551. (b) Tanaka K Ushio H Suzuki H (1989) J Chem Soc Chem Commun 1700. (c) Corey EJ Yuen PW Hannon FJ Wierda DA (1990) J Org Chem *55:* 784

51 (a) Riediker M Duthaler RO (1989) Angew Chem, Int Ed *28:* 494. (b) Duthaler RO Herold P Lottenbach W Oertle K Riediker M (1989) Angew Chem, Int Ed *28:* 495. (c) Bold G Duthaler RO Riediker M (1989) Angew Chem, Int Ed *28:* 497

52 Riediker M Hafner A Piantini U Rihs G Togni A (1989) Angew Chem, Int Ed *28:* 499

53 Matteson DS (1988) Accounts of Chem Res 294

54 Ikeda N Arai I Yamamoto H (1986) J Am Chem Soc *108:* 483

55 Yamamoto Y Nishii S Maruyama K Komatsu T Ito W (1986) J Am Chem Soc *108:* 7778

56 Masamune S Choy W Petersen IS Sita LR (1985) Angew Chem, Int Ed *24:* 1

57 Hoffmann RW Dresely S (1989) Chem Ber *122:* 903

58 Genet JP Kopola N Juge S Ruiz-Montes J Antunes OAC Tanier S (1990) Tetrahedron Lett *31:* 3133

59 Blystone SL (1989) Chem Rev 1663

60 Tomiak K (1990) Synthesis 541

61 Izawa H Shirai R Kawasaki H Kim H Koga K (1989) Tetrahedron Lett *30:* 7221

62 (a) Duhamel L Duhamel P Lannay JC Plaquevent JC (1984) C Bull Soc Chim Fr *II:* 421. (b) Duhamel L Plaquevent JC (1978) J Am Chem Soc *100:* 7415. (c) Duhamel L Plaquevent JC (1980) Tetrahedron Lett *21:* 2521. (d) Duhamel L Fouguay S Plaquevent JC (1986) Tetrahedron Lett *27:* 4975
63 Fehr C Galindo J (1988) J Am Chem Soc *110:* 6909
64 Ojima I Clos N Bastos C (1989) Tetrahedron *45:* 6901
65 (a) Eder U Sauer G Wiechert R (1971) Angew Chem *83:* 492. (b) Hajos ZG Parrish DR (1974) J Org Chem *39:* 1612
66 Winterfeldt E (1984) In: Bartmann W Trost BM (eds) Selectivity a Goal for Synthetic Efficiency. Verlag Chemie Weinheim
67 Knowles WS Sabacky MJ Vineyard BD Weinkauf DJ (1975) J Am Chem Soc *97:* 1975
68 Brunner H (1983) Angew Chem *95:* 921
69 Noyori R (1989) J Chem Soc Rev *18:* 187
70 Brunner H Becker R Riepl G (1984) Organometallics *3:* 1354
71 (a) Corey EJ Chen CP Reichard GA (1989) Tetrahedron Lett *30:* 5547. (b) Corey EJ Link JO (1989) Tetrahedron Lett *30:* 6275
72 Leutenegger U Madin A Pfaltz A (1989) Angew Chem *101:* 61
73 Narasaka K Iwasawa N Inoue M Yamada T Nakashima M Sugimori J (1989) J Am Chem Soc *111:* 5340
74 Corey EJ Imwinkelried R Pikul S Xiang YB (1989) J Am Chem Soc *111:* 5493
75 (a) Hayashi T (1988) Tetrahedron *44:* 5253. (b) Togni A Pastor SD Rihs G (1989) Hel Chim Acta *72:* 1471
76 Ward RS (1990) J Chem Soc Rev *19:* 1

8 Transition Metal Chemistry and Optical Activity – Werner-Type Complexes, Organometallic Compounds, Enantioselective Catalysis

H. Brunner

8.1 Werner-Type Complexes

Preparative transition metal chemistry was already well established in the last century. Thus, *cis*-$PtCl_2(NH_3)_2$, today worldwide the most powerful drug in the chemotherapy of human cancers, was discovered in 1844 [1]. The existence and separation of isomers, mainly *cis/trans* isomers in square planar or octahedral complexes, was the basis for Alfred Werner's theory of coordination, acknowledged by the Nobel Prize 1913. Optical activity in transition metal chemistry came into play in 1911, when Alfred Werner resolved the octahedral complex $[Co(en)_2(NH_3)Cl]^{2+}$ (Scheme 8.1) using bromocamphorsulfonate as the counterion [2].

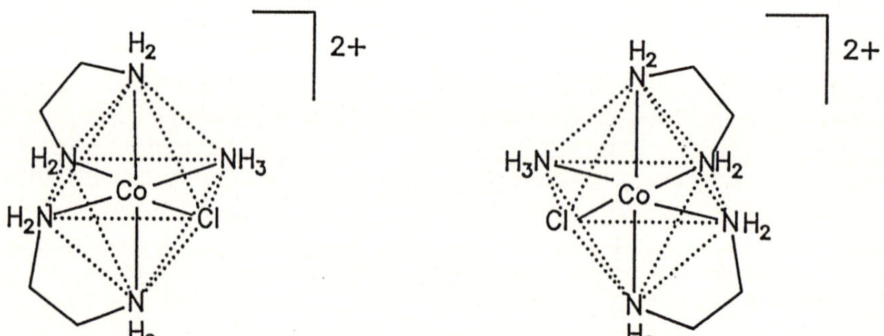

Scheme 8.1.

In the following decades numerous chiral transition metal complexes have been resolved, especially metal trischelates. At present, the metal trischelate type $M(LL)_3$ is the most celebrated chiral structure after the asymmetric carbon atom C(a,b,c,d).

Depending on the metal and the chelate ligand, $M(LL)_3$ compounds may be configurationally stable at the metal center, e.g. $[Co(en)_3]^{3+}$, or racemize by ligand dissociation, chelate ring opening or twist mechanisms, e.g.

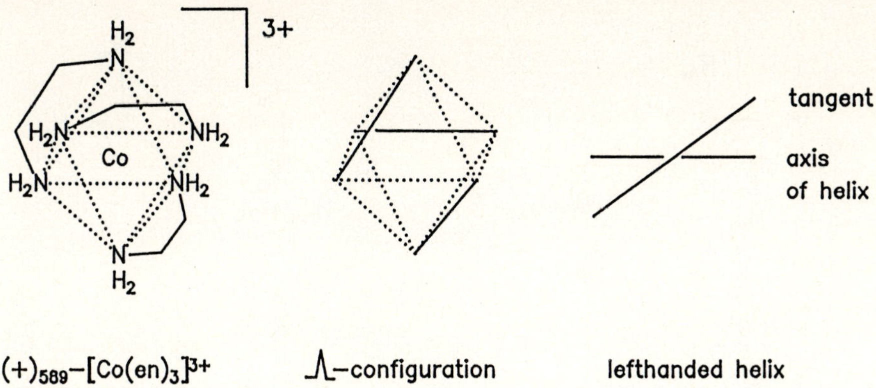

Scheme 8.2.

$[Fe(bipy)_3]^{2+}$ and $[Fe(phen)_3]^{2+}$ [3]. With the advent of modern NMR spectroscopy, the change of the metal configuration could be monitored by diastereotopic probes, e.g. benzyl or isopropyl groups within the ligands. This method does not require optical resolution and allows to measure racemization reactions which are too fast for chiroptical methods.

$(+)_{589}$-$[Co(en)_3]^{3+}$ (Scheme 8.2) was one of the first complexes, the absolute configuration of which was determined by anomalous X-ray scattering, establishing the correlation between chiroptical properties, such as the sign of the optical rotation, and the absolute configuration at the Co atom [4–6]. As the application of the Cahn–Ingold–Prelog rules for the specification of the configuration is not straightforward for metal trischelates, a Δ/Λ-nomenclature has been introduced [7]. Two of the chelate ligands, represented by the edges which they span, form two skew lines, considered to be axis and tangent of a helix. Following the helix away from the observer, righthandedness defines a Δ-configuration and lefthandedness a Λ-configuration. Scheme 8.2 shows that $(+)_{589}$-$[Co(en_3)]^{3+}$ has the Λ-configuration at the Co atom.

X-ray structure determinations demonstrated that five-membered M(en) rings definitely are puckered. Following the conformational analysis of cyclohexane in organic chemistry, a conformational analysis of M(en) complexes was initiated by Corey and Bailar [5, 6]. In the following paragraph, the conformational analysis of $[Co(en)_3]^{3+}$ is discussed.

The puckered M(en) unit is a chiral entity (Scheme 8.3). The line connecting the two N atoms and the C-C bond form a pair of skew lines which define a helix. The chirality of this helix, designated by the small Greek letters δ and λ, is used to specify the conformation of the M(en) system. As a consequence of the puckering, the N-H and C-H bonds are differentiated into axial and equatorial as indicated in Scheme 8.3.

$[Co(en)_3]^{3+}$ contains 4 different elements of chirality: the configuration at the cobalt atom and the 3 chiral conformations of the en rings. According to the 2^n rule, 4 elements of chirality entail a total of 16 possible isomers, 8 belonging to the Δ-configuration and 8 to the Λ-configuration. However, the

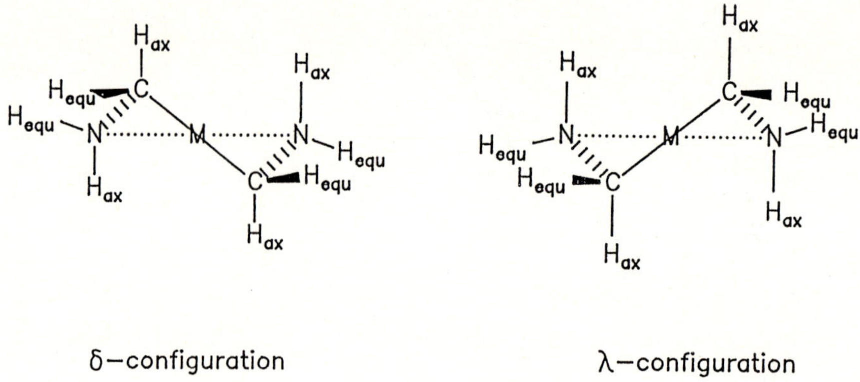

Scheme 8.3.

3 combinations, respectively, with either two δ-rings and one λ-ring or one δ-ring and two λ-rings are identical. This reduces the number of isomers to 8, 4 pairs of mirror image isomers as shown in the first line of Scheme 8.4.

Possible isomers	$\Delta(\lambda\lambda\lambda)$	$\Delta(\lambda\lambda\delta)$	$\Delta(\lambda\delta\delta)$	$\Delta(\delta\delta\delta)$	$\Lambda(\delta\delta\delta)$	$\Lambda(\delta\delta\lambda)$	$\Lambda(\delta\lambda\lambda)$	$\Lambda(\lambda\lambda\lambda)$
Degenerate forms		$\Delta(\lambda\delta\lambda)$ $\Delta(\delta\lambda\lambda)$	$\Delta(\delta\lambda\delta)$ $\Delta(\delta\delta\lambda)$			$\Lambda(\delta\lambda\delta)$ $\Lambda(\lambda\delta\delta)$	$\Lambda(\lambda\delta\lambda)$ $\Lambda(\lambda\lambda\delta)$	
Probability factor	1	3	3	1	1	3	3	1
Isomer ratio	10	30	10	1	10	30	10	1

Scheme 8.4

Due to this three-fold degeneracy, isomers with different δ- and λ-conformations get a probability factor of 3 compared to isomers with only one type of δ- or λ-conformation.

In the 4 different isomers of Δ-$[Co(en)_3]^{3+}$, N-H_{ax}, N-H_{equ}, and C-H_{ax} give intramolecular interactions within one M(en) ring and also with different M(en) units. Only C-H_{equ}, oriented to the outside, does not take part in these intramolecular repulsions. The intramolecular interactions together with the entropy factors 1 and 3 for the respective isomers cause the stability series indicated in the last line of Scheme 8.4. Thus, for the Δ-configuration the diastereomer $\lambda\lambda\delta$ (factor 3) is the most stable isomer. The diastereomers $\lambda\lambda\lambda$ (factor 1) and $\lambda\delta\delta$ (factor 3) have about the same free energy, whereas the diastereomer $\delta\delta\delta$ (factor 1) is the least favored by far. Kinetically, the cobalt configuration is stable, the conformations of the en rings, however, change rapidly. Thus, in a racemic mixture of $[Co(en)_3]^{3+}$ for each Co-configuration there are 4 interconverting diastereomers in approximate ratios 30 : 10 : 10 : 1 (Scheme 8.4).

The situation changes dramatically on substitution of the en ligands. Replacement of a C-H bond by a C-methyl group increases the intramolecular interactions for an axial position to such an extent that only an equatorial arrangement is possible. For $[Co(bn)_3]^{3+}$, bn = (SS)-(+)-1,2-diaminobutane, the energy difference between the diaxial and the diequatorial arrangement is about 4 Kcal/mol per Co(bn) moiety (Scheme 8.5), implying a distribution of δ/λ-conformations way above 99.9 : 0.1. That means, exclusively δ-conformations are possible and on complex formation, only the two isomers $\Lambda(\delta\delta\delta)$: $\Delta(\delta\delta\delta)$ in the ratio of roughly 95 : 5 have to be taken into account.

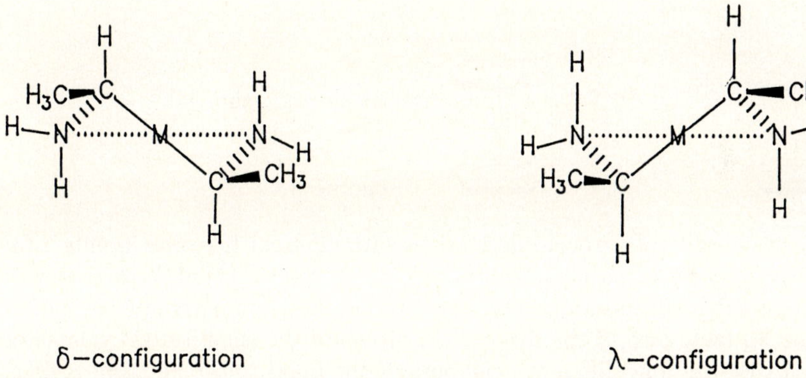

Scheme 8.5.

In the preceding sections the problems of configuration and conformation in $M(en)_3$ complexes were discussed, which are also involved in more complex systems. The part on configuration was kept simple and short, while the part on conformation was relatively broad and much more detailed. May be the reader has received the impression that conformational analysis in such metal complexes is a nice playground, but a playground without any relevance. This might have been true for the 1960s when this type of conformational analysis was developed. In the late 1970s, however, it became apparent that the conformational analysis of M(en) complexes was a prerequesite for the understanding of the chirality transfer in the enantioselective catalysis of organic reactions with transition metal compounds as outlined in the third part of this review.

8.2 Organometallic Compounds

The first organometallic compound, Zeise's salt $K[PtCl_3C_2H_4]\times H_20$, was discovered as early as 1827 [8]. However, it was not until 1953 that it was recognized as a π-complex in which the ethylene ligand is η^2-bonded to the Pt atom [9, 10]. Together with the simultaneous discovery of ferrocene this stimulated the development of organometallic chemistry of the transition elements. Early in this development, optical activity came into play, first in 1959, when chiral ferrocene derivatives with different substituents in 1, 2-positions were resolved [11]. An example is shown in Scheme 8.6, in which the two substituents, which must be different, are methyl and carboxyl.

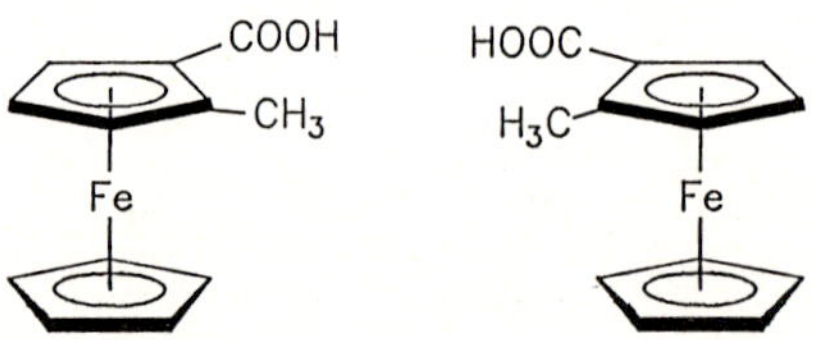

Scheme 8.6.

The chirality of ferrocene derivatives with different 1, 2-substituents is frequently referred to as planar chirality [12]. 1, 2-substituted ferrocenes with different substituents can also be visualized as containing asymmetric carbon atoms. In fact, each of the five carbon atoms of the substituted cyclopentadienyl ring has four different neighbors, if the bond to the Fe atom is taken into account. Planar chirality is also due to 1, 3-disubstituted ferrocenes with different substituents and to other 1, 2- and 1, 3-cyclopentadienyl complexes [12].

As ferrocene derivatives behave like organic compounds, the methods established for optical resolution in organic chemistry can be applied. Hundreds of optically active ferrocene derivatives have been obtained. In some of them, the unique geometry of the ferrocene skeleton is utilized, e.g. in derivatives with interconnected rings [12].

The planar chirality typical for unsymmetrically disubstituted ferrocene is also found in benchrotrene derivatives (benzenetricarbonylchromium derivatives) provided different substituents are present in the *ortho-* or *meta-*positions. The enantiomers of the tricarbonylchromium complex of *o*-methyl benzoic acid are shown in Scheme 8.7.

It is evident that on π-complex formation, the symmetry plane present in free *o*-methyl benzoic acid is removed giving rise to planar chirality. Obviously, the phenomenon of planar chirality in benchrotrene complexes is restricted to *o-* and *m*-derivatives. An arene containing two different substituents in *p*-positions, irrespective of a π-complexation by a metal-carbonyl fragment, has a plane of symmetry.

In benchrotrene complexes the $Cr(CO)_3$ fragment is easily removed, giving the free arene. With this strategy, the synthesis of optically active natural

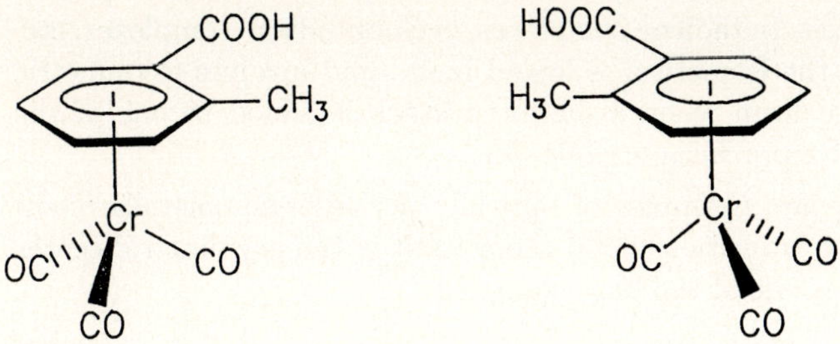

Scheme 8.7.

products has been carried out taking advantage of the planar chirality and the different reactivity of arenes in benchrotrene complexes compared to free arenes.

Ethylene is a highly symmetrical molecule having the symmetry elements 3σ, $3C_{2v}$ and i. On introduction of a substituent, e.g. a methyl group, all the symmetry elements disappear except one symmetry plane. Therefore, monosubstituted olefins, such as propylene, are prochiral with respect to π-complex formation. On complexation, the symmetry plane is removed and the substituted carbon atom becomes an asymmetric center in the same way as in the ferrocene and benchrotrene derivatives, discussed above. In Scheme 8.8, the platinum complex of propylene, the higher homologue of Zeise's salt, is shown, which can be resolved by replacing one of the Cl^- ligands by an optically active amine, such as 1-phenylethylamine. With similar methods stereochemical problems in olefin chemistry, e.g. the resolution of the chiral olefin *trans*-cyclooctene, could be achieved on the basis of platinum complexes [13].

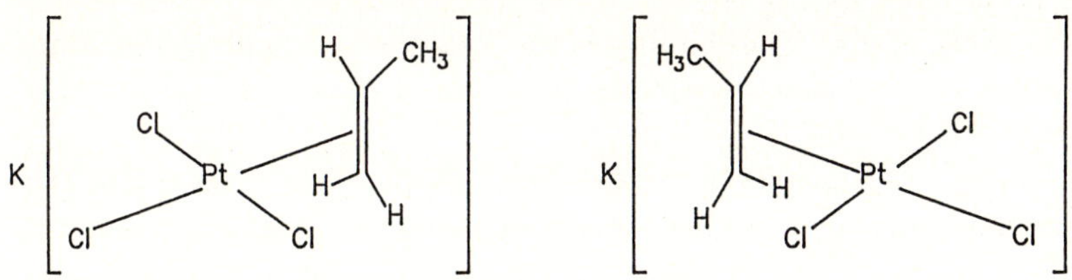

Scheme 8.8.

Whereas the optically active ferrocene and benchrotrene derivatives are configurationally stable, olefin complexes may racemize in solution. Racemization involves a decomplexation of the olefin. On readdition, the olefin may bind with its front or back, usually called *re*- and *si*-faces, leading to racemization.

It is straightforward that the optical isomerism discussed for ferrocene, benchrotrene, and olefin complexes can be extended to other π-complexes, such as allyl complexes, butadiene complexes, cyclobutadiene complexes, etc. In these complexes, the isomerism is ligand based and involves asymmetric carbon atoms, which form when a metal rucksack is added to one of the enantiotopic faces of a prochiral organic unit.

Since 1969 there are examples of optically active organometallic compounds, in which the transition metal atom itself is the asymmetric center [14]. Scheme 8.9 gives one of the first examples.

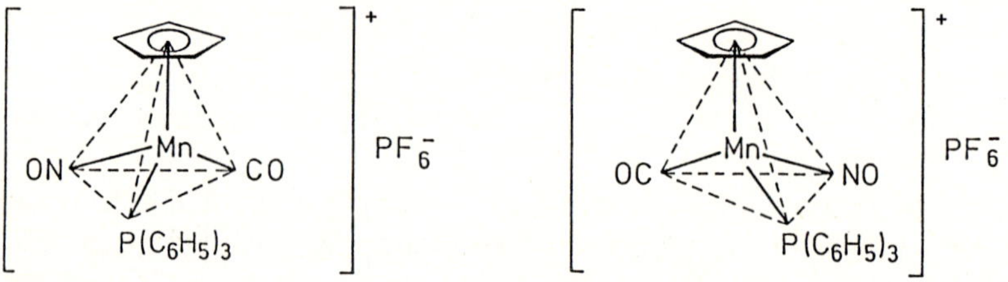

Scheme 8.9.

Although surrounded by four different ligands, the geometry around the Mn atom is approximately that of an octahedron with the cyclopentadienyl ligand occupying three *cis*-positions. This is corroborated by bond angles close to 90° between the ligands other than cyclopentadienyl. In addition, there are truly tetrahedral and octahedral compounds [15]. For a variety of such compounds the absolute configuration has been determined by X-ray crystallograpy [15].

The pair of compounds in Scheme 8.10 contains the same optically pure ligand and differs only in the configuration at the metal atom. Their electronic absorptions are dominated by the metal chromophore. Due to the opposite metal chirality, the CD spectra of the diastereomers are almost mirror images of each other. The chirality in the ligand usually makes only minor contributions in the visible part of the spectrum of the colored compounds, as shown in Scheme 8.10 [15].

Some of the organometallic compounds, chiral at the transition metal atom, are configurationally stable, others racemize or epimerize in solution with respect to the metal configuration. For the compounds in Scheme 8.11 – square pyramidal if the cyclopentadienyl ligand is taken as the top of the pyramid – the epimerization is an intramolecular process. For other compounds it may occur by ligand dissociation [16].

Configurationally stable compounds can be used to elucidate the stereochemistry of substitution reactions at the metal center. With the help of the chiral label at the metal atom the same mechanistic information can be acquired as for organic compounds with asymmetric carbon atoms and a va-

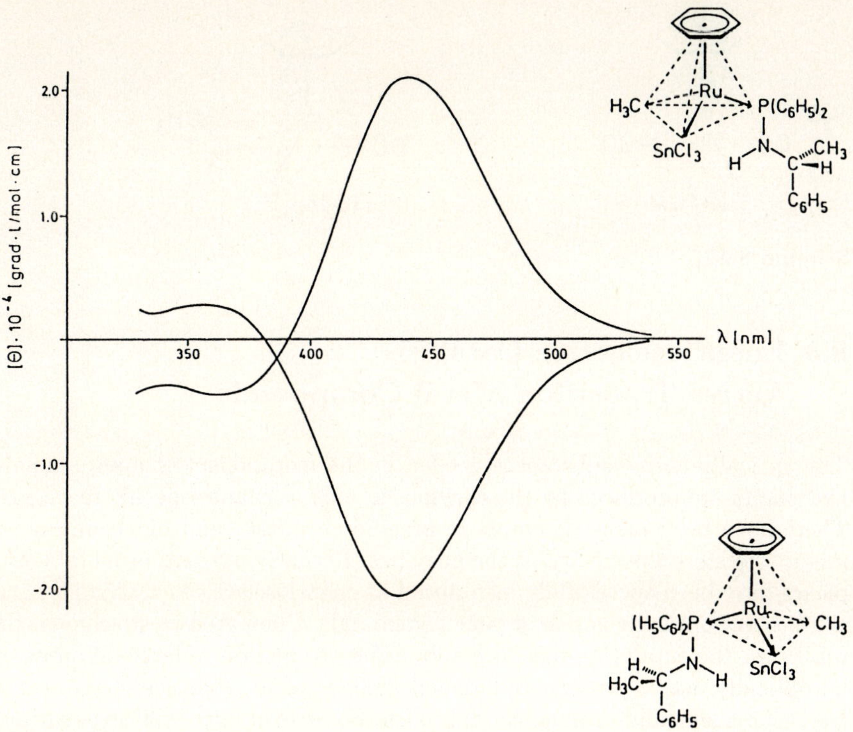

Scheme 8.10.

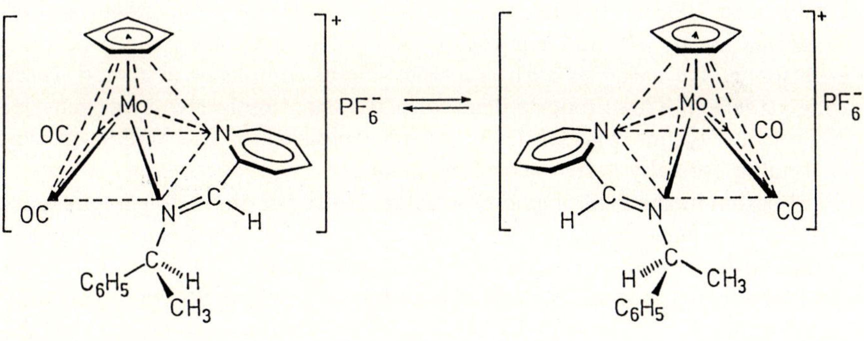

Scheme 8.11.

riety of different mechanisms has emerged [15]. Extended investigations have been carried out in the series of the Fe compounds shown in Scheme 8.12.

A question still open is how the concept of chirality at a transition metal atom can be brought to work in enantioselective catalysis, an important topic outlined in the next section.

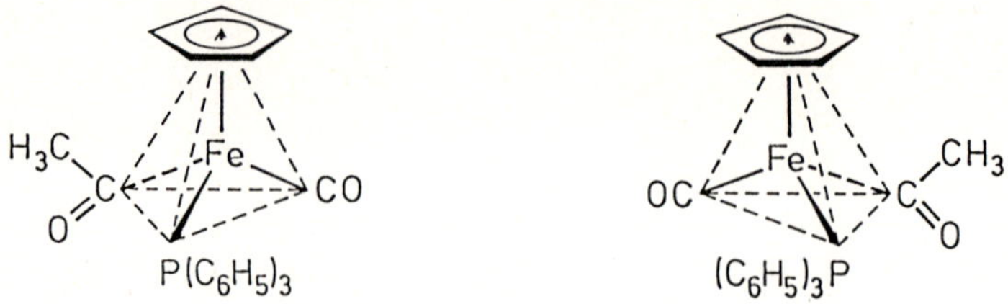

Scheme 8.12.

8.3 Enantioselective Catalysis Active Transition Metal Compounds

The optically active substances needed in the metabolism of man, animals and plants are produced by the enzymes in mostly enantiospecific reactions. There have been many attempts in organic chemistry and biochemistry to mimic the stereospecificity of the enzymes. Recently, transition metal complexes have been successfully introduced as enantioselective catalysts for the synthesis of optically active organic compounds. Compared to stoichiometric reactions, the catalytic procedure allows the production of large amounts of an optically active product by a small amount of an enantioselective catalyst, an elegant and economical approach. Some examples will demonstrate the relevance of the use of optically active transition metal compounds.

The first optical inductions, which were clearly outside the limits of error, were reported for heterogeneous catalysts in 1956 [17] and for homogeneous catalysts in 1966 [18]. However, it was not until the independent reports of Horner et al. [19] and Knowles et al. [20] in 1968 that the concept of enantioselective catalysis with transition metal complexes attracted worldwide attention. These reports dealt with the enantioselective hydrogenation of prochiral olefins with optically active complexes of the Wilkinson type containing optically active phosphine ligands.

Progress in the field of enantioselective hydrogenation of olefins included the choice of dehydroamino acids as prochiral substrates and the development of new optically active phosphines. Both aspects are demonstrated in Scheme 8.13. The hydrogenation of the dehydroamino acid (Z)-α-acetamidocinnamic acid to give N-acetylphenylalanine is a frequently used model reaction. Rh complexes of Diop, a chelate phosphine derived from natural tartaric acid, give optical inductions of up to 81% ee [21].

The success of Diop stimulated the synthesis of several hundreds of optically active phosphines for application in enantioselective transition metal catalysts [22]. Scheme 8.14 shows four representative examples, to which, similar to Diop, short acronyms have been assigned. In Dipamp, the two phosphorus atoms are the asymmetric centers, whereas in Prophos and Norphos the PPh_2 groups are connected by a chiral carbon skeleton, for Prophos derived from optically active lactic acid and for Norphos from an optical resolution. Binap is a compound belonging to the class of axial chirality.

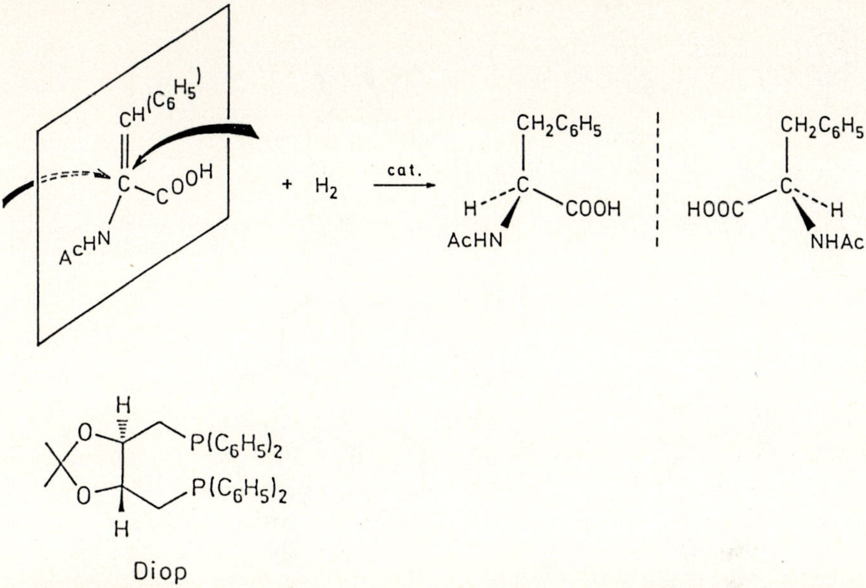

Scheme 8.13.

Dehydroamino acids, such as (*Z*)-α-acetamidocinnamic acid in Scheme 8.13, proved to be especially suitable substrates. They bind as chelate ligands to the catalyst i) with the oxygen atom of the *N*-acetyl group and ii) with the olefinic double bond to which hydrogen is transferred within the catalyst during the hydrogenation reaction. The first industrial application of the concept of enantioselective catalysis with transition metal complexes was the synthesis of the amino acid *L*-Dopa (3, 4-dihydroxyphenylalanine), a drug against Parkinson's disease. The highly enantioselective hydrogenation of the corresponding dehydroamino acid with a Rh/Dipamp catalyst is known as the Monsanto amino acid process [23].

The enantioselective hydrogenation of prochiral olefins can be carried out with isolated Rh(I) complexes, such as [Rh(cod)Prophos]PF_6, or more favorably with in situ catalysts. Such in situ catalysts consist of two independent parts, the so-called procatalyst and cocatalyst, both of which are usually commercially available. A celebrated example of a procatalyst is $[Rh(cod)Cl]_2$, a chloro-bridged rhodium(I) compound, shown in Scheme 8.15.

Cocatalysts are the optically active phosphines or other optically active ligands. In solution, procatalyst and cocatalyst combine to give the actual catalyst. Thus, no synthetic work for the preparation of the catalyst is required prior to the catalytic reaction, which has to be carried out [24].

In the 1970s, the enantioselective hydrogenation of prochiral olefins dominated the field of asymmetric catalysis with transition metal compounds. Increasingly, however, new reactions are made accessible to control by optically active transition metal catalysts. The following two examples will demonstrate this point.

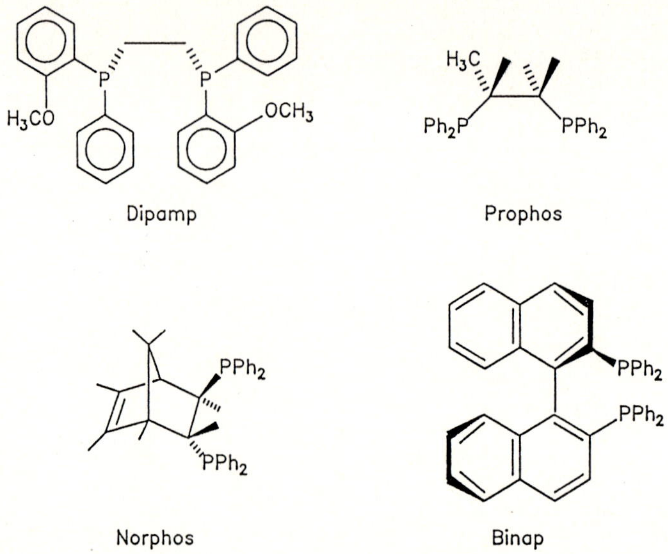

Scheme 8.14.

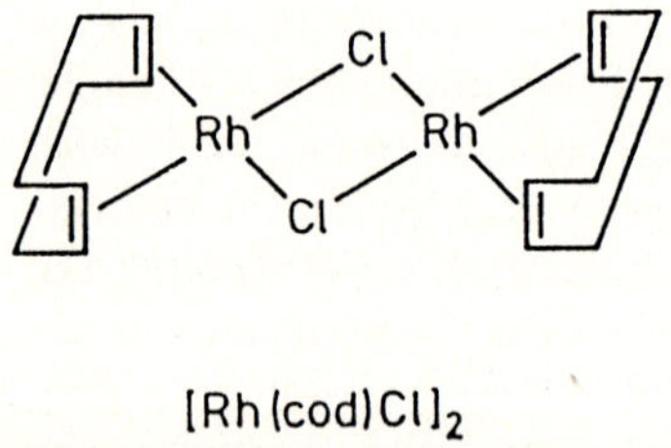

Scheme 8.15.

Enantioselective hydrosilylation of acetophenone with diphenylsilane (Scheme 8.16): In this reaction, a Si-H bond is added to the C=O bond of acetophenone. The hydrogen atom is attached to the prochiral C atom and the oxophilic $SiHPh_2$ fragment to the O atom of acetophenone. This reaction is catalyzed by Rh complexes. The primary product is a 1-phenylethylsilylether which on subsequent hydrolysis of the O-Si bond gives 1-phenylethanol as the ultimate product (Scheme 8.16). With optically active nitrogen ligands, e.g. the pyridinethiazolidine shown in Scheme 8.16, optical inductions close to 100% are achieved [25].

Enantioselective allylamine isomerization (Scheme 8.17): This reaction type is exemplified for a commercially important reaction. Starting material is diethylgeraniolamine, the NEt_2 derivative of geraniol. Its isomerization via a $1,3$-hydrogen shift is catalyzed by Rh complexes. With a Rh/Binap catalyst the optical induction with $> 99\%$ ee is virtually perfect. The enamine obtained in the isomerization can be converted into menthol. At present, 1000

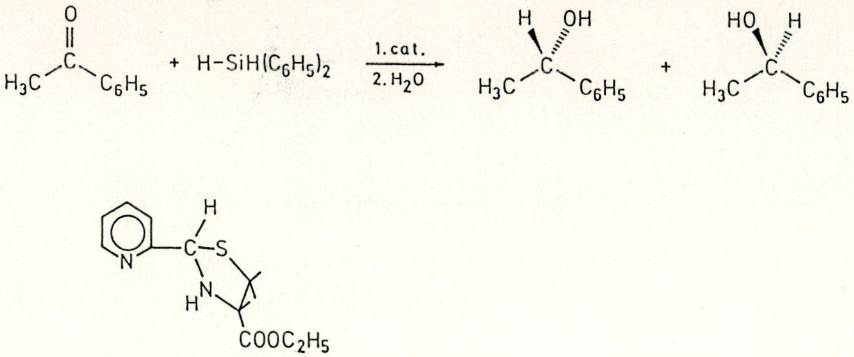

Scheme 8.16.

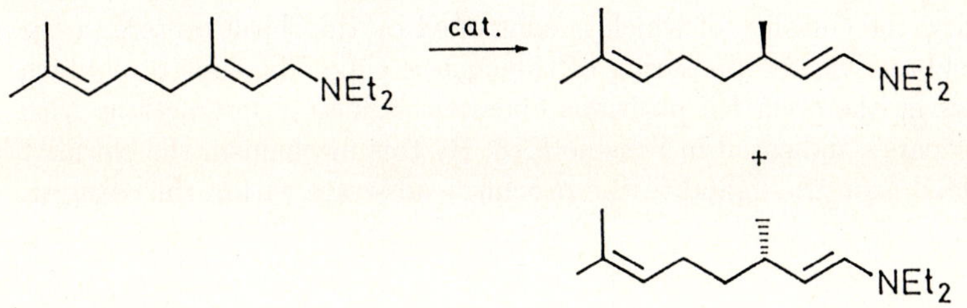

Scheme 8.17.

tons per year of menthol, approximately one third of the world production, is prepared by this Takasago process [26].

The mechanism of the enantioselective hydrogenation of dehydroamino acid derivatives is known in detail [22]. For other reactions the mechanisms are unknown. First ideas are developing to understand how the chiral information present in the optically active ligand is transferred within the catalyst to the coordination positions, where the prochiral substrates are converted into the optically active products. This explanation refers to the first part of the present article, in which the puckering of five-membered rings of the type M(en) has been discussed. The same puckering holds for five-membered chelate rings M(PP) of bidentate phosphines PP, such as Prophos or Norphos. In Scheme 8.18 it is demonstrated for (S,S)-Chiraphos. Similar to ethylenediamine complexes, the carbon substituents in M(PP) rings are differentiated into axial and equatorial substituents. The tendency of a methyl group to occupy an equatorial position controls the puckering of the chelate ring in a M(Chiraphos) complex, imposing a λ-conformation on the chelate ring.

Additionally, the puckering of the five-membered M(Chiraphos) ring differentiates the two phenyl substituents at the phosphorus atom, one becoming axial and the other equatorial (Scheme 8.18). Furthermore, the two phenyl rings usually arrange almost perpendicular to each other, one being called "face-exposed" and the other "edge-exposed". Thus, the PPh_2 moiety is a

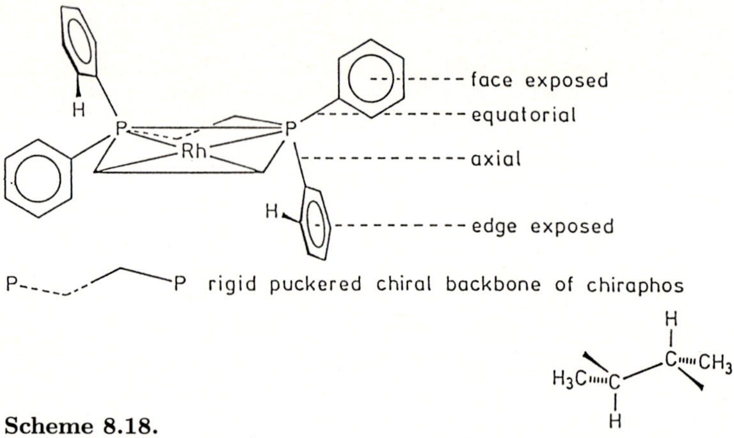

Scheme 8.18.

chiral entity, the chirality of which is controlled by the chiral centers in the ligand backbone via the puckering of the chelate ring. The substrate which coordinates at the open Rh positions experiences mainly interactions with the PPh_2 "ears", indicated in Scheme 8.18. By this mechanism the chirality is transferred from the ligand to the prochiral substrate within the catalyst.

8.4 References

1 Peyrone M (1844) Justus Liebigs Ann Chem *51:* 1
2 Werner A (1911) Ber dtsch chem Ges *44:* 1887
3 Basolo F Pearson G (1967) Mechanisms of inorganic reactions. Wiley New York p 300
4 Saito Y Nakatsu K Shiro M Kuroya H (1955) Acta Crystallogr *8:* 729
5 Hawkins CL (1971) Absolute Configuration of Metal Complexes. Wiley New York
6 Saito Y (1979) Inorganic Molecular Dissymmetry. Springer Berlin
7 (1970) Inorg Chem *9:* 1
8 Zeise WC (1831) Pogg Ann *21:* 497
9 Dewar MJS (1951) Bull Soc Chim Fr *C71*
10 Chatt J Duncanson LA (1953) J Chem Soc 2939
11 Thomson JB (1969) Tetrahedron Lett 26
12 Schlögl K (1967) Top Stereochem *1:* 39
13 Herberhold M Komplexe mit mono-olefinischen Liganden. Elsevier Amsterdam, part 1 1972, part 2 1974
14 Brunner H (1969) Angew Chem *81:* 395; (1969) Angew Chem Int Ed Engl *8:* 382
15 Brunner H (1980) Adv Organomet Chem *18:* 151
16 Brunner H (1979) Acc Chem Res *12:* 250
17 Akabori S Sakurai S Izumi Y Fujii Y (1956) Nature *178:* 323
18 Nozaki H Moriuti S Takaya H Noyori R (1966) Tetrahedron Lett 5239
19 Horner L Siegel H Büthe H (1968) Angew Chem *80:* 1034; Int Ed Engl *7:* 942
20 Knowles WS Sabacky MJ (1968) Chem Commun 1445
21 Kagan HB Dang TP (1972) J Am Chem Soc *94:* 6429
22 Morrison JD (1985) Asymmetric Synthesis, vol 5. Academic Orlando

23 Knowles WS (1986) J Chem Educ *63:* 222
24 Brunner H (1988) Top Stereochem *18:* 129
25 Brunner H Becker R Riepl G (1984) Organometallics *3:* 1354
26 Tani K Yamagata T Akutagawa S Kumobayashi H Taketomi T Takaya H Miyashita A Noyori R Otsuka S (1984) J Am Chem Soc *106:* 5208

9 Strategies for Liquid Chromatographic Resolution of Enantiomers

W. Lindner

9.1 Background of Basic Chromatographic Terms

Classic liquid–liquid chromatographic separation relies basically on the distribution of a compound between two immiscible phases by which one is moving (the mobile phase) with respect to the stationary phase. However, in the same way, similar processes occur in the chromatography by classifying the stationary phase as adsorbent which might be a chemically modified surface (e.g. with chiral compounds). The chemical and physico-chemical nature of the phases may vary to a great extent and leads to the various modes of chromatography together with their technical translation. The heart of every chromatographic system is the column which contains the (modified) particles whose surface serves as stationary phase and where the separation of a mixture of compounds (in the following for instance a mixture of stereoisomers or enantiomers) depending on the mobile phase chosen takes place. During the passage of the compounds through the packed column (over the stationary phase) the formation of chromatographic bonds with a concentration profile according to a Gaussian distribution curve takes place, and when the individual sorption isotherms of each component are non-identical, the compounds will become separated. This is the idealistic case, further details on chromatography may be found in more specific textbooks.

The position of the sorption isotherms and their relative difference for a given pair of compounds, (e.g. stereoisomers) is driven by physico-chemical difference of the separated compounds in the given chromatographic system, whereas the range is dominantly influenced by column parameters and the particle size of the adsorbent which is coated with the stationary phase. We speak about column efficiency and describe it by the number of theoretical plates N and the plate height H. Both terms can be easily calculated from a chromatogram according to the formula

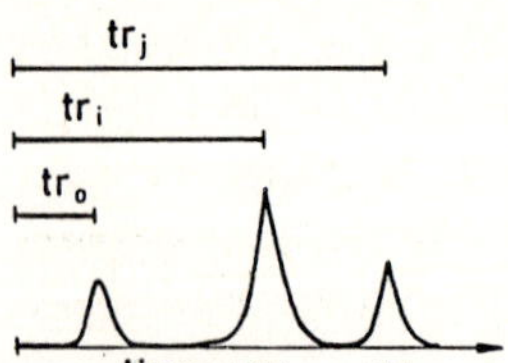

$$N_{\text{theor.}} = 16\left(\frac{tr_i}{w_i}\right)^2 \qquad H = \frac{L}{N} \qquad (9.1)$$

L = column length
tr_i = retention time of compound i
w_i = baseline peak width of comp. i
t_0 = retention time of non-retained compound

The capacity ratio (k') is defined as:

$$k'_i = \frac{tr_i - t_0}{t_0} \tag{9.2}$$

and the chromatographic separation of the two components i and j is described by the separation factor (α) which means also chromatographic selectivity.

$$\alpha = \frac{k'_j}{k'_i} \tag{9.3}$$

However, α does not reflect whether or not the bands (in chromatography usually termed as peaks) are completely separated, since the column efficiency has not yet been taken into account. This will be done by the resolution (R_s) of two peaks expressed as:

$$R_s = \frac{2(tr_j - tr_i)}{(w_i + w_j)} = \frac{2\Delta tr}{(w_i + w_j)} \tag{9.4}$$

Equation (9.4) can be transformed by expressing R_s as a function of α, k', and N.

$$R_s = \left(\frac{\alpha - 1}{4\alpha}\right)\left(\frac{k'_j}{1 + k'_j}\right)\sqrt{N} \tag{9.5}$$

With $R_s = 1$ a 98% separation of the two peaks and with $R_s = 1.5$ a complete separation is performed. From Eq. 9.5 it becomes clear that we have basically two possibilities of generating baseline resolution: a) by increasing the chromatographic selectivity α, or b) by tuning up the plate number of a column. The latter is limited by technological conditions whereas α deals with chemical parameters and phenomena which can be quite effectively influenced. To separate stereoisomers, particular stereochemical viewpoints have to be taken into account and this is the fascinating field of the various stereoselective chromatographic techniques.

9.2 Strategies to Separate Enantiomers by Chromatographic Techniques

In order to discriminate enantiomers (chiral analytes, selectands, SA) of each other by means of physico-chemical principles it is necessary to provide a chiral source (chiral handle, chiral selector, SO) with which the SAs come into contact by forming quasi diastereomeric transient complexes or molecule associates (see generalized reaction Scheme 9.1).

To get enantioseparation these diastereomeric complexes must differ in their free energies (ΔG) of formation and differ in their stability constants due to sterical facts derived from the spatial conformation of the SO and SA molecules and the strength and possibility of intermolecular "interactions" which might be attractive but also repulsive. We speak of "chiral recognition"

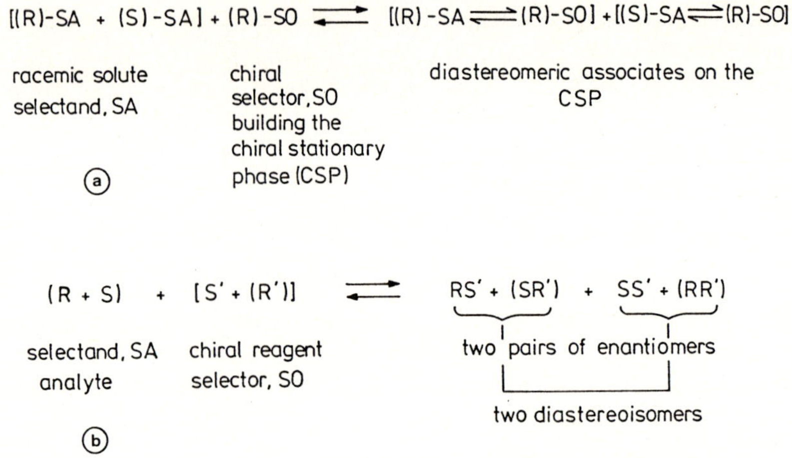

Scheme 9.1. Reaction Scheme of the formation of **a**) diastereomeric molecule associates, **b**) diastereoisomers via derivatization with an optically active derivatizing reagent

also in chromatographic terms and mean that one enantiomer, e.g. $SA_{(R)}$, binds more strongly to the chiral SO molecule than $SA_{(S)}$ assuming that these SOs have been immobilized onto the surface of silica particles used in liquid chromatography and serve as such as chiral stationary phases (CSP).

Chiral recognition and the enantioselective adsorption, respectively, is based on "multipoint" interactions (see later) between SO and at least one of the SA enantiomers. And at least one of these interactions has to be stereochemically dependant. What means that a particular chiral selector molecule or selector region (see later at the paragraph describing chiral polymers and proteins as chiral SOs) fits spatially "ideal" one SA enantiomer in comparison to a "non-ideal" fit of the other. Considering only one simple isolated SO moiety with e.g. only one chiral center a stereoselective binding model as illustrated in Fig. 9.1 may be conceivable.

However, in reality the SOs might be clustered and lined up by (semi) crystalline structures and might generate in this form differently spatially shaped selector regions or "selector surfaces" which might act synchronically to the isolated SO–SA chiral recognition model and mechanism, but not necessarily. This can become particularly evident using SO polymers as polysaccharides and proteins with their particular secondary and tertiary structure on top of the shape of the chiral subunits, as for instance one amino acid amide element within a peptide chain.

The nature and the number of intermolecular SO–SA interactions necessary for chiral recognition have to be more specific. A distinction between single-point and multi-point interactions should be tried, but all are based on complementary binding forces. Ion–pair formation via coulomb attraction, hydrogen bonding and dipole–dipole interaction are single-point. However, if you consider the formation of a dipole you have electron rich and elec-

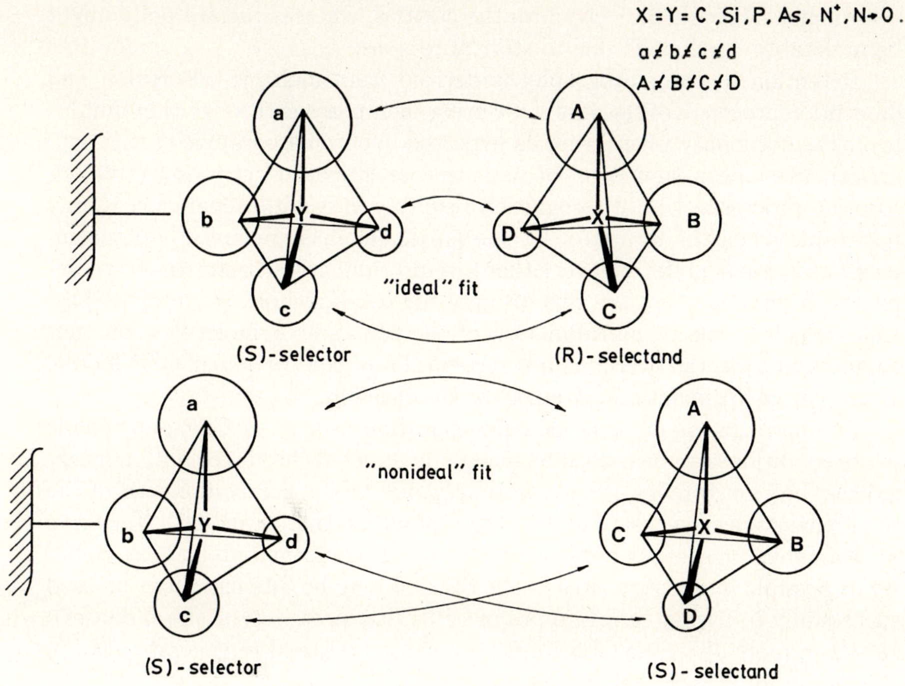

Fig. 9.1. Stereoselective interaction (binding) model between chiral selector and racemic selectand molecules

tron deficient centers within one functional group thus creating a two-point attachment region for a complementary dipole. Along these lines are the interactions between linear and planar functions. As just mentioned, dipole–dipole stacking and the π–π interactions between electron rich and electron deficient planes are multipoint in nature.

To explain chiral recognition it is the best to follow the "three point interaction" rule derived from Ogsten's [1] and later Dalgliesh's [2] "three point binding" theory which conceptually means that for spatial and chiral recognition a minimum of three simultaneous interactions between the SO and one SA enantiomer are required where at least one of them is stereochemically dependant.

To define a plane in space, three interaction sites of SO and SA must exist and be reciprocally approached throughout the chiral recognition process. However, "interactions" does not necessarily mean, that all of them have to be attractive bindings, one or two bindings as driving forces for the formation of the diastereomeric associate can be sufficient, together with stereochemical attractions or repulsions of space filling groups within the chiral SO and SA compounds, to perform chiral recognition. Their conformation under given chromatographic conditions is of importance, particularly when chiral cavities are considered as a chiral SO source for chiral recognition, one SA enantiomer might fit perfectly into the cavities, whereas the antipode might be preferably "excluded" due to sterical reasons.

Retention in chromatography is derived from multiple adsorption and desorption processes of the analytes during their passage through the columns (over the stationary phase) and as expected, a chromatographic enantioseparation must be a sum total of non-stereoselective and stereoselective adsorption processes. Chiral recognition resulting in enantio separation is only noticeable when the formation of one particular diastereomer SO–SA complex formation is preferred over other less discriminative diastereomeric complexes. Assuming there are two discrete SO–SA$_{(R)}$complex formations possible, which is due to multifunction of the SO and SA molecules, but not collinear in sign, the overall chiral recognition of the particular CSP for the given pair of SA enantiomers might be diminishing.

Summarizing, poor or no enantioseparation of a given CSP and mobile phase conditions can be caused by unequally directed chiral recognition mechanisms, but much more often it is simply due to the non-compliance of the requirements necessary to fulfill a "three point interaction model". Variation of the mobile phase thus influencing the overall conformation and excessibility of possible interaction sites of the SO and may be SAs can often be used successfully to trigger certain types of CPSs (e.g. protein type chiral phases). However, more often the CPS itself also has to be varied to generate enantio separation for a given pair of antipodes.

9.3 Thermodynamic and Kinetic Considerations for Chromatographic Enantioseparation

As pointed out earlier the chromatographic parameters retention time tr_i/tr_j, the capacity factors k'_i, k'_j, and the (enantio)selectivity $\alpha_{i,j}$ are thermodynamically controlled. k' is related to equilibrium constants and applying the Gibbs–Helmholtz equation to the molar free energy differences $\Delta\Delta G^0$ between the SO–SA equilibrium constants of the (R)- and (S)-enantiomers in an enantioselective chromatographic system, Eqs. 9.6 and 9.7 can be expressed:

$$-\Delta\Delta G^0 = RT \ln \frac{k'_{(S)}}{k'_{(R)}} = RT \ln \alpha(k'_{(S)} > k'_{(R)}) \tag{9.6}$$

$$\ln \alpha = -\frac{\Delta\Delta H^0}{RT} + \frac{\Delta\Delta S^0}{R} \tag{9.7}$$

The molar free energy difference ($\Delta\Delta G^0$) is associated by the observed α-value and can be easily calculated from chromatographic data. Peak shapes, plate height and the plate number expressed as efficiency of a chromatographic system are influenced by kinetics of the mass transfer, packing factors of the column and particle size of the packing material on which the chiral selector (SO) is immobilized. An α-value of 1.05 reflects a $\Delta\Delta G^0$

value (cal/mole) of 29 which is rather small. And according to Eq. (9.5) one needs about 15 000 theoretical plates to generate a complete resolution (R_s = 1.5) of the pair of enantiomers. Assuming a highly efficient column quite small energy differences may be sufficient for a successful chromatographic enantioseparation, whereby the observed α-value, as expressed earlier, is the "weighted time average" of all possible adsorbed and desorbed diastereomeric SO–SA molecule associates and of which the stability constants in terms of chiral recognition might be added in an agonistic or antagonistic way. In liquid chromatography α-values between 1.1 and 2.0 are usual, but also α-values up to $\alpha = 100$ have been noticed in special cases [3] reflecting an extremely well steric fit of one enantiomer to the SO compared to the others. Enhancing the temperature usually leads to decreased enantiodifferentiation as illustrated in Fig. 9.2, where $\ln \alpha$ is plotted as function of $1/T$ according to Eq. (9.7).

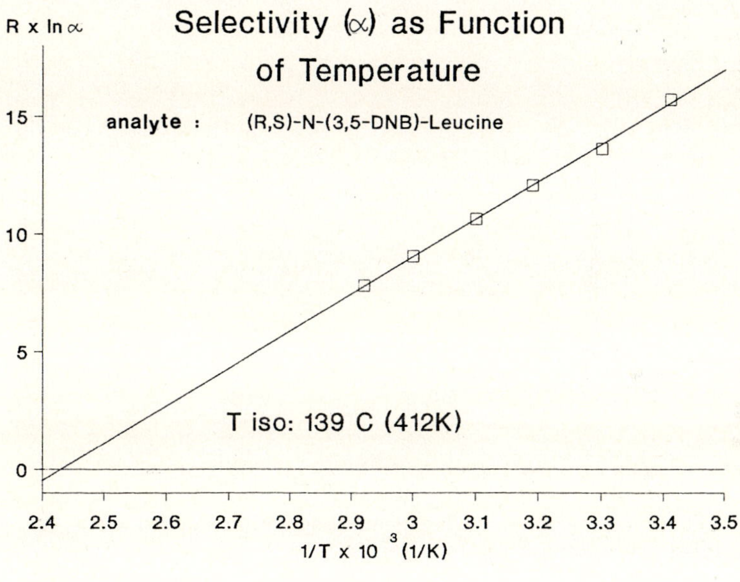

Fig. 9.2. Decrease of enantioselectivity with increasing temperature. The plots permit calculation of enthalpy and entropy contributions according to Eq. (9.7). (Reprinted, with permission, from Ref. [4])

Extrapolating the line to $\ln \alpha = 0$ one gets the temperature where the enthalpy and entropy contributions of the chromatographically observed chiral recognition process cancel each other. However, by increasing the temperature usually the overall column efficiency increases and thus the resolution might decrease much slower than α, R_s might even increase and go through a maximum [4]. From this it becomes clear, that measuring and comparing α-values without controlling and specifying the working temperature is problematic, particularly when chiral recognition models are derived. The working temperature is certainly a parameter to optimize the overall chromatographic conditions to resolve a given pair of enantiomers; for analytical purposes a resolution R_s of about 1.5 at a minimum of analysis time, and for preparative scale resolutions maximum R_s values to increase the column loadability is optimum. However, at elevated temperature the risk of thermal racemization, particular of atropisomers, by changing the molecule conformation by rotating along a bonding axis is enhanced. In this case the chiral recognition leading to a chromatographic resolution might be accompanied by a continuous partial racemization and the phenomenon is termed "peak coalescence", as depicted in Fig. 9.3.

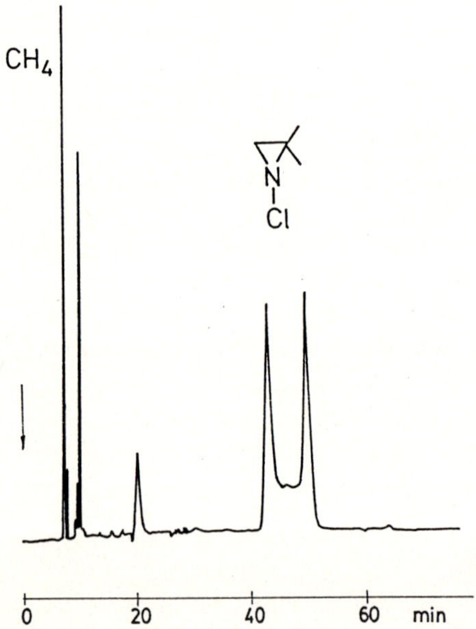

Fig. 9.3. On-column racemization during enantioseparation noticed on a GC-column. (Reprinted, with permission, from Ref. [5])

Such phenomena have been noticed in enantioselective gas-chromatography where base- or acid-catalysed racemization reactions also have to be considered depending on the mobile phase conditions used. Classical examples in LC are benzodiazepines under aqueous conditions which tend to racemize quite easily (Fig. 9.4).

In order to avoid such reactions the overall chromatographic conditions have to be chosen carefully, consequently one should never forget the chemical behavior of the SO and the SAs in the environment of the mobile phase as an indispensable partner in a liquid chromatographic system.

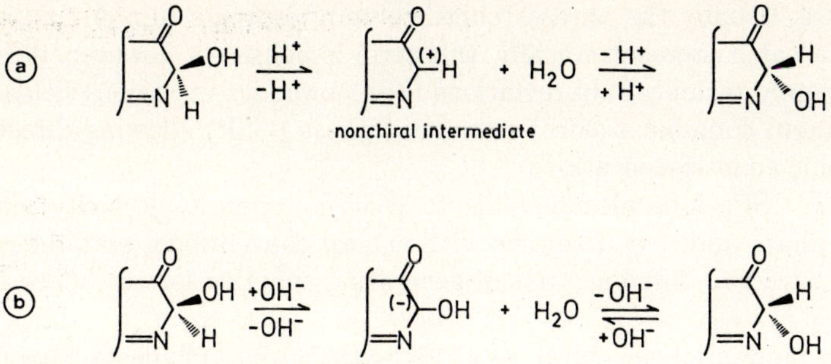

Fig. 9.4. Scheme of **a**) acid-catalyzed and **b**) base-catalyzed reaction yielding racemization

9.4 Enantioselective Liquid Chromatography

As pointed out earlier, the key for enantioseparation is the discriminative formation of diastereomeric transient molecule associates between a chiral selector (SO) and chiral selectands (SAs). Techniques following such principle strategies are usually termed "direct", in contrast to enantioseparation techniques which involve the covalent formation of diastereomeric molecules using an optically active derivatizing reagent (CDR) as the chiral selector. Such approaches are termed "indirect" enantioseparation techniques by which the pairs of diastereomers formed (see generalized reaction Scheme 9.1) should be separable per se on a non-chiral stationary phase, since diastereomers are physicochemically non-equivalent therefore different from each other; they are two distinct compounds and may be resolved by any conventional separation technique. Pros and cons of this technique will be discussed in a separate section.

9.5 Direct Enantioseparation by Liquid Chromatography

Over the last few years the concepts of generating chiral recognition in a "direct" fashion and adapting these to chromatographic systems have been expanded substantially and the number of publications in this field including applications is booming and was well over 1000 by the end of 1989. In order to structure this research area by methods and applications several review articles [6–8] have been written and recently also books on enantioseparations have appeared on the market [12–15]. However, it is the aim of this article to select the most promising techniques to date, to discuss their chiral recognition principles together with possibilities of modulating them to a certain extent via the mobile phase and to give some practical examples. Up to now more than 45 enantioselective stationary phases (CSPs) and HPLC columns, respectively, are commercially available and many more have been described in the literature. Usually the phrase "chiral column" is found in many publications; looked at it stereochemically, this term is nonsense, however, it is one of these sloppy technical abbreviations used nowadays in various fields. A "chiral" column contains a chiral stationary phase (CSP) allowing direct chromatographic enantioseparations.

Besides the CSPs it is also possible to generate enantioselectivity via chiral mobile phase additives together with achiral conventional stationary phases, so we have two different ways of generating enantioselective LC systems:
a) via covalent binding of the chiral selectors on the surface of the backbone of the packing material or on the walls of capillary columns or
b) via dynamic coating of the surface, which might be non-chiral premodified, with a chiral selector through the mobile phase which contains defined amounts of SO.

Both ways have been applied successfully by using essentially the same type of chiral selector, however, there are some differences noticeable in enantioselectivity and elution order of the resolved pair due to the steric influence of the surface itself, but the principal characteristics of the SOs remain normal to a great extent.

The general methods mentioned above deal with the technical generation of CSPs and the mode of enantioselective chromatographic systems, but the key is the type of chiral selector, its stereochemical structure and whether it is a monomeric compound or a chiral polymer, which leads to a rough classification of CSPs (see Table 9.1).

9.6 Chiral Phases Using Polymers as Chiral Selectors

From an historic point of view the naturally occurring polymers (polysaccharides), such as wool, starch, or cellulose, served first as chiral selectors, as for instance in paper chromatography some racemic amino acids could be resolved [16–18]. These polymeric particles of e.g. cellulose as chiral selector were used as such in non-immobilized form and were of crystalline structure. Crystalline cellulose is readily available and inexpensive, but the columns filled with such material are only modestly efficient but sufficiently so for preparative separations. However, Hesse and Hagel [19] introduced (1973) microcrystalline triacetyl-cellulose (MCTA) as enantioselective particles and CSP, respectively, and created the term chiral "inclusion chromatography", since they assumed, that the chiral recognition must be due to some type of chiral cavities within the original chiral structure of the cellulose particles which was not ruined during the solid state acetylation process. As soon as the MCTA material was dissolved and recovered, but in an amorphous form, the original enantioselectivity was greatly diminished or the observed elution order of the resolved enantiomers was even reversed [20]. But nevertheless,

Table 9.1. Main chiral stationary phases (CSPs) used for direct enantioseparation by LC

CSP	Chiral Selector	Column Efficiency Application
Chiral Polymer	*polymers of natural origin*	*moderate to low*
	polysaccharides	preparative
	polysaccharide derivatives	analytical and preparative
	proteins, polypeptides	mainly analytical
	Synthetic polymers based on chiral monomers	
	helically coiled polymers *"imprinted" chiral cavities in a polymer matrix*	analytical and preparative
Chiral Monomer	*cyclodextrines*	*moderate to high*
"brush type"	cyclodextrine derivatives *amino acid derivatives* *alcaloides* *various synthetic mono- and multifunctional compounds*	analytical and preparative

MCTA with a particle size of about 10 μm to 25 has proved to be an excellent CSP to resolve efficiently, for analytic but mainly preparative purposes, chiral compounds of a broad range of polarity and of conformational shape including rotamers. They usually contain a ring system (most often of aromatic character) as substituent. Mannschreck and co-workers, in particular, have revived MCTA and made extensive use of it [21]. An example of preparative resolution using MCTA column is depicted in Fig. 9.5. Okamoto and his co-workers [23] followed another approach of using polysaccharides and particularly cellulose as chiral selectors. This group derivatized the free OH groups with varius reagents to esters or carbamates (see examples in Table 9.2), dissolved the new chiral polymers and immobilized (coated) them onto a silica gel surface of wide pore particles. By this protocol numerous polysaccharide based CSPs have been prepared, many of them have come onto the market and are sold by Daicel.

The most likely chiral recognition process involves some type of chiral inclusion phenomena combined with hydrogen bonding, dipole–dipole and/or π–π interactions between the coated SO layer which is presumably liquid crystalline to some extent, and the SAs. Examples of separations are given in Fig. 9.6. Due to the coating rather than covalently binding of the polymeric SO, it can be washed off with strong eluting solvents, as for instance chloroform. The selection of mobile phases is therefore somewhat restricted.

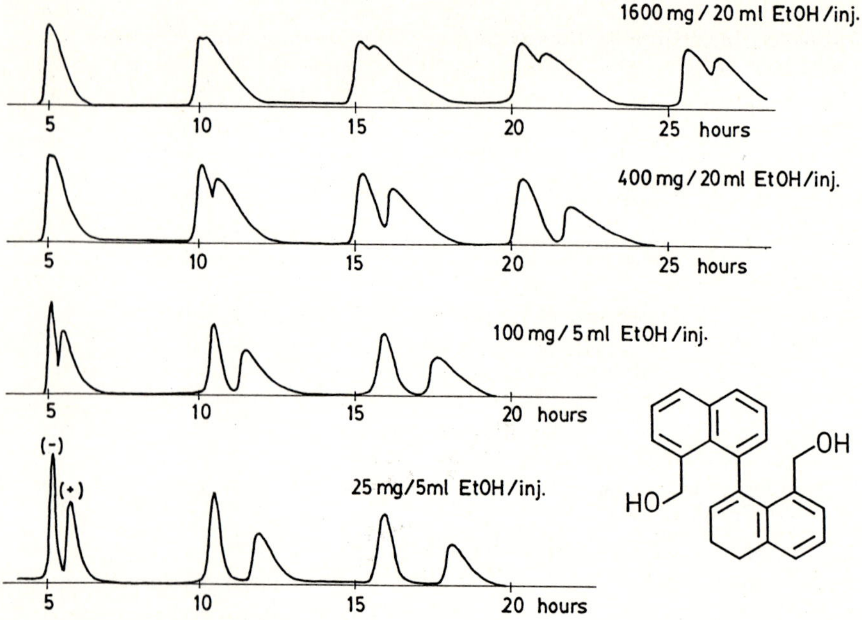

Fig. 9.5. Evaluation of loadability onto a preparative MCTA column used in recycling mode. Exact conditions see Ref. [22]. (Reprinted, with permission, from Ref. [22])

However, the spectrum of chiral SAs which have been successfully resolved is wide and to my knowledge the Chiralcel OD and OJ phases of Daicel (see Table 9.2) are the most broadly applicable ones. But there is little knowledge of if a given pair of SA can be resolved (will fit to the chiral recognition region). The existence of an aromatic ring in the SA molecule seems crucial.

9.7 Chiral Stationary Phases Using Proteins (Polypeptides) as Chiral Selectors

These types of natural biopolymers have unique primary, secondary, and tertiary structures depending on the amino acid sequence as well as on the degree and type of glycosyclation. As we, for instance, know from high affinity type receptor proteins for drugs, they act mostly quite remarkably stereospecifically [13, 25] and the reason must be an area, "pocket", "cleft", or "bay region" with a pronounced chiral recognition capability due to the unique conformation of the particular peptide chain surrounding the cleft and which is supported by the multiple bonding and steric interactions of the more or less lipophilic amino acid side chains. But not only receptor type proteins express stereoselectivity, also many others may do the same, as for instance human plasma transport proteins for small compounds like the

Table 9.2. Chiral stationary phases based on cellulose

Substance	Trade name	Supplier
Microcrystalline cellulose triacetate	CHIRALCEL CA-1 art 16362, 16363 Chiral Triacel	Daicel E. Merck Macherey-Nagel
Cellulose triacetate on silica gel	CHIRALCEL OA	Daicel
Cellulose tribenzoate on silica gel	CHIRALCEL OB	Daicel
Cellulose trisphenyl-carbamate on silica gel	CHIRALCEL OC	Daicel
Cellulose tricinnamate on silica gel	CHIRALCEL OK	Daicel
Cellulose tris(3,5-dimethylphenyl carbamate) on silica gel	CHIRALCEL OD	Daicel
Cellulose tris(4-chlorophenyl carbamate) on silica gel	CHIRALCEL OF	Daicel
Cellulose tris(4-methylphenyl carbamate) on silica gel	CHIRALCEL OG	Daicel
Cellulose tris(4-methyl-benzoate) on silica gel	CHIRALCEL OJ	Daicel

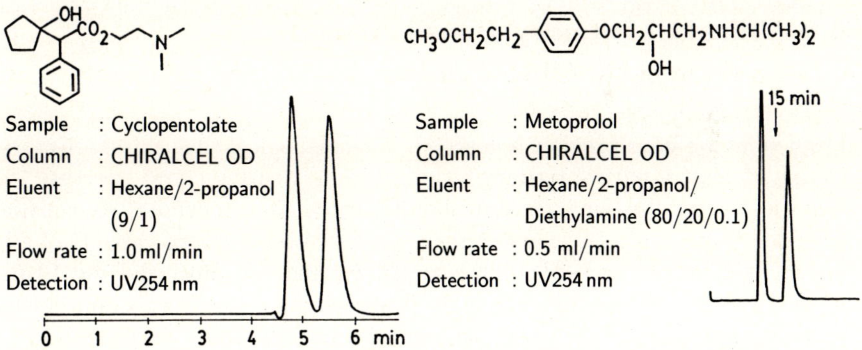

Fig. 9.6. Examples of enantioseparations on a cellulose derivative type column, Chiralcel OD (Daicel). (Reprinted, with permission, from Ref. [24])

alpha-1-acid glycoprotein (AGP). However, it was first Allenmark who, basing on his work on early findings using BSA as enantioselective mobile phase additives [26], used BSA immobilized covalently onto a silica gel surface and succeeded with this CSP in the resolution of a variety of racemates [12, 27].

Hermansson turned his interest to acidic AGP which is claimed to be a main cationic binding protein and should therefore bind well and stereospecific chiral drugs with an amino function. He immobilized AGP successfully [28] and with the first generation columns of this AGP-type CSP, called EnantioPack, he obtained remarkable enantioselectivity for basic drugs as well as for neutral but somewhat polar and acidic compounds, depending on mobile phase conditons (pH and buffer strength) and additives (proteolytic and non-proteolytic), which obviously alter the overall conformation of the immobilized protein probably expressing different enantioselective "pockets".

Schill and Wainer [29] started to study these phenomena in depth to use them as tools for optimizing and controlling the enantioselectivity for protein type bonded CSP, and which are now common in practice. It turned out that the EnantioPack columns and the CSP, respectively, were not sufficiently stable and by modifying the bonding procedure Hermansson came up with a much more rugged second generation of an AGP–CSP (Chiral-AGP) [30a]. Such new protein type CSPs will definitely become quite popular in the future, like we see with the ovomucoide type CSP [30b], since the source of highly selective proteins seems unlimited. This approach might also open up e.g. protein–drug binding type studies by chromatographic means, but also of many other applications in biomedical research areas. That reversible conformational changes of the immobilized proteins occur as has been shown to a certain extent by CD and fluorescence spectroscopy. However, more "insight" information is needed for better characterization of the stereochemical aspects of the enantioselective "pockets".

9.8 Chiral Stationary Phases Based on Synthetic Chiral Polymers

From what we have previously learned, it seems logical that chiral selectors, having at least a partial tertiary structure thus creating chiral "pockets" or "cavities" accessible for inclusion, should express stereochemical recognition provided that within the cavities there are functional groups capable of e.g. intermolecular hydrogen bonding between SO and SA. Since the mid 1970s, Blaschke and co-workers [31] have dealt successfully with such a concept by polymerizing monomeric (meth)acrylamides derived from chiral amines as e.g. (S)-phenylalanine methylester (see Fig. 9.7).

Recently this type of polymerization was also performed in situ together with modified silica gel resulting in covalently immobilized chiral polymers. Such CSPs showed remarkable enantioselectivity for several drugs as exemplified in Fig. 9.8. The predictability of chiral recognition of these CSPs seems poor. However, inclusion type phenomena seem likely due to the claimed but speculative partial helical structure of the linear polymer. The loadability of such CSPs is high and therefore suitable for chromatographic enantioseparations on a preparative scale. Besides the way of generating chiral polymers

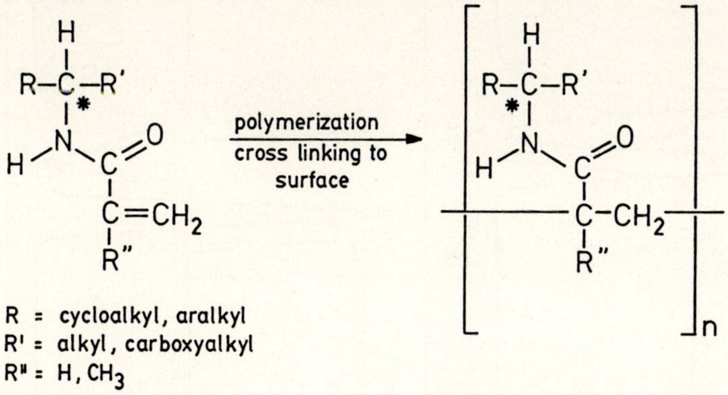

Fig. 9.7. Reaction scheme for the synthesis of chiral polyacrylamides

due to the chirality of the monomeric subunits it is also possible to generate truly helical polymers by isotactic polymerization of non-chiral monomers and chiral ionic radical starters. The relevant literature in polymer science is full of such chiral polymers, however, to date only Okamoto and his research group [33] have applied this concept for preparing chiral helical polymers for chromatographic purposes by coating these materials onto wide pore silica gel. The concept of preparing such phases is depicted in Fig. 9.9 together with a chromatogram from a Daicel brochure, the company who sells these columns. Due to the coating procedure, strong eluting solvents have to be avoided and the predictability of chiral recognition has not been rationalized yet, but it seems that particularly molecules with aromatic rings have a good chance of being retained enantioselectively. One should think that in the future more of this type of helical polymer will be evaluated for their chromatographic applicability.

9.9 Chiral Stationary Phases Based on "Brush Type" Immobilization of Small Selector Molecules

This type of CSP should be schematically visualized according to Fig. 9.10, whereby the size and the conformation of the chiral selector group may vary, but it is assumed that the SO molecule can be chemically characterized fully prior to its "brush type" and covalent immobilization onto a surface. It is obvious from a literature survey, that "brush type" CSPs build the majority within the total spectrum of enantioselective chromatographic systems. In the following only some variations of chiral recognition models within this CSP group will be discussed together with recent and illustrative examples. At first glance the mass of chiral selectors and CSPs seems difficult to classify, but it can be grouped roughly in the following way:

Fig. 9.8. Enantioseparation of racemic drugs on polyacrylamide coated silica gel (reprinted with permission from Ref. [28])

Enantioseparation of racemic drugs on polyacrylamide coated silica gel (reprinted with permission from ref. [28]). The chromatographic system:
Column: LiChrospher 100 Diol, 7 μm, modified with poly-N-acryloyl-L-phenylalanine ethylester, dimensions: 250 × 4.00 mm I.D.;
Detection: UV 254 nm in serie with Perkin-Elmer polarimeter 241 with a 100 cm cell (302 nm or 365 nm);
Sample volume: 10 μl (sample 1 mg/ml);
Eluent: n-hexane/dioxan/2-propanol (1 ml/min) as shown in the chromatogram.

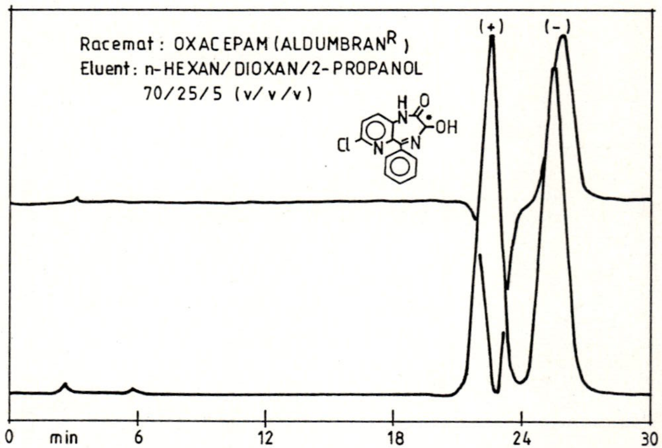

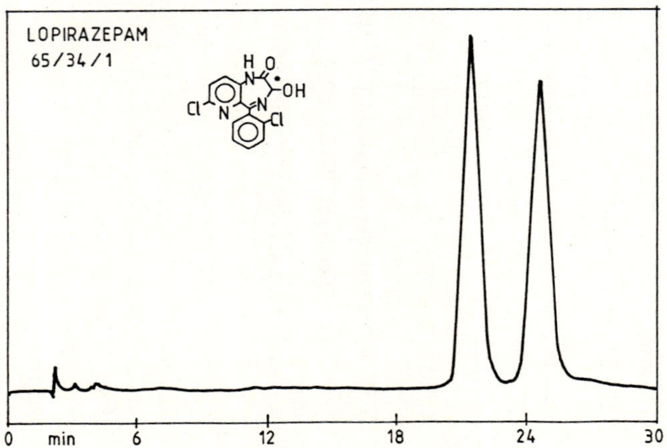

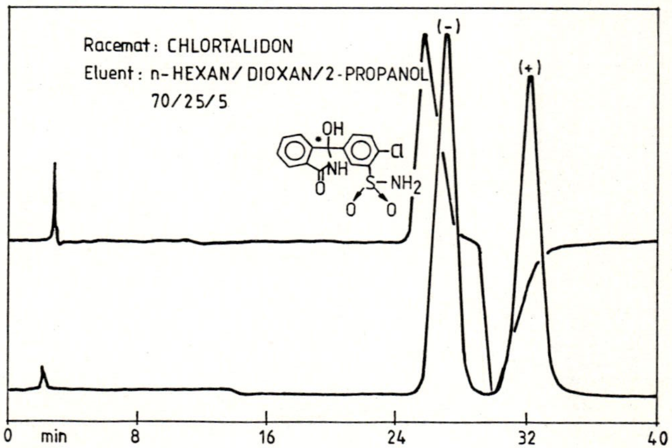

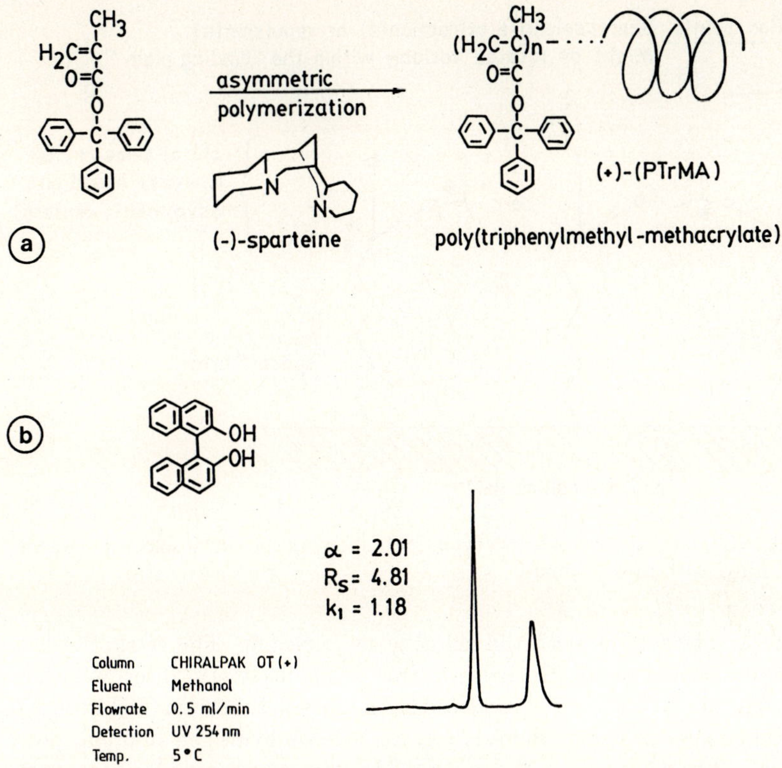

Fig. 9.9. a) Reaction scheme for synthesis of isotactic chiral polymer phase and silica gel coating. **b**) A typical chromatogram

a) inclusion type selectors such as cyclodextrines and crown ethers
b) selectors involving predominantly π–π interactions
c) selectors involving predominantly hydrogen bondings
d) selectors capable of chelating transition metal ions
e) selectors expressing charged groups thus acting as ion exchangers.

However, within these groups overlapping is likely, which means that many SOs express "mixed mode" interaction sites. This might be advantageous for chiral SA compounds, as e.g. drugs which are not designed to fit in a well-defined way the proposed and classified enantioselective binding models.

Type a) As we have learned from the polymeric CSPs, chiral recognition by inclusion phenomena is an effective way of enantioseparation.

Cyclic polysaccharides as cyclodextrins CD (depicted in Fig. 9.11) are cone shaped chiral molecules based on 6 to 12 cycled glucose units; β-CD contains 7 units and the diameter at the mouth of the cone is about 8 Å. The inner sphere of the cone is lipophilic and expresses interactions with particular aromatic rings which fit into the cone. At the mouth of the cone, the hydrogen bonding groups (CH_2-OH) are located and their conformation

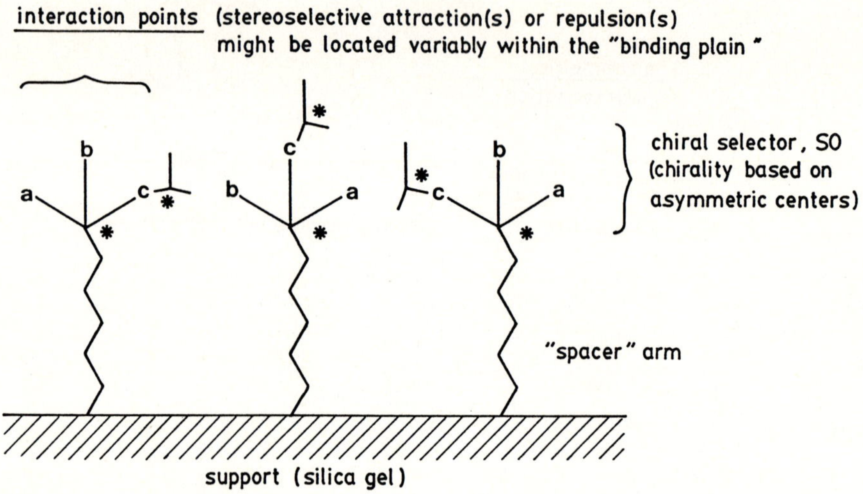

Fig. 9.10. Schematized view of brush type CSPs showing the chiral selector towards the space filled with mobile phase

allows steric interactions with the chiral guest molecule. The steric SO–SA interactions become only effective in terms of enantiomeric discrimination when the chiral center(s) of SA are close to the interaction sites (OH-groups) and of the course, provided there is e.g. an SO–SA hydrogen bonding possible, which depends also on the size of SA. Armstrong prepared the first stable cyclodextrin type CSPs based on silica gel and these CSPs have also been commecialized as "Cyclobond" columns [34]. According to restrictions given by the size of the bacterially produced product, e.g. β-CD, one tried to modify it by derivatizing the hydroxyl groups thus generating "extended" cyclodextrins. After successful attempts by König and others generating new "methylated" cyclodextrins, Armstrong and co-workers [36a] recently introduced "hydroxypropyl" extended CD by reacting the majority of the CD-hydroxyl groups with optically pure propylene oxide to the corresponding hydroxy ethers. Other types of derivatization strategies have been attempted recently [36b]. Finally the extended CDs were covalently immobilized onto a silica gel surface (see Fig. 9.11).

These new SOs have a much broader spectrum of enantioselectivity also to larger SA molecules, and a new generation of inclusion type CSPs has been developed and characterized including structure elucidations by X-ray crystallography. A further derivatization of the OH groups, for instance acetylation, leaded to a new enantioselective stationary phase used in capillary gas chromatography. The cyclodextrin rings may host more or less a medium sized SA molecule via inclusion.

In contrast, the crown ether rings like to host particularly small cations or ammonium ions, due to the hydrogen accepting capability of the ether bridges. A space filling model of a diastereomeric complex is depicted in

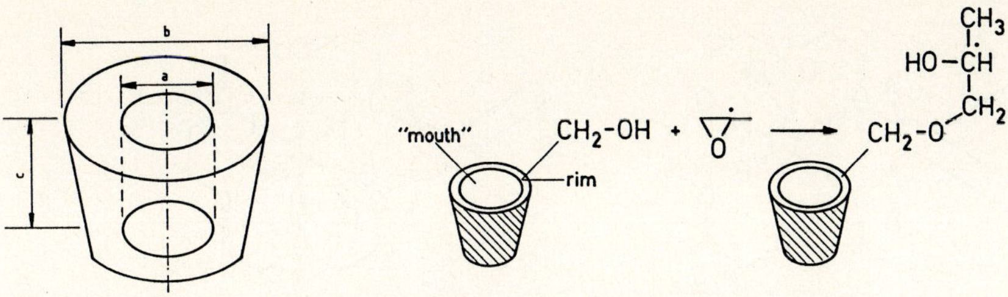

CD	Dimensions Å			Cavity volume $Å^3$	Molecular mass	Specific optical rotation $[\alpha]_D25$	Solubility in water at 25°C (g 100 ml^{-1})
	a	b	c				
α	5.7	13.7	7.8	174	972	+150	14.5
β	7.8	15.3	7.8	262	1135	+162	1.8
γ	9.5	16.9	7.8	427	1297	+177	23.2

Fig. 9.11. Reaction scheme for preparing extended cyclodextrines (only one hydroxyl group shown)

Fig. 9.12 which becomes possible with "chiral barriers" within the chiral host crown ether. It took many years after the pioneer work of Cram [35] for chiral crown ethers to be immobilized onto a silica gel surface (Fig. 9.12). However, such CSPs are now available from e.g. Daicel and show remarkable enantioselectivity for protonized chiral primary amines, particular amino acid derivatives.

Type b) Historically seen, the various donor–acceptor including π–π interaction "brush type" CSPs described in the literature derive from Pirkle's early work in this field and his constant enthusiastic and successful effort to defining chiral recognition models which allow one to predict for certain classes of compounds stereoselective expressions of interaction regions. Pirkle and Pochabsky [6] have summarized their work and that of others concerning the π–π donor–acceptor CSPs in an excellent review and try to explain the possible chiral recognition models in depth. When the SO contains only one chiral center, the models seem straightforward and are also well supported by numerous SO structure variations together with chromatographic results and spectroscopic measurements. However, SOs containing a number of chiral centers and a variety of potential interaction cites with SAs are mechanistically more difficult to study. An example of possible ways of interpreting chromatographic results is given in Fig. 9.13.

Summarizing, Pirkle's work and the Pirkle-type CSPs played a major role in the rapid development of the whole field of liquid chromatographic enantioseparation, although it seems that there is a trend towards the use of protein and inclusion type CSPs, particular in pharmaceutical analysis, since

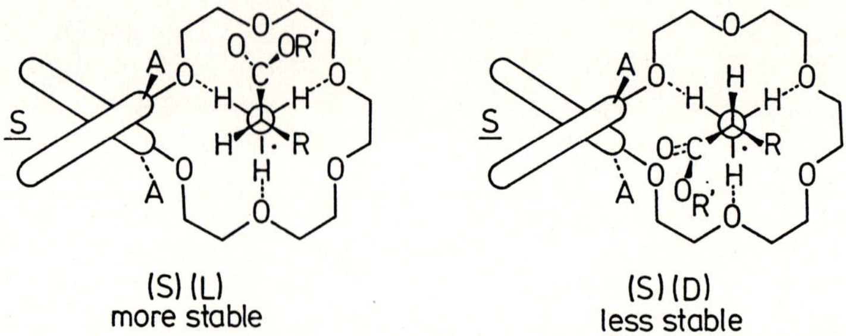

Fig. 9.12. Diastereomeric complexes between 3,3′-substituted monolocular chiral crown ether and amino acid esters as SA. (Reprinted, with permission, from Ref. [35])

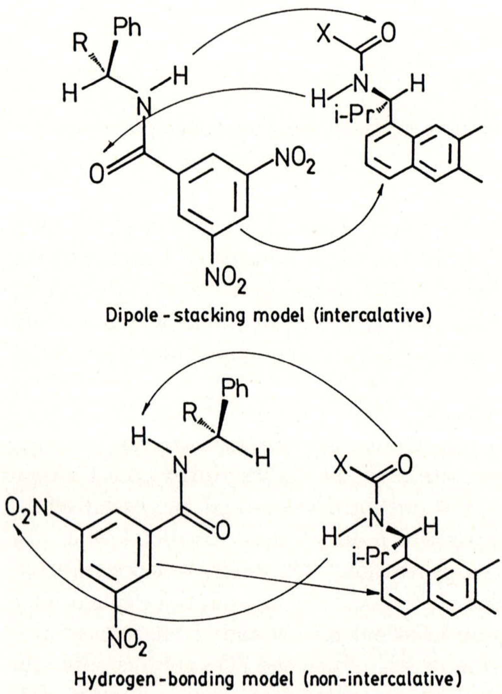

Fig. 9.13. Scheme of two competing mechanisms for chiral recognition involving π–π interactions and dipole-stacking and hydrogen bonding, respectively. (Reprinted, with permission, from Ref. [6])

these techniques mostly do not require a frequent non-chiral functionalization of the SA molecules in order to fit the donor–acceptor type CSPs.

Type c) Donor–acceptor type selectors based predominantly on multiple SO–SA hydrogen bondings are of considerable interest, since such CSPs have served for a long time as enantioselective GC phases [36]. Also in LC and

SFC the use of hydrogen donor–acceptor type chiral selctors can be of great interest, particularly for studying chiral recognition phenomena. For instance the tartaric acid diamide phases developed by Lindner [37] and Hara [38] belong to this group. Along this line are also the results of using tartaric acid esters as chiral selectors. First observed by Prelog and co-workers and later on adapted for enantioselective LC [39] the chiral esters were used as mobile phase additives leading to new dynamically generated CSPs. However, it is not the aim of this article to go more deeply into this specific and interesting approach.

Type d) Chiral ligand exchange chromatography (CLEC) is, along with chiral inclusion chromatography, a well-established discipline developed by Davankov and Rogazhin in 1971 [15]. The principle of CLEC is based on the reversible formation of mixed diastereomeric complexes between metal ions, the chiral selector (SO), and the chiral selectand (SA) ligand. In this case for a conformationally fixed SO–SA chelation, two bifunctional molecules are needed, each having two or more functional groups within the molecule at a distance to each other so that the most stable five, six or seven membered rings can be formed together with a central Cu^{2+}, Ni^{2+}, Zn^{2+} or Cd^{2+} ion. A proposed chiral recognition model which might reflect to some extent the conformation of a covalently immobilized chiral ligand exchange selector (CLE-SO) together with a chelating SA is shown in Fig. 9.14.

Over the years several CLE-SOs, almost all of them derived from alpha-amino acids but also some from tartaric acid [40], have been used and immobilized onto support surfaces (organic polymer or silica gel) by different chemical procedures and spacing groups [15]. These various SO molecules, together with the underlying support surface participate in the chiral recognition processes and control the overall observed enantioselectivity in CLEC. Mobile phase additives and conditions as type of buffer ions and molarity, pH, type and amount of organic modifier, type of chelating ion, may have an effect on the stabilities od the SO–SA chelate complexes and influence the overall retention and stereoselectivity. Due to the number of interactions and "orientations" involved in forming the mixed bidentate or multidentate complexes, also entropy effects may be considered, expressing a slight increase of enantioselectivity by raising the temperature. The rather complex CLEC systems need to be studied carefully by potential applicators; an excellent guide which helps to understand CLEC is a recent book on this subject by Davankov [15]. CLEC has been successfully applied to resolve alpha-amino acids, derivatives thereof, alpha-hydroxy acids, some amino ethanols, but also imide type drugs, as shown in Fig. 9.15.

Type e) Covalently bonded polypeptides, glycosylated (proteins), as for instance the acidic AGP-CSP, might act to a certain extent as cation exchanger, exposing but embeded in a chiral environment, e.g. free carboxylic groups excessable for coulomb attraction with amines, thus forming reversible diastereomeric ion-pairs under pH and buffer controlled mobile phase conditions.

Chiral Ligand Exchange Chromatography (CLEC)

Assumptions: Accomplishment of metal complexes of the chiral stationary phase (selector) and the compounds (selectands) to be resolved.

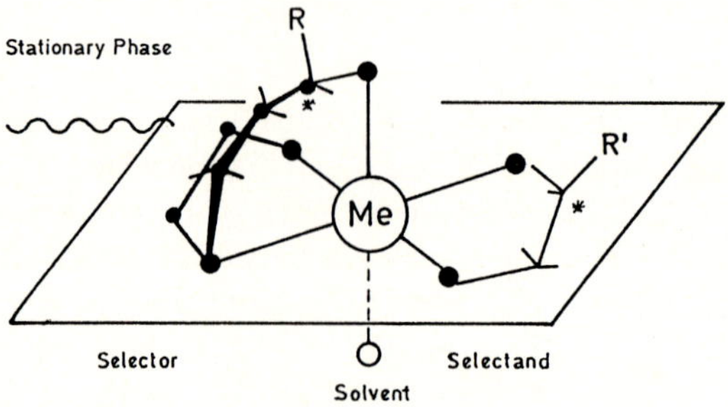

Parameters reflecting chelate stabilities:

a) Nature of ligand atoms (N, O, S, P)
b) Functionality of the ligand atoms
c) Size of chelate rings (4,5,6 or 7 membered rings)
d) Number of formable chelate rings
e) Nature of the chelated metal ion
f) Nature and concentration of competing ligands of the mobile phase (solvents, ions, buffers)
g) Temperature

Fig. 9.14. Model of chiral recognition during ligand exchange chromatography

The enantioselective ion-pair chromatography with much simpler structured anions or cations as chiral selctors (always used as mobile phase additives) follows this principle as elegantly demonstrated by Pettersson [13, 42]. As discussed earlier, at least one second intermolecular SO–SA interaction is necessary together with additional conformational orientation barriers including the non-flexible support surface.

By immobilizing e.g. rigid and chiral amines (but small molecules rather than polypeptides) onto a support material an enantioselective anion exchange in the classic sense should be generated. As seen in Fig. 9.16 this concept works indeed remarkably well [43]. The rigid tertiary amino function of the quinocludin ring of quinine or quinidine is obviously surrounded by chiral substituents which support the chiral recognition process based on the driving coulomb attraction with anions (e.g. N blocked amino acid derivatives together with π–π and hydrogen bonding areas) resulting in α-values up to 6. A representative example is shown in Fig. 9.16. The development of this

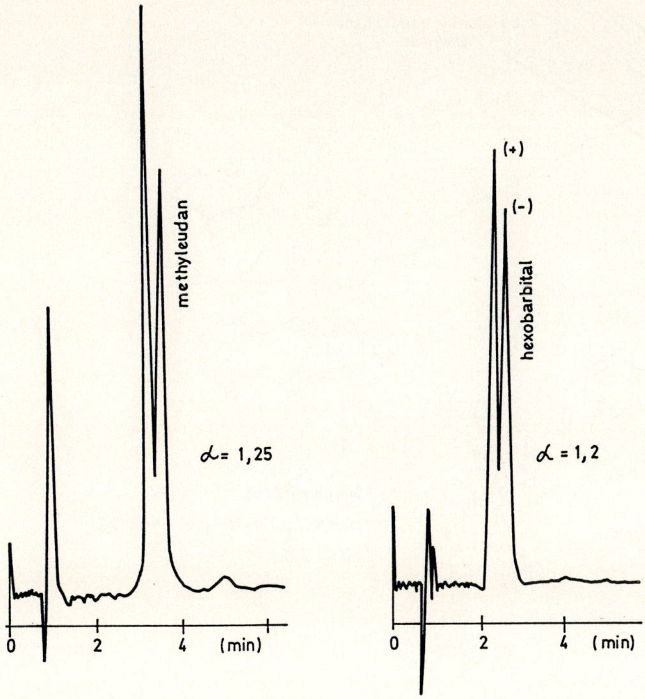

Fig. 9.15. Direct enantioseparation of racemic barbiturates on a L-Pro-amide CLEC system. (Reprinted, with permission, from Ref. [41])

special field of CSPs seems at the beginning and furthermore, tailor-made chiral ion exchangers might come up in the future.

9.10 Final Remarks on Brush Type and Inclusion Type CSPs

By summarizing the various CSPs, they express quite different types of stereotopic contact areas (planes) and it becomes obvious, that various chiral selectands (SA) might be resolvable on more than just one CSP. However, in order to make a successful estimation of which CSP should best express chiral recognition and retention for a given SA, it is essential for the user to learn to think stereochemically and also to focus on the chemical characterization of the functional groups of SA and SO. The rational models for predicting chiral recognition work for some classes of compounds, but there are no clear rules yet to be followed. Computer modeling with docking experiments of SO and SA is at its beginning, but should have great potential for the future. To date, the user of CSPs relies still more on his stereochemical imagination, but it is hoped that the chiral recognition models of the main

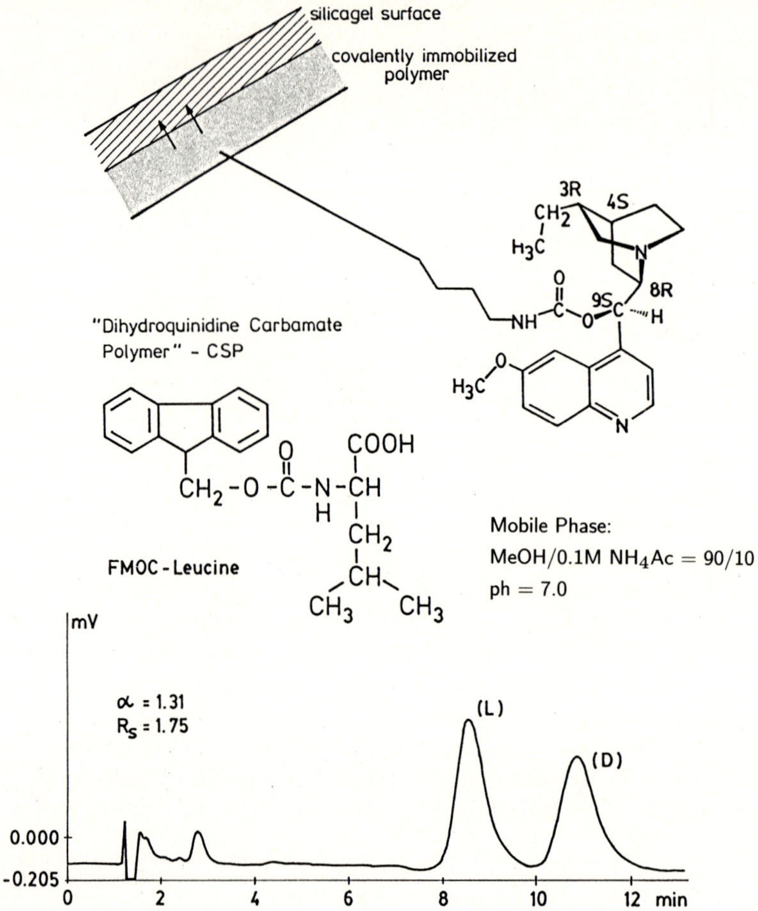

Ref.: S. Rauch-Puntigam, F. Reiter, W. Lindner; J. Chromatogr. (1990)

Fig. 9.16. *Above.* Structure of immobilized dihydroquinidine carbamate serving as anion-exchanger. *Below.* Chromatogram showing the resolution of D,L-FMOC-Leu. (Reprinted, with permission, from Ref. [43])

CSPs discussed throughout the chapter will support substantially and limit the trial and error approach to a minimum.

9.11 Indirect Enantioseparation

As pointed out in the introduction, this technique deals in principle with the formation of covalently formed diastereomers by derivatizing the mixture of SAs with an optically pure chiral derivatizing reagent (CDR). (See Reaction Scheme 9.1). These compounds should now be resolvable on conventional non-chiral stationary phases. This "indirect" principle has been applied for a long time to generate enantioseparation and seems to be less attractive these days compared to the "direct" methods. However, there are still applications which are easier and faster to perform "indirectly", particularly in the biopharmaceutical area. To study more deeply the pros and cons of this technique and to get an overview of the structures of chiral reagents, their reactivity and conform shape and features, it should be referred to a recent article by Lindner [44] and to the references in [14].

9.12 Final Remarks

With this article I have tried to explain simple chiral recognition models in the various modes of enantioselective (liquid) chromatography. Only the main CSPs selected from an individual point of view have been mentioned explicitly and numerous but not less important papers could not be cited adequately due to self-imposed limits in the size and content of this article. I have tried to demonstrate that to date a great number of successful chromatographic resolutions have been performed; however, important questions concerning a deeper understanding of their diverse separation mechnisms and the prediction of resolvability of given analytes, are waiting for an answer not only from a scientific point of view, but also for many practical reasons. In the future, many more new CSPs will appear within the framework of the search for mostly broad but also specifically applicable enantioselective chromatographic systems; their success will be judged critically by (analytical) chemists working in diversified areas but concerned with stereochemical aspects.

9.13 References

1 Ogsten AG (1948) Nature *162:* 963
2 Dalgliesh C (1952) J Chem Soc 3940
3 Pirkle W Pochapsky T (1986) J Chromatogr *369:* 175
4 Ruach-Puntigam S Erni F Lindner W (1990) J Chromatogr, paper submitted
5 Burkle W Karfunkel H Schurig V (1984) J Chromatogr *288:* 1
6 Pirkle W Pochapsky T (1989) Chem Rev *89:* 347–362
7 Lindner W (1987) Chromatographia *24:* 97–107
8 Däppen R Arm H Meyer V (1986) H Chromatogr *373:* 1
9 Lindner W Pettersson C (1985) In: Wainer J (ed) Liquid chromatography in pharmaceutical Developments: An introduction. Springfield, p 62
10 Schurig V (1983) In: Morrsion J (ed) Asymmetric Synthesis, vol 1, Analytical Methods. Academic New York
11 Pirkle W Finn J (1983) In: Morrison J (ed) Asymmetric Synthesis, vol 1, Analytical Methods. Academic New York
12 Allenmark S (1988) In: Chromatographic Enantioseparation: Methods and Applications. Ellis Horwood John Wiley New York
13 Krstulovic A (1989) In: Chiral Separation by HPCL. Ellis Horwood John Wiley New York

14 Lough W (1989) In: Chiral liquid chromatography. Blackie, Chapman and Hall New York

15 Davankov V Navratil J Walton H (1988) In: Ligand Exchange Chromatography. CRC Press Inc Boca Raton Florida USA

16 Dent C (1948) Biochem J *43:* 169

17 Dalgliesh C (1952) Biochem J *52:* 3

18 Contractor S Wragg J (1965) Nature *208:* 71

19 Hesse G Hagel R (1973) Chromatographia *6:* 277

20 Shibata T Okamoto Y Ishii K (1986) J Liq Chromatogr *9:* 313

21 Mannschreck A Kolber H Wernicke R (1985) Kontakte (E Merck Darmstadt FRG) 40 and citations therein

22 Werner A (1989) Kontakte (E Merck Darmstadt FRG) 50

23 Okamoto Y Kawashima M Hatada K (1984) J Am Chem Soc *106:* 5357

24 Application Guide for chiral column selection. Daicel Chemical Industries (1989)

25 Ariens E (1984) Eur J Clin Pharmacol *26:* 663

26 Stewart K Doherty R (1973) Proc Natl. Acad Sci USA *70:* 2850

27 Allenmark S Bomgren B (1982) J Chromatogr *252:* 297

28 Hermansson J (1983) J Chromatogr *269:* 71

29 Schill G Wainer I Barkan S (1986) J Chromatogr *265:* 73; (1986) J Liq Chromatogr *9:* 641

30a Hermansson J Schill G (1988) In: Brown PA Hartwick R (eds) High Performance Liquid Chromatography. Wiley New York

30b Erlandsson P Marle J Hansson L Isaksson R Pettersson C Pettersson G (1990) J Am Chem Soc *112:* 4573

31 Blaschke G (1974) Chem Ber *107:* 237

32 Blaschke G Frankel W Kinkel J (1987) Kontakte (E Merck Darmstadt FRG) 3

33 Okamoto Y Hatada K (1986) L Liq Chromatogr *9:* 369 and citations therein

34 Ward T Armstrong D (1986) J Liq Chromatogr *9:* 407

35 lingenfelter D Helgeson R Cram C (1981) J Org Chem *46:* 393

36 Armstrong D Stalcup A Hilton M Duncan J Faulkner J Jr. Chang S-H (1990) Anal Chem *62:* 1610

37 Lindner W Hirschböck I (1984) J Pharmac a Biol Anal *2* (2): 183

38 Dobashi A Hara S (1987) J Org Chem *52:* 2490

39 Prelog V Mutak S Kovacevic K (1983) Helv Chim Acta *66:* 2279

40 Lindner W Hirschböck I (1986) J Liq Chromatogr *9:* 551

41 Lindner W (1983) In: Lawrence JF Frei RW (eds) Chemical Derivatization in Analytical Chemistry, vol 2. Plenum Press New York

42 Pettersson C Schill G (1988) In: Zief M Crane L (eds) Chromatographic Chiral Separations. Chromatographic Science Series vol 40. Marcel Deccer Inc New York, p 283–313

43 Rauch-Puntigam S Reiter F Lindner W (1990) J Chromatogr, paper submitted

44 Lindner W (1988) In: Zief M Crane L (eds) Chromatographic Chiral Separations. Chromatographic Science Series. Marcel Dekker New York, p 91–130

10 The Nucleoproteinic System

S. Hoffmann

In memory of Jiři Beránek who made essential contributions in Prague in the years between (1968–1989) to the sense of scientific community and the progress of science among nucleic acid people, and dedicated to the memory of Richard Altmann who coined the term "nucleic acids" in 1889 in Leipzig

10.1 Introduction

Our roots reach back to the depths of the past. The grand process – at least within our view of space and time – seems to have endeavored over a period of 10–20 billion years to gain a certain consciousness and understanding of itself. Together with the universe, life patterns originated in their early infancy from an alien phase transition between nothingness and existence in the incomprehensible beginning. In all our insufficiencies, we were a part of these patterns at the very beginning, and we will share their final termination.

10.2 The Chiral Message

Partial freezing of the originally unified forces, accompanied by corresponding symmetry breaks, seem to have mediated the primarily symmetric grand unification of our universe into the diversifications of its present appearance (Fig. 10.1).

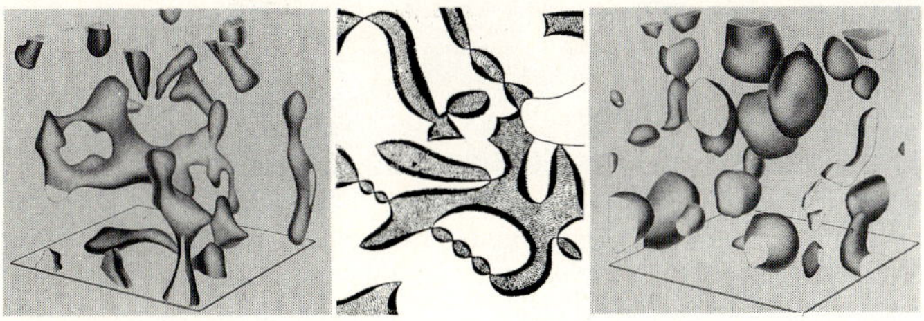

Fig. 10.1. "Phase and phase transition patterns" – Lehmann's first painting of a liquid crystal schlieren texture (*middle*) in comparison to computer simulations of the early universe (*left* and *right*) – modified from [1, 2]

By the subsequent freezing of gravity, strong and electroweak interactions, the grand process evolved through the GUT (grand unification theories), the electroweak, the quark, the plasma and elementary particles eras into the still lasting period of atoms [1, 3, 4]. A multitude of heavier atoms, burnt in the hell-fires of stars and liberated in their catastrophes, engaged in quite different chemical interactions and in this way created hierarchical patterns of increasing complexity [3, 4–7].

The chaos, however, appears to be predetermined by a strange message. Among the four forces (gravity, strong, electromagnetic and weak interactions) governing the world of elementary particles, the weak interaction and its unification with the electromagnetic interaction to the electroweak force exhibit an unusual characteristic. Contrary to the other forces which display parity-conserving symmetries, the electroweak force – mediating by $W^{\pm}$ and Z^0-bosons both weak charged and neutral currents – gives the whole process a chiral, parity-violating, asymmetric component (Figs. 10.2–5) [1, 8–18].

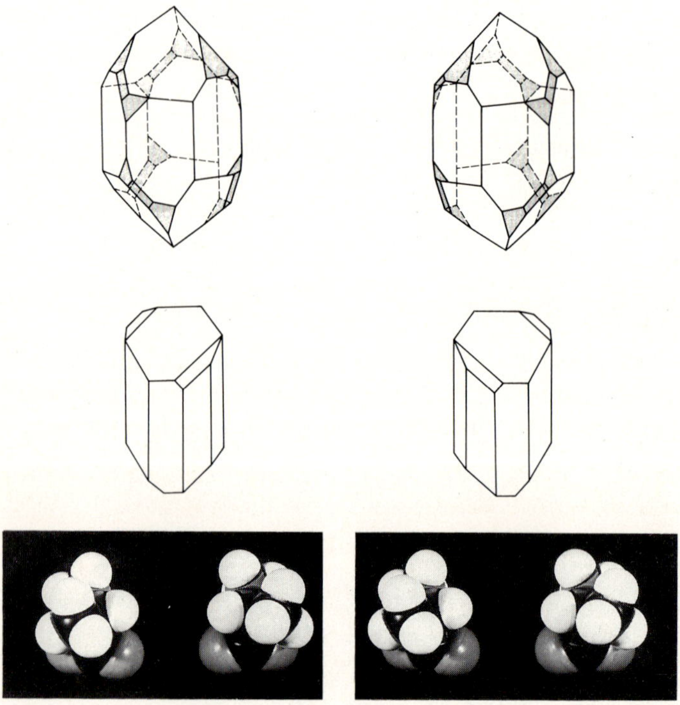

Fig. 10.2. Landmarks in chirality recognition (*left* to *right* and *top* to *bottom*): Biot's, Herschel's and Haüy's left- and right-handed quartz crystals – distinguished by the screw patterns of minor crystal facets – as inorganic solid state representatives; Pasteur's [8] (+)- and (−)-enantiomers of sodium ammonium tartrate – likewise solid phase representatives of chirality of molecular species in the liquid phases of solutions; CPK-illustration of L- and D-alanine – as common representatives of chirality characteristics of life pattern constituents

Not only atomic nuclei, but also atoms and molecules as well as their multifarious aggregations are sensitized to the special message. Static and

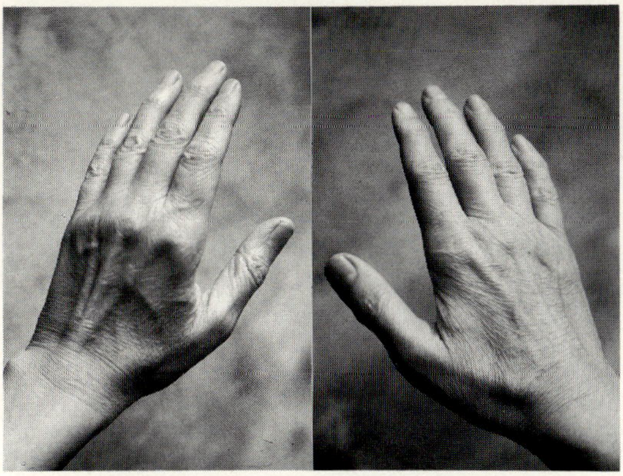

Fig. 10.3. $\chi\varepsilon\iota\varrho$ – Kelvin's intriguing term for symbolizing "chir" ality patterns

dynamic states of enantiomeric species are distinguished from the beginning by a minute but systematic preference for one enantiomer and discrimination against its mirror-image isomer [17, 18]. Amplification mechanisms, travelling long evolutionary roads, elaborated the first weak signals into dominant guiding patterns.

The freezing of strong chemical interactions at the interfaces of phase boundaries – and by this also the freezing of the special characteristics of their spatio-temporal coherence into the individual chirally affected mesophases – liberated the richness of their folds into the directionalities of chirally ordered dynamics. By the freezing of self-reproduction and chiral self-amplification conditions within nucleation trajectories, the systems appearing gained the abilities of information adaptation, storage, processing, transfer and, finally, optimization (Fig. 10.4)

Based on the unique chiral amphiphilic designs of their constituents, the biomesogen patterns evolving from there developed complex structure-motion linguistics and forwarded their more and more homochirally based contents in synergetic regulations. Within their adaptational and spatio-temporal universalities, biomesogens retraced the impetus of the early dynamics and reflected within the developing consciousness of their chirally structured organismic organizations the grand unifications that dominated their origins. Their creativity, however, somehow aims at beyond early limitations.

10.3 The Evolution of the Chiral Amphiphilic Patterns

Understandable only as the last and most highly sophisticated derivatives of the universe, life patterns included and lived in their growing complexity all the facilities from which they originated and that contributed to their further development. Between the theses of solid order and fluid disorder (Fig. 10.4), facilities for optimizable function and information processing seem to have been opened to phase systems (Fig. 10.5) that could develop flexible and adjustable structure-function correlations by the multitude of their transient,

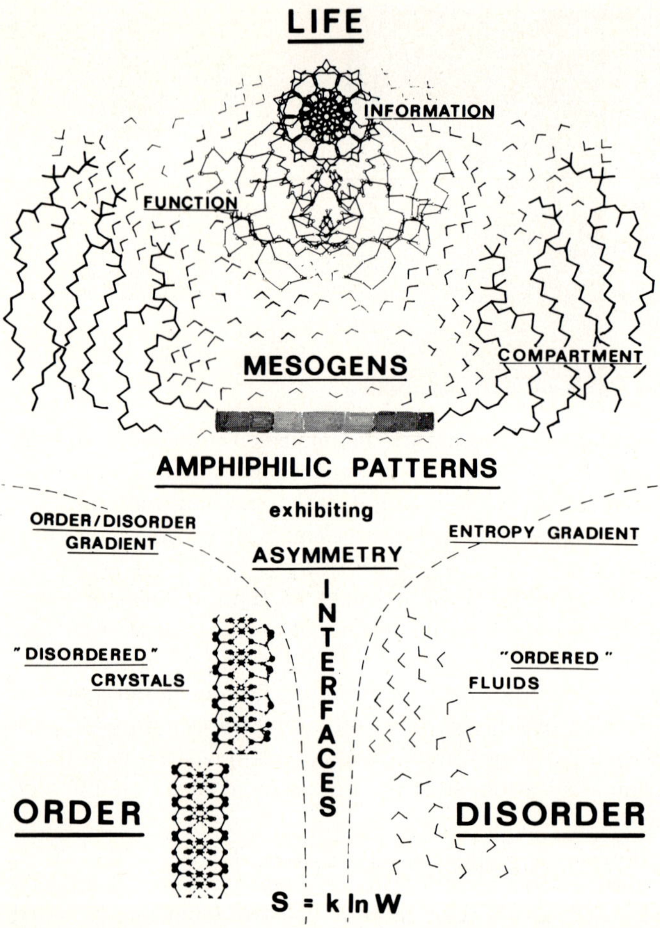

Fig. 10.4. Evolution of the chiral amphiphilic patterns

interchanging phase-domain organizations. The highly advanced life patterns of today learnt their first lessons, while endeavoring to blueprint their maternal matrix patterns, and they have been further educated in the trials to follow up their environmental changes. They gained individuality in the successful handling of dialectical syntheses out of all these fertile contradictions.

10.3.1 Darwinian Selection for Chiral Information-Processing Patterns

Among the molecular species screened by evolution in Darwinian selection for suitable constituents of first dynamic reality-adaptation and, later on, reality-variation and -creation patterns, amphiphiles with specific hydrophilic-hydrophobic and order-disorder distributions, sensitized to the chiral message of

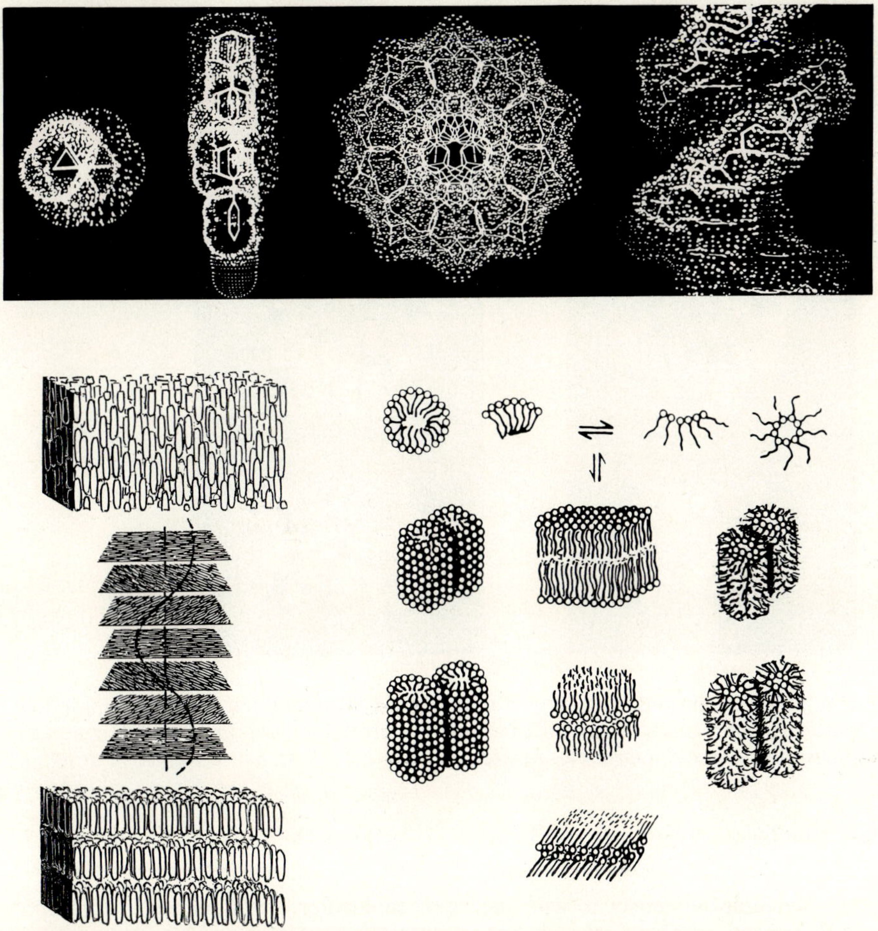

Fig. 10.5. (Bio)mesogenic patterns and their constituents (*left* to *right* and *top* to *bottom*): relationships between artificial mesogens and native biomesogens for unifying conceptions – computer graphics of 1-[*trans*-4-(alk-3-en-1-yl)-cyclohex-1-yl]-4-cyano-benzene [19] and B-DNA [20] (courtesy of Laurence H. Hurley); abstraction schemes of thermotropic (nematic, cholesteric, smectic) phases in comparison with lyotropic phase arrangements

the electroweak force, were rendered preferred survivors of the grand process (Figs. 10.6–8).

Born within the tensions between the realms of crystalline order and fluid disorder, molecularly imprinted by the very theses and antitheses of their origins, informed by the chiral message of their catalytical birth zones, experienced in the creative dialectics of interchangeable values between chirally enriched "disordered" crystal surfaces and their partially "ordered" water shells, and later on in the fluctuations of the conquered areas of their increasingly complex and permanently endangered world – mesogens, inflicted with a chiral impetus and endowed with dynamic order, managed the grand

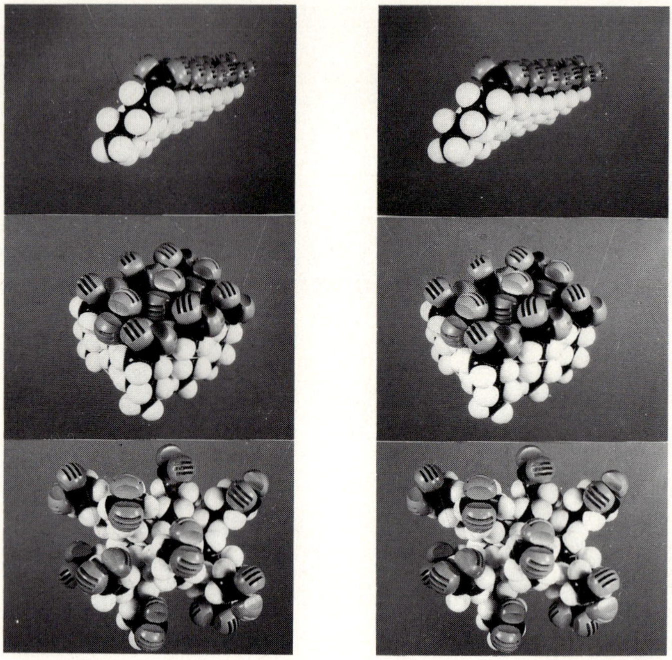

Fig. 10.6. Evolutionary strategies according to the dimensionalities of amphiphilic patterns (*top* to *bottom*): one-dimensional: information; two-dimensional: compartmentation; three-dimensional: function

escape forward from the sterile extreme futilities that dominated the boundaries of their extended birth zones.

By preintelligently forwarding their molecular asymmetries in dynamic directionality, the chiral mesogenic constituents of the developing chiral amphiphilic patterns succeeded in a fertile and creative synthesis. Avoiding hyperstatics and hyperdynamics – the disadvantages of the extreme states that contradicted their origins – they developed the creative meso-positions of ongoing ordered dynamics. Optimizable free-energy strategies on the basis of their molecularly imprinted affinity patterns selected, by preintelligently handling the entropic order-disorder gradients, patterns of chiral mesogenic backbone structures (Figs. 10.7, 8).

In the beginning, a rather omnipotent mesogenic biopolyelectrolyte pool evolved – dependent on the phase dimensionalities of its outsets (Fig. 10.6) – by division of labor into preferably informational, functional and compartmental components. Their interaction facilities together with their aptness for cooperation created by non-linear dynamics the richness of dissipative structures far from thermal equilibria and forwarded – breaking symmetry – the transition of the whole process from its mainly racemic prebiotic period into the optimizable biotic patterns of homochirality. The grand process, however, remained subjected to the dialectics from which it originated: the

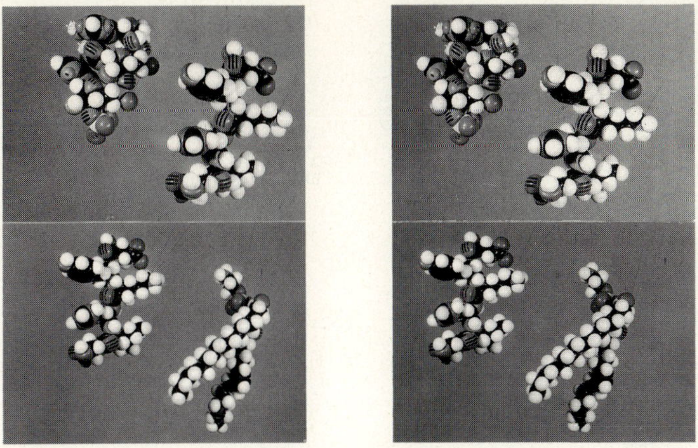

Fig. 10.7. Chiral biomesogenic backbone structures as survivors of a Darwinian selection for optimized homochiral order-disorder distributions (*left* to *right* and *top* to *bottom*): stacked nucleic acid single strand, extended protein single strand, lecithin with an only small, but nevertheless chiral backbone

general chiral mesogenic approach between order and disorder and the permanent renewal and achievement of forward-directed path-finding out of the contradictions.

It had been the singular usefulness of optimizable chiral backbone arrangements (Figs. 10.7, 8), that allowed for a division of labor development into the specializations of information, function and compartmentation, conserving, however, beneath the skin of their specific adjustments, the continued primitive universalities of their origin. Thus, while a first sudden glance might connect the structural features of nucleic acids with information, that of proteins with function and the remaining characteristics of membrane components with compartmentation, a nearer and more detailed intimacy with the three dominants of biopolymeric and biomesogenic organization nowadays reveal much broader ranges of different abilities, unravelling, for instance, functional capabilities of nucleic acids in the widespread landscape of catalytic RNA-species, offering informational ambitions of proteins in their instruction of certain old-protein production lines and recognizing within the complex instrumentary of functional membranes additional informational and functional potencies.

By this, the classical views of interacting structural individuals submerge into the new qualities of transient mosaics of mutual domain cooperativities, where chirally instructed stereoelectronic patterns of individual representatives of the grand triad anneal into the spatio-temporal coherences of newly achieved biomesogenic domain organizations. It is within this dual view of biomesogens, that both the structural and the phase-domain aspects will

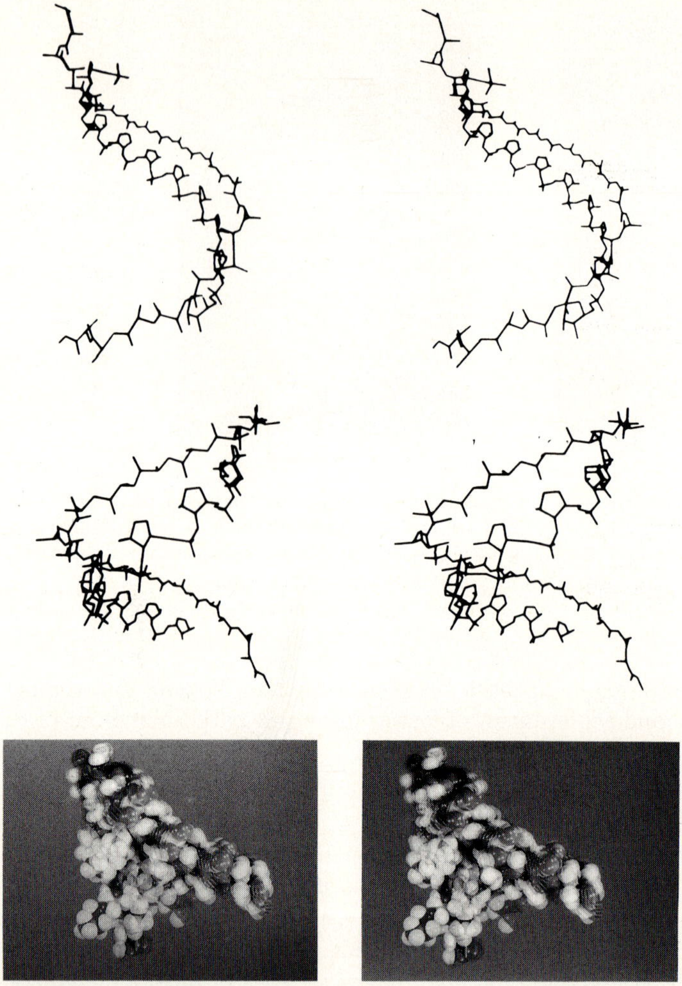

Fig. 10.8. Chiral intimacies of nucleic acids and proteins in the nucleation of the nucleoproteinic process (*top* to *bottom*): computer simulations of mutual backbone arrangements [21] and CPK-illustration of nucleoproteinic dynamics – montmorillonites being capable of catalyzing both polypeptide and polynucleotide formation [22]

contribute intriguingly to a consistent picture of life patterns and their operational modes [4, 6, 7].

10.3.2 Basal Geometries of Chiral Nucleoproteinic Constituents

Miescher's prophetic view of his "Nuclein" to be the genetic material [23, 6] could not have been valued at a time, when proteins, the $\pi\rho\tilde{\omega}\tau o\varsigma$ (greek: the first), had been the beloved species of all people engaged in "early life sciences". And though some early intuitive perceptions appear convincingly

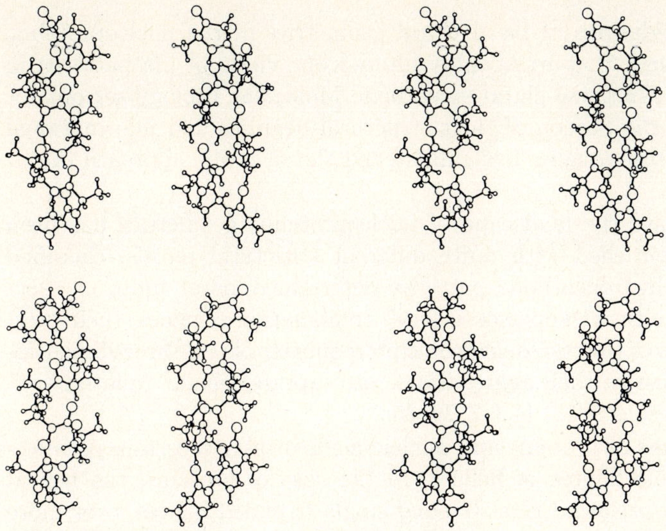

Fig. 10.9. Enantiomeric and diastereomeric expressions of L- and D-amino acids integrated into superior chirality patterns of α-helical arrangements (*left* to *right* and *top* to *bottom*): left- and right-handed α-helices of L-amino-acid constuents; right- and left-handed α-helices of D-amino-acid enantiomers – some details taken from [31]. Electroweak force and energy minimizations of side-chain instrumentary selecting for L-α-amino acids, preferentially adopting right-handed helices

modern [24], the real structural designs of the patterns of life remained obscure for nearly a century.

Somehow indicative of the "optical disposition" of the species "homo sapiens", it was not the admirable achievements of the classical period of natural products chemistry, starting together with Fischer at the turn of the century [25] and highlighted perhaps by the work of Eschenmoser and Woodward [26], it was also not the lonely prophecy of a physicist for an aperiodic crystal to be the genetic material [27], and it was not even the mystic secret-formulae of Chargaff [28] concerning the base-pairing schemes of nucleic acids and anticipating the whole story to be further elucidated, that caused the dramatic scientific "phase transition" in the 1950s. It was the perception of the two "holy" chiral structures: Pauling's protein α-helix [29, 30] and Watson and Crick's (as well as Wilkins and Franklin's) DNA double-helix [32, 33] (Figs. 10.9, 10) that introduced a new age.

The first elucidations of basal chiral geometries of the nucleoproteinic system was an event that irrevocably altered our views of life patterns in nearly all aspects. In these chiral structures, Molecular Biology set out for far horizons, and the beacon of their structural beauties and informational and functional transparencies has enlightened the scientific approach of our time.

Since these days the landscape of nucleoproteinic geometries has been more and more enriched with quite different structural motifs, classified into the mainly hydrogen-bond-patterns determined basal units of "secondary" structure, their "supersecondary" combination schemes, their "tertiary" adaptations to the stereoelectronic prerequisites of real three-dimensional entities and, finally, their aggregations into supramolecular "quaternary" organizations (Figs. 10.9–15) [4, 6,7, 29–35].

Both the realms of proteins and nucleic acids display a certain predilection for the flexible filigree of helices. In the case of proteins, the helical designs vary from preferred right-handed single-stranded helices, over more restricted right- and left-handed single- and double-stranded arrays to special forms of left-handed helix triples (Figs. 10.9, 11, 12) [4, 6, 7, 29–31, 35–42]. For nucleic acids, the structural landscape is dominated by right-handed antiparallel double-helix motifs, enriched by differently intertwisted helical triples and quadruples and contrasted by alien versions of left-handed antiparallel double helices (Figs. 10.10, 13–15) [4, 6, 7, 32–34, 42–52].

The antitheses of more rigid basic secondary-structure motifs appear in the parallel and antiparallel β-sheets of proteins and will find some correspondence in the cylindrically wound-up double-helix design of the nucleic acid A-families as well as in the so far somewhat dubious versions of suprahelical Olson-type arrangements (Figs. 10.10–14) [4, 6, 7, 29–35, 43, 48, 49].

In the rivalry between the D- and L- enantiomers of protein and nucleic acid pattern constituents as well as in their chiral amplifications into larger supramolecular motifs, the "more unfortunate" enantiomers, energetically discriminated by the chiral instruction of the electroweak force, fought a losing battle from the beginning of the grand evolutionary competition. Ab-initio calculations of the preferred aqueous-solution conformations of some α-amino acids and glyceraldehyde, the classification-ancestor of all sugar-moieties, are indicative of a stabilization of L-α-amino-acids and D-glyceraldehyde to their discriminated mirror-imaged enantiomers by some 10^{-14} Jmol^{-1} [14–16]. The same seems to hold true for the α-helical and β-sheet patterns of L-α-amino acids in comparison to the structural mirror-images of their D-antipodes. Absolute weighing by the electroweak force, further selections in different amplification mechanisms, as well as the rejection of mirror-image species in building up supramolecular chiral arrangements, all this taken together with energy minimizations of the side-chain instrumentary both in proteins (the 20-1 proteinogenic side-chain versions) and nucleic acids (the four-letter alphabet of the nucleobases) selected, finally, for the preferentially appearance of L-α-amino acids' right-handed protein-α-helices and right-handedly twisted parallel and antiparallel protein-β-sheets [6, 14–

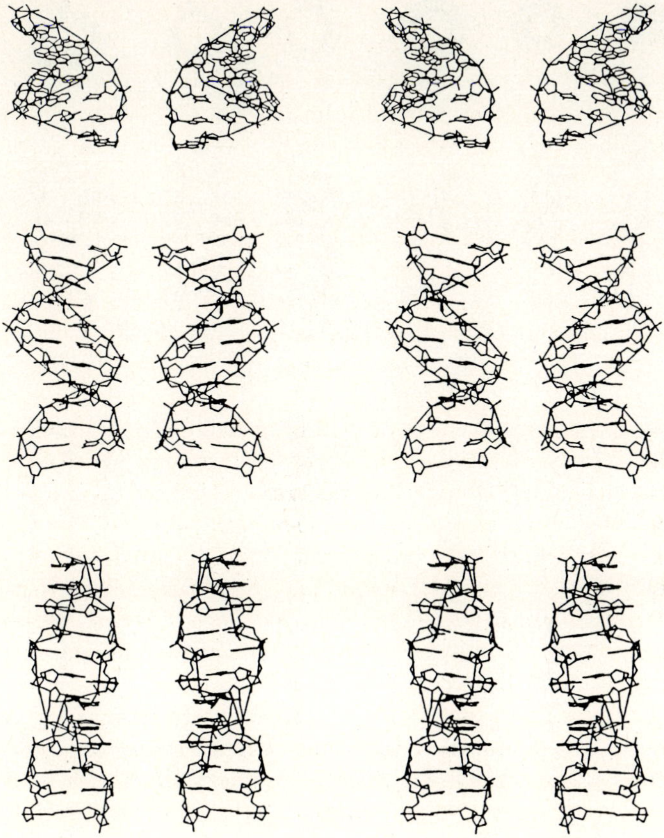

Fig. 10.10. Enantiomeric versions of nucleic acid patterns in antiparallel double helical arrangements (*left* to *right* and *top* to *bottom*): A-type left- and right-handed versions of L- and D-ribose and 2′-deoxy-L- and D-ribose moieties, respectively; B-type left- and right-handed arrays of 2′-deoxy-L- and -D-ribose cycles, respectively; Z-type right- and left-handed double helices of 2′-deoxy-L- and D-ribose enantiomers. Electroweak force and energy-minimizations of base-stack instrumentary selecting for D-ribose and its 2′-deoxy-derivative, adopting right-handed double helices in A- and A′-type RNAs as well as in A-, B-, C-, and D-type DNAs, and creating left-handed double helical varieties for Z-type DNAs and RNAs – some details taken from [34]

18, 29–31, 35, 52], and favored (2′-deoxy)-D-ribose moieties' right-handed antiparallel double helices of nucleic acids A-, B-, C- and D-families [6, 14–18, 31–34, 52]. And it is also due to those very selection and optimization pro-

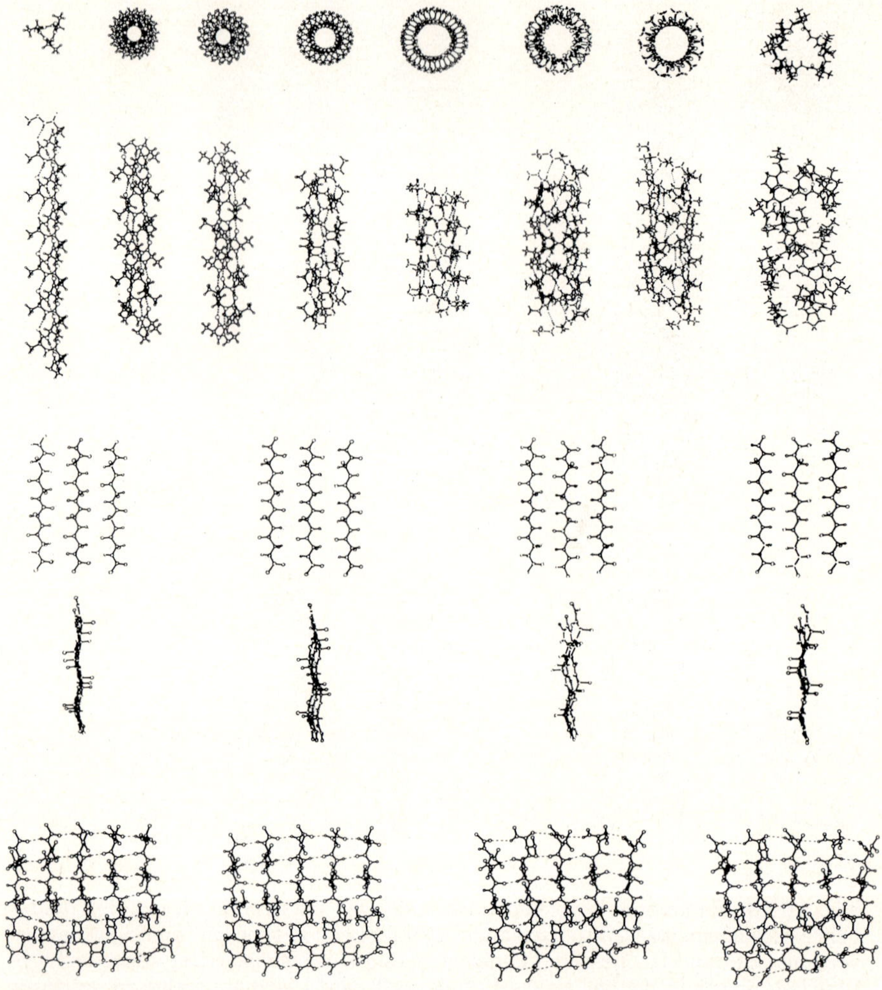

Fig. 10.11. Basal chiral protein geometries between helices and sheets in secondary structure (*left* to *right* and *top* to *bottom*): polypeptide helical designs between single- and double-stranded, right- and left-handed, parallel and antiparallel, helix, spiral- and channel-expressions [36, 37]; parallel and antiparallel β-sheets in plane and right-handed twisted forms [38, 39] – modified from the cited references

cedures, that poly-L-proline backbones might be forced into left-handed helical versions, and special alternating sequences of nucleic acids designs tend to adopt the strange double-helical left-handed Z-motif [6, 14–18, 31, 34, 49, 52] (Figs. 10.9–15).

The beauties of all these chiral structural standards and the respective families around them in proteins and nucleic acids represent, however, only the main building blocks which evolution has been operating on. The colored loop- and knot-stretches and all the more disordered parts of protein and

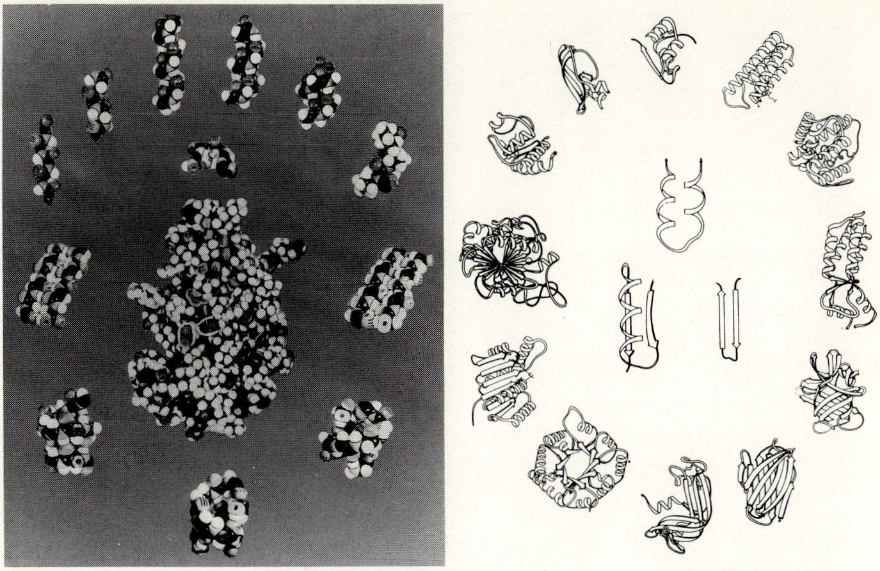

Fig. 10.12. Basal chiral motifs of protein secondary, supersecondary and tertiary structures: *Left plate* (central part and surrounding circle clockwise): ribonuclease A; α-structures of 2_7-band, right-handed $2{,}2_7$-, 3_{10}-, α-, and π-helices, left-handed $(\text{Pro})_n$-helix and β-turn, parallel β-sheet, parallel and antiparallel right- and left-handed double-helical β-strands, left-handed single-stranded β-helix, antiparallel β-sheet. *Right plate* (central part and surrounding circle clockwise): $\alpha\alpha$-, $\beta\beta$- and $\beta\alpha\beta$-supersecondary structure motifs; (circle starting $1^{\circ\circ}$ all α-(antiparallel α)-proteins: cytochrome b_{562}, thermolysin (domain 2), tobacco-mosaic virus protein; all-β-(antiparallel β)-proteins: soybean trypsin inhibitor, immuno-globulin (V_L-domain), ribonuclease A; regular α/β-proteins (containing often $\beta\alpha\beta$-motifs): triosephosphate isomerase, lactate dehydrogenase (domain 1), carboxypeptidase; (small) irregular α- and/or β-proteins: cytochrome c, pancreatic trypsin inhibitor (BPTI), insulin – all strongly modified from [40, 41]

nucleic acid structural designs are of primary importance both for the evolutionary aspects of informational and functional processings in our current life patterns (Figs. 10.12, 14, 15) [4, 6, 34, 35, 40–42, 49, 52].

It fits nicely into the picture of a dual structure-phase view of biomesogenic organizations, that "rod-like appearances" as for instance the little world of the tobacco mosaic virus as well as protein and nucleic acid helical expressions are typical mesophase builders in the sense of classical liquid crystal phase expectations (Figs. 10.5, 15) [4, 6, 7, 52–60]. It is within the frame of its close connections with the preinstruction of biomesogenic species by the chiral message of the electroweak force, that the transformation of chiral molecular units to the potencies of "supra-chiral" macromolecular patterns and from here to the still amplified chiral expression of supramolecular phase organizations provided a useful sequence for the savage and elaborations of an early message of our universe from the noise of its first hidden appearances up to the homochirality as a matter of course nowadays.

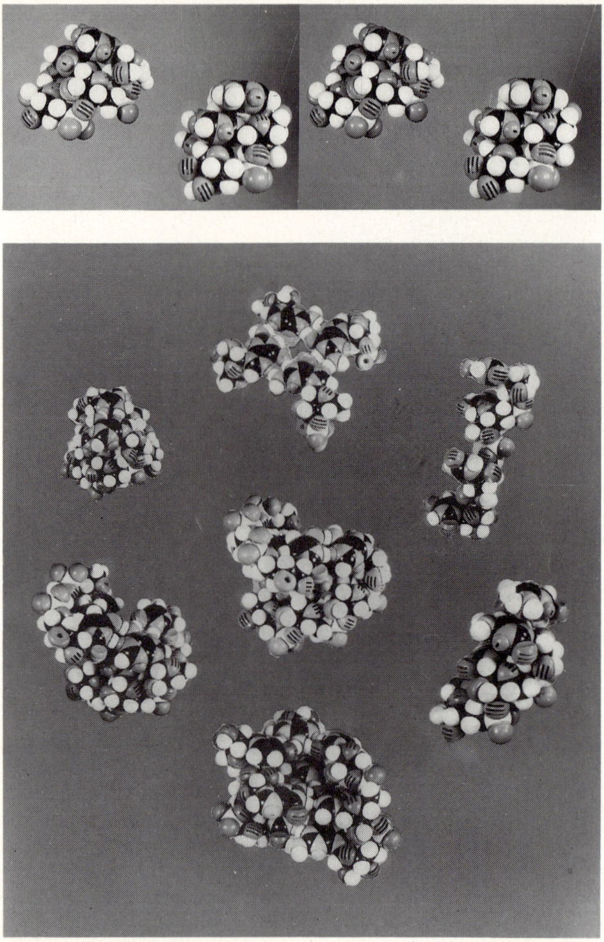

Fig. 10.13. Basal chiral nucleic acid geometries (*upper part*: stereo-presentation, *lower part*: central motif and clockwise arrangement) between secondary and tertiary structures: 3′-5′- and 2′-5′-strand arrangements; A-type double-stranded RNA $(U)_n \cdot (A)_n$; random coil of single stranded $(U)_n$ – unstructured at ambient temperatures; single stranded stacked $(C)_n$ at ambient temperature; uncommon base triples in hypothetical triple-stranded complex of self-splicing Tetrahymena pre-rRNA (see for details Fig. 10.18); triple-stranded arrangement of $(U)_n \cdot (A)_n \cdot (U)_n$, displaying both Watson-Crick- and Hoogsteen-pairing; double-stranded parallel version of $(A)_n \cdot (A)_n$ and $(AH^+)_n \cdot (AH^+)_n$, respectively; parallel quadruple-stranded versions of $(G)_n \cdot (G)_n \cdot (G)_n \cdot (G)_n$

10.3.3 The DNA-RNA-Protein Triad

The "Central Dogma", proposed by the creators of the DNA-double helix model in order to channel their considerations of the information flow in biological systems, probably had been inverted in historical sequence. Its α and ω determined the evolutionary action scheme: variation in the information con-

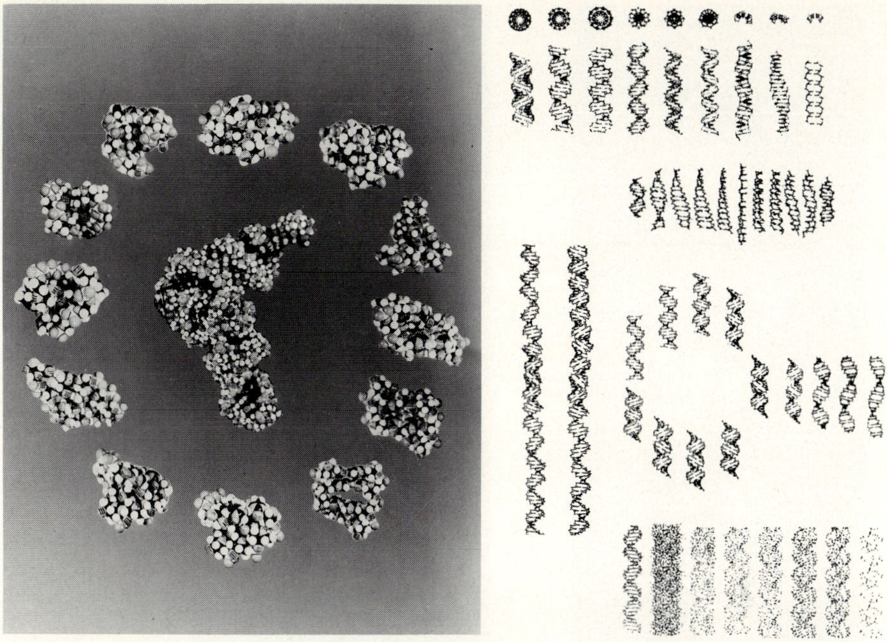

Fig. 10.14. Nucleic acids chiral secondary and tertiary structures and parts of their operation modes [4, 6, 7, 34, 42–53]. *Left table* (centre and clockwise circle/1°°): yeast tRNA^{Phe}; alternating B-DNA, Sussman-bend B-DNA, Crick- and Sobell-kink B-DNA, intercalation geometry B-DNA, B-like D/RNA-hybrid, A-D/RNA, Olson-D/RNA, left-handed A- and B-DNA, Z-D/RNA, B-DNA. *Right table* (*left* to *right* and *top* to *bottom*): nucleic acid polymorphs – fibrous polynucleotides unwinding right- to left-handed helices (Olson- and Z-families omitted) [47–49]; computer simulation of the transition from right- to left-handed DNA [49]; A-B-transitions and intercalation-geometry generations mediated by Sobell-kinks, presumably based on non-linear excitations and soliton formations [50]; B-DNA counterion and water cover as resulting from Monte-Carlo simulations [51] – partially taken from the references cited

tent of the genotype (DNA) and subsequent functional selection in the phenotype (protein). "Dynamic" proteins and "static" nucleic acids competed as the "hen-egg" problem for historical evolutionary priority [4, 6, 7, 53, 61–63]. But while the original alternative found its solution in both the informational and functional "hypercycle" [61–63], and while dual structure-phase views – contrary to classical perceptions – foresaw much closer spatio-temporal coherences for the evolving mesophase systems [4, 6, 7, 52, 53], while RNA turned out to be in some cases DNA-informative [64, 65], the old hen-egg problem offered a further surprising aspect for the "meso" positions of the so far unfortunate and disregarded RNAs: a view on what might be called an early RNA-world [66–74].

The evolutionary main roads in the development of the DNA-RNA-Protein triad, however, seem to have been preceded by a peculiar intermezzo. The ambitious enterprise of proteins to cover self-consistently both informa-

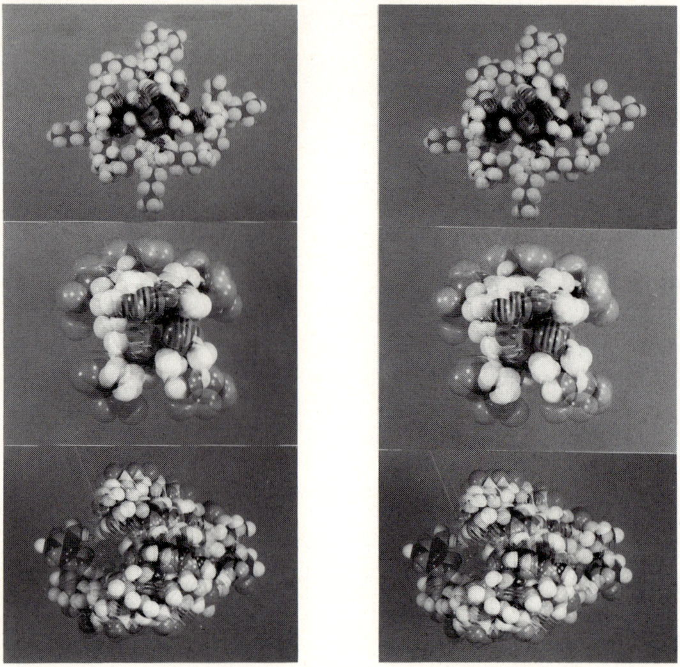

Fig. 10.15. Dynamics of chiral polymeric thermotropics and lyotropics – models and native standards (*top* to *bottom*): poly-(γ-alkyl-L-glutamate) α-helical array – a thermotropically designed original lyotropic [56]; lyotropic poly(L-glutamic acid) [57, 58]; lyotropic DNA/RNAs [58–60] – all chiral monomers chirally amplificated in secondary structure helical and double helical designs and in subsequent superhelical cholesteric phase arrays

tional and functional capacities survived up to our days in some alien, but nevertheless within all their structural restrictions admirable molecular appearances. A group of depsipeptide and peptide carriers and channels as well as their possible subunit structures – produced on old, both informational and functional protein lines – display enchanting achievements in the skillful handling of biomesogenic operation modes (Fig. 10.16) [4, 6, 7, 75–78]. Interestingly, all this was brought about without any preference of a special homochirality. On the contrary, it is just the surprising selection of differently alternating chirality patterns from the prebiotic racemic pool and the careful and intelligent designing of rather small molecular entities with adjusted alternating chiral codes, that made such beautiful arrangements as, for instance, the "bracelet" of valinomycin work (Fig. 10.16). The ambitious enterprise of proteins, the "first", to create by the power of their own legislative and executive intelligent "bichiral" patterns, ended up, notwithstanding all the convincing achievements that match not only intriguingly later-to-be-developed operation modes of the DNA-RNA-Protein triad but also our own intelligence, in an evolutionary impasse.

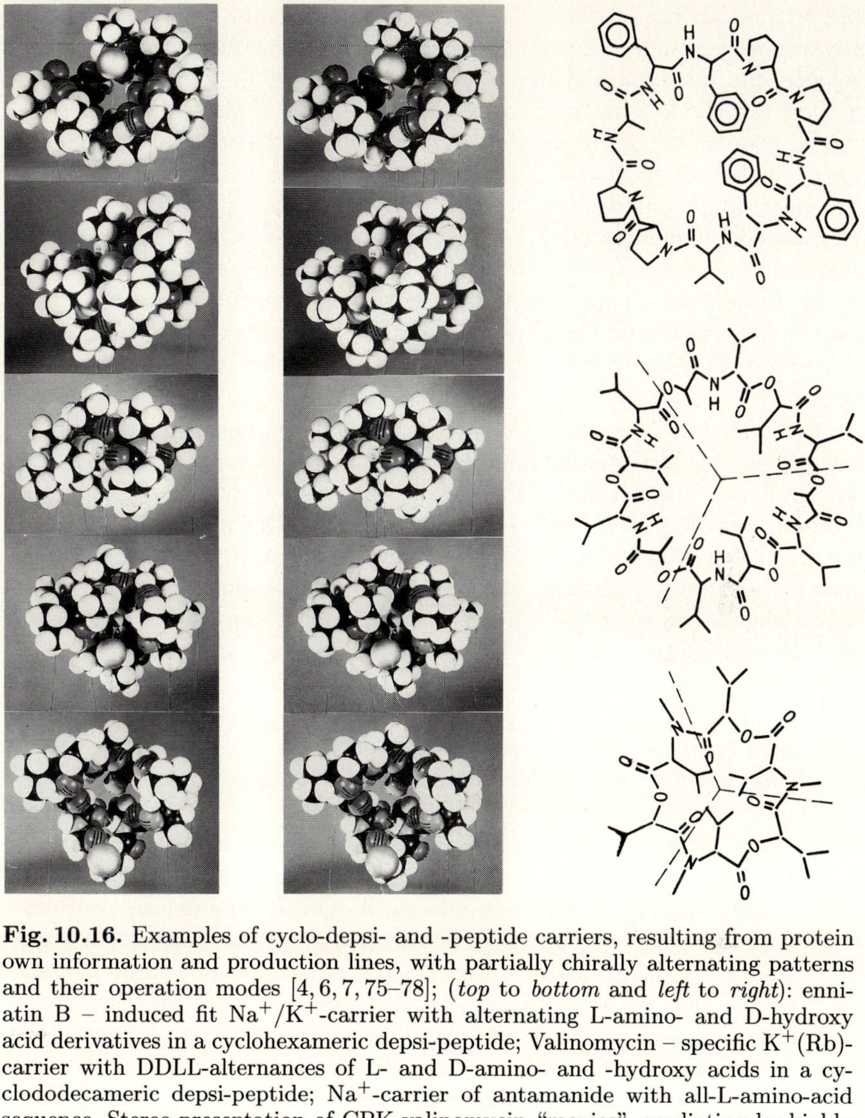

Fig. 10.16. Examples of cyclo-depsi- and -peptide carriers, resulting from protein own information and production lines, with partially chirally alternating patterns and their operation modes [4, 6, 7, 75–78]; (*top* to *bottom* and *left* to *right*): enniatin B – induced fit Na^+/K^+-carrier with alternating L-amino- and D-hydroxy acid derivatives in a cyclohexameric depsi-peptide; Valinomycin – specific K^+(Rb)-carrier with DDLL-alternances of L- and D-amino- and -hydroxy acids in a cyclododecameric depsi-peptide; Na^+-carrier of antamanide with all-L-amino-acid sequence. Stereo-presentation of CPK-valinomycin "movies", mediating by highly sophisticated biomesogenic interplays a K^+-ion membrane passage

While proteins thus failed in their omnipotency, the obviously less qualified RNAs seem to have built up a preliminary, mediating and forwarding RNA-world (Figs. 10.17, 18). "A tRNA looks like a nucleic acid doing the job of a protein", Crick's pensive considerations [6, 32, 33] became reality [4, 6, 7, 66–74]: mRNAs – the differentiating blueprints of the DNA-information store, tRNAs – the structurally disrupted mediators of nucleoproteinic interrelationships and rRNAs – the long disregarded "structural support" of ribosome protein-synthesis machinery, they all developed not only a more and more interesting "eigenleben", RNAs, moreover, proved to be both informational and functional [4, 6, 7, 52, 66–74]. They were among the first to provide with their special characteristics of molecular hystereses a basal understanding for memory records, oscillations and rhythm generations in biological systems [4, 6, 7, 52, 53, 70–74], and they advanced – starting with their catalytical facilities in self-splicing and ending up with general "ribozyme" characteristics [66–71] – successful predecessors and competitors of proteinic enzyme potencies both in basic and applied research. RNAs – combining in their early appearances both genotypic and phenotypic aspects – seem to have mediated the grand DNA-RNA-protein triad into its present expressions and survived up to now as the first inherent principle of life patterns.

The final ways for utmost complexity in informational and functional processings are illustrated by impressive landmarks (Figs. 10.17, 19–20). The early prebiotic interactions of the shallow groove of sheet-like right-handed A-type RNAs (Fig. 10.17) with the right-handed twisted antiparallel β-sheet of a peptide partner (Fig. 10.19) might have established the first fertile intimacies between nucleic acids and proteins [6, 79]. Within this hypothetical first productive organization, the archetypic protein-β-sheet with its aptness for easily sorting polar–apolar and hydrophilic-hydrophobic distributions as well as the archetype A-RNA with its partially complementary informational and functional matrix patterns could "live" first vice versa polymerase facilities, and thus mutually catalyze early reproduction cycles, connected with chiral amplifications and informational and functional optimizations.

The last step for liberating all the inherent potencies of these early nucleoproteinic systems of RNA and protein was found in the dual conformational abilities of the small 2′-deoxy-D-ribose cycle to account for a conformationally amplificative switch between A- and B-type nucleic acid versions (Fig. 10.17, 19–20) [6, 31–34]. The detection of the information store of DNA enabled the systems to separate the replication of a more densely packed DNA-message from the informational and functional instrumentary needed in the hitherto existing nucleic-acid-protein-interaction schemes and forwarded by this transcriptional and translational operation modes with far-reaching maintainance of so far elaborated RNA-protein cooperativities. The new qualities had been brought about by the structurally deepening of the shallow RNA-groove into the minor groove of DNA that allowed the continuation of the successful RNA-protein contacts together with newly to be established small effector regulations, and by the opening of the so far hidden huge information content of the RNA-deep-groove into the much more exposed major groove of B-type DNA. All these newly developed and achieved operational facilities advanced a dramatic transition from the unspecific contacts between proteins and nucleic acids in their shallow and minor grooves, respectively, to specific recognition and functional information processing, mostly exemplified by protein helices in the B-DNA major groove (Fig. 10.20) [6, 31–34, 52, 80–85].

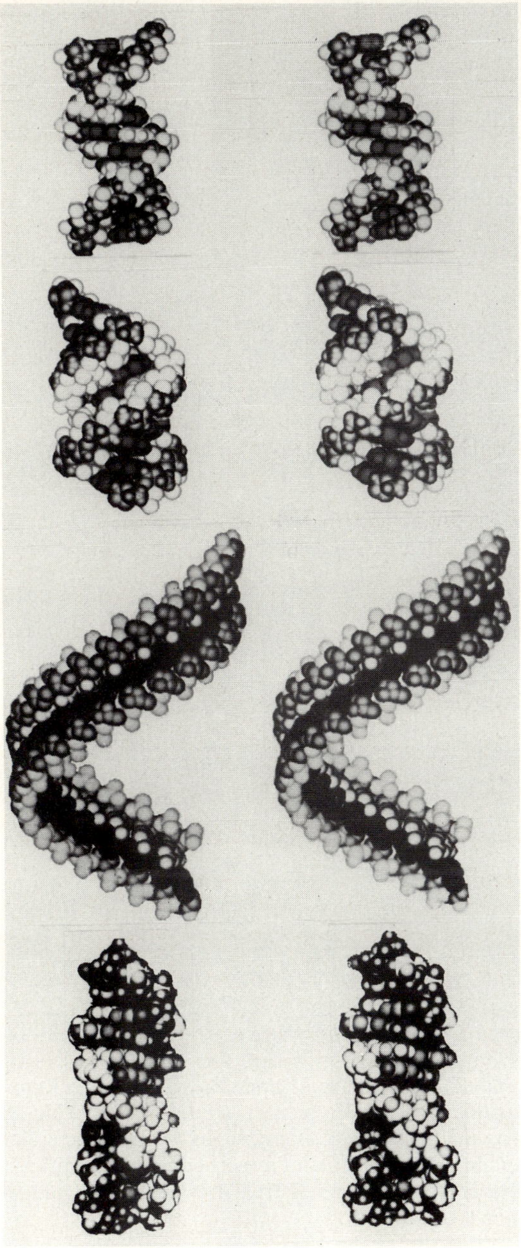

Fig. 10.17. Stereo-visualizations of nucleic acids A/B/Z- and Olson-arrangements (*top* to *bottom*): B-type DNA exhibiting minor and major groove design; A-type DNA and RNA, respectively, with minor-groove corresponding shallow groove and major-groove derived deep groove; superhelical expression of Olson-type DNAs and RNAs, Z-type DNAs and RNAs, dramatically deepening minor- and shallow-groove design, respectively, and exposing former major and deep groove of B-type DNAs and A-type DNAs and RNAs – modified from [76]

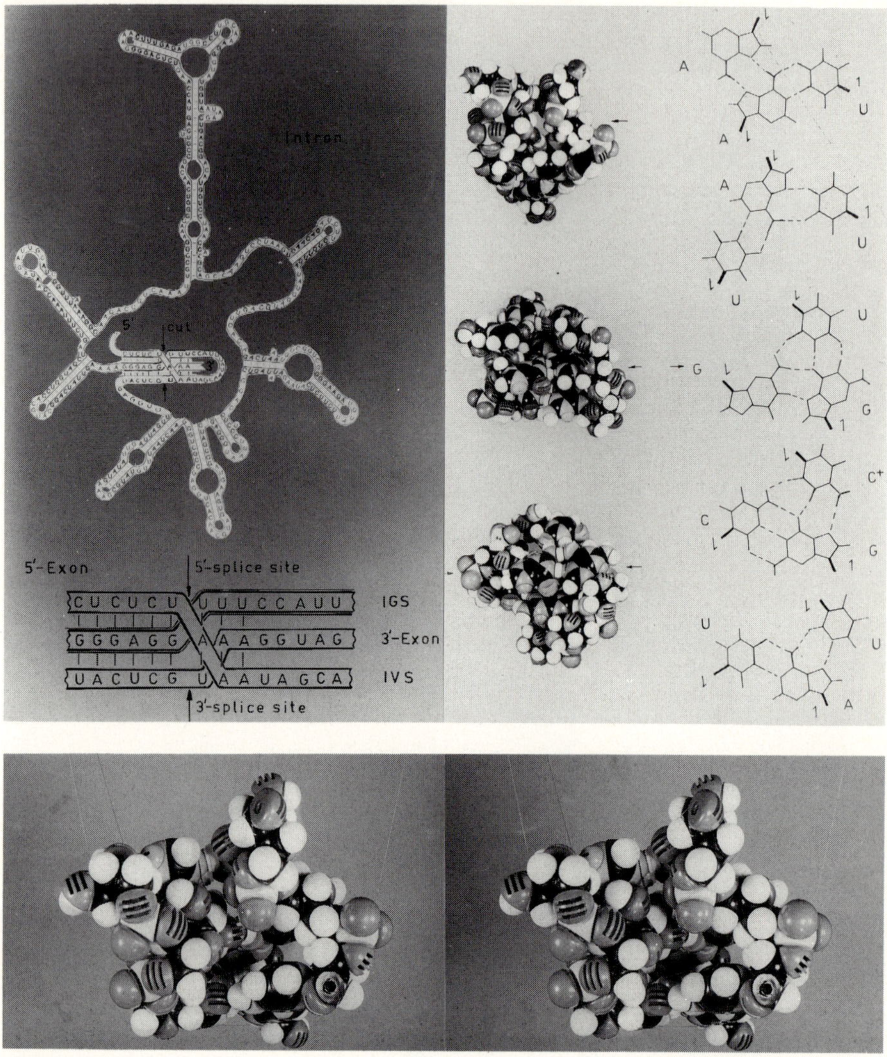

Fig. 10.18. Self-splicing RNAs [66–71] (*left* to *right* and *top* to *bottom*): hypothetical triplex formation in self-splicing of Tetrahymena pre-rRNA, mediating by an all-purine guide pattern a constraint triple helix arrangement around the splice sites and forwarding by this the splicing process; cuts of hypothetical mini-triplexes around the (circled) splice sites; stereo presentation of triple-stranded arrangement of the splicing complex – for details see [70]

The resulting DNA-RNA-Protein triad – mediated by intelligently handled clusters of water molecules – allowed the development of universally applicable informational and functional patterns, that advanced not only their order-disorder distributions into optimizable operation modes of biomesogenic systems, but responded, moreover, intelligently to the early chiral mes-

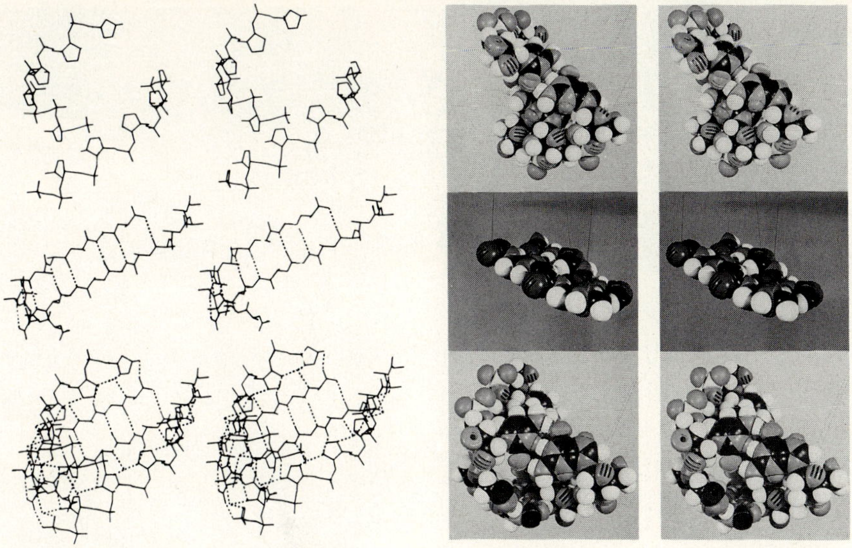

Fig. 10.19. Chiral recognition motifs between nucleic acid and protein backbones, modelling an early vice-versa polymerase activity between A-RNA and a chirally twisted protein β-sheet in skeletal and space-filling presentation [79] (*top* to *bottom*): recognition pattern of antiparallel double helical A-RNA shallow groove; structural complementary pattern of right-handed twisted antiparallel protein β-sheet; informational and functional processings in an early nucleoproteinic system, envisaging the fits of hydrogen-bonding patterns – modified in part from [79]

sage from the inherent qualitities of the universe and its translation into "molecular creativity".

10.4 Stabilization Within the Dynamics

Derivatives of cholesterol, a small chiral biomolecule, initiated, more than 100 years ago, the scientific history of liquid crystals (Figs. 10.5, 15, 21–25) [2, 4, 6, 7, 24, 52–60, 86–100]. The colored, playful movements of chiral cholesteric mesophases delighted the first researchers and prompted Lehmann [87, 88] to call these mysterious creations "seemingly living crystals" [88]. Liquid crystal aspects in general and chiral mesophase behavior in particular caused Lehmann, the grand "romanticist" of early mesogen views and coiner of the very term in Karlsruhe, to outline his visions on "Liquid crystals and the theories of life" [87, 88] (Fig. 10.22). Chiral supramolecular directions of mesophases – an early signal amplification model – puzzled Vorländer and his scientific school in Halle during the first "academic" period in the field (Fig. 10.21) [2, 4, 6, 7, 24, 89, 90]. Chiral mesophases continued to attract our attention up to the present day when ferroelectrics [93–96] and all the other "advanced material" aspects not only appear as a playground for artificial

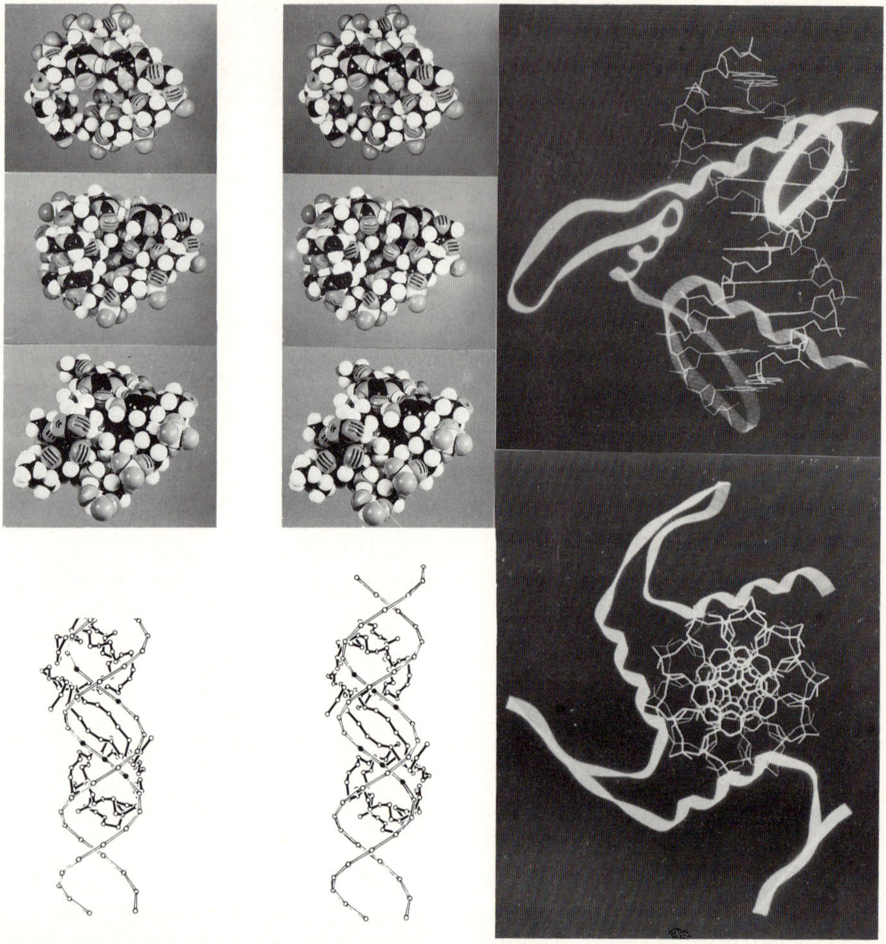

Fig. 10.20. Extensions of early RNA-protein recognition motifs to DNA-protein interactions [4, 6, 7, 79–85] (*top* to *bottom* and *left* to *right*): hydrogen-bonded antiparallel protein β-sheet in the minor groove of B-DNA; hypothetical regulation facilities of small molecule effectors (steroids a.o.) within this unspecific recognition complex [81]; specific recognition between protein α-helix and B-DNA major groove, as envisaged by cro-repressor-operator interactions [82, 83]; specific and unspecific recognitions via cro-α-helices/operator-major-groove and cro-β-sheets/minor-groove contacts [82]. Alternatives exemplified, for instance, in the Zn-finger-proteins/B-DNA interactions [84] – partially modified from cited references

technological ambitions, but also as fundamentals in biomesogenic operation modes [4, 6, 7, 55].

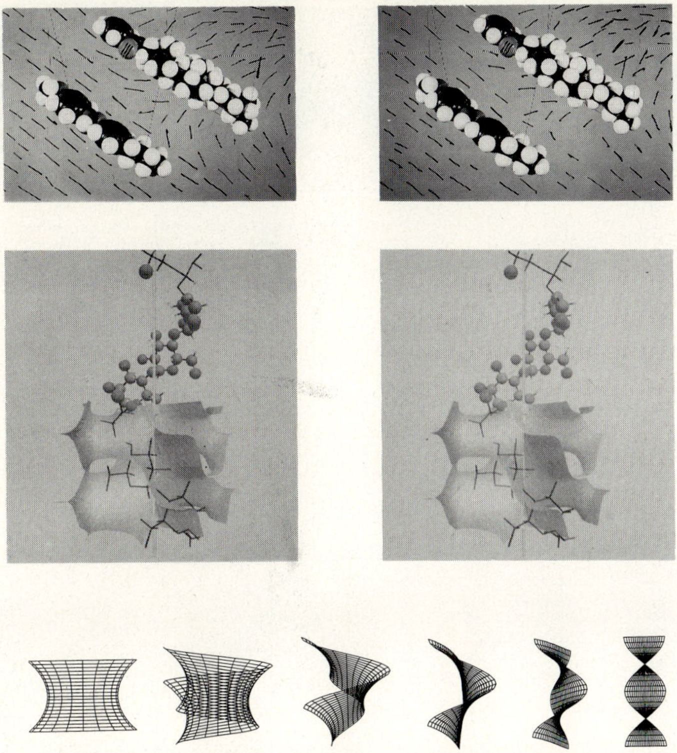

Fig. 10.21. Chiral guiding patterns in biomesogen phase-domain transition strategies (*top* to *bottom*): Reinitzer's chiral cholesterylbenzoate [86, 87] and Kelker's MBBA [2], elucidating Vorländer's "zirkulare Infektion" of nematic phases by chiral doping compounds [89, 90] – an early biomimetic thermotropic hormone model, simulating biomesogenic amplifications of molecular signals in biomesogenic surroundings; space-partitioning patterns as chiral guides from inorganic quartz to organic α-amylose [101]; chiral guiding patterns both relevant for inorganic materials and for highly sophisticated biological chromatin rearrangements as visualized by Bonnet transformations [102] – partially modified from cited references

Since these early days of fruitful dialectics between Karlsruhe and Halle, the field has been developed by mutual stimulation of Molecular Biology and Liquid Crystals [2, 4, 6, 7, 24, 52, 53] and now seems capable of creative dialectical syntheses of so-far dominating thesis-antithesis tensions. The unique spatio-temporal coherences of chiral biomesogen systems advance self-consistent evolutionary views and seem to afford an intriguing concept for the stabilizations of the grand process.

Even very simple and primitive mesogens (Fig. 10.5) seem to differ from non-mesogenic species in that they bear some sort of rudimentary intelligence [4, 6, 7]. At present, however, we appear to be faced with difficulties in scientifically treating this gleam of preintelligent behavior. Our theoretical approaches so far prefer rather abstract and etheric shadows of real entities [4, 6, 7, 52–54, 91–93, 97–100]. Changing from here to complex artifical

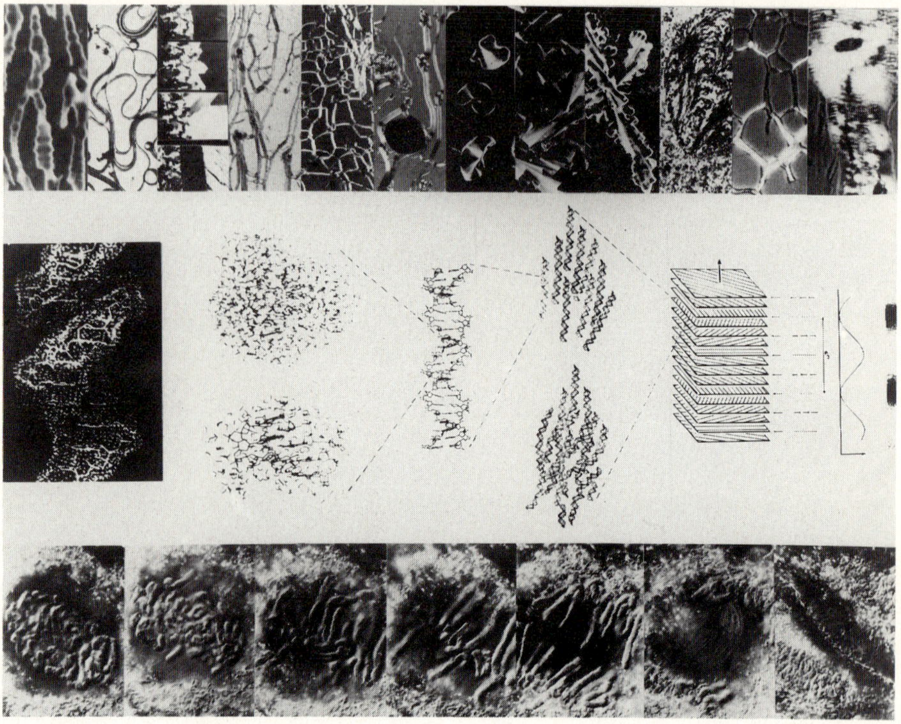

Fig. 10.22. "Developing" textures, preliminary phase interpretations and morphogenetic implications (*left* to *right* and *top* to *bottom*): Ammonium oleate – Lehmann's first "scheinbar lebender Kristall", original photograph as a kind gift of Hans Kelker [2, 87, 88]; Lehmann's "worm-like appearances" [88]; Halle-rediscovery of phase-chemistry by a miscibility rule for promesogenic imidazole derivatives in the 1950s [4, 6, 7]; cholesteric streak textures of thermotropically designed steroid hormone derivatives (three plates) [4, 6, 7, 43, 103, 104]; textures of 146bp-DNA fragment (three plates) and cholesteric organization of Dinoflagellate chromosome [59, 60]; cholesteric oily streak and chevron textures of high-molecular-mass chicken-erythrocyte DNA (two plates) [105, 106]. Preliminary interpretation of DNA cholesteric phase organizations, ranging from chiral secondary structure basic geometries built up from chiral residues, over water shell cover and counterion clouds, to biophysical texture interpretations with special emphasis on order-disorder distributions within superhelical chiral twist of cholesteric phase arrangements. Speculations for morphogenetic extensions, exemplified in case of the course of mitosis in Haemanthus cells (modified from [107])

and even native (bio)mesogen organizations (Figs. 10.5, 15, 21–25) amplifies the problem for quantitative treatment [4, 55]. Therefore the constraints are growing for qualitative inferences – if one is not tempted to speculate.

Though we developed our mesogens in terms of thermotropic and lyotropic characteristics, biomesogens both blurr and dialectically combine these extreme positions within the cooperative network hierarchies of interdependent chiral complexity patterns and live a permanent dialectic synthesis from the driving forces of complex thesis-antithesis tensions. The combina-

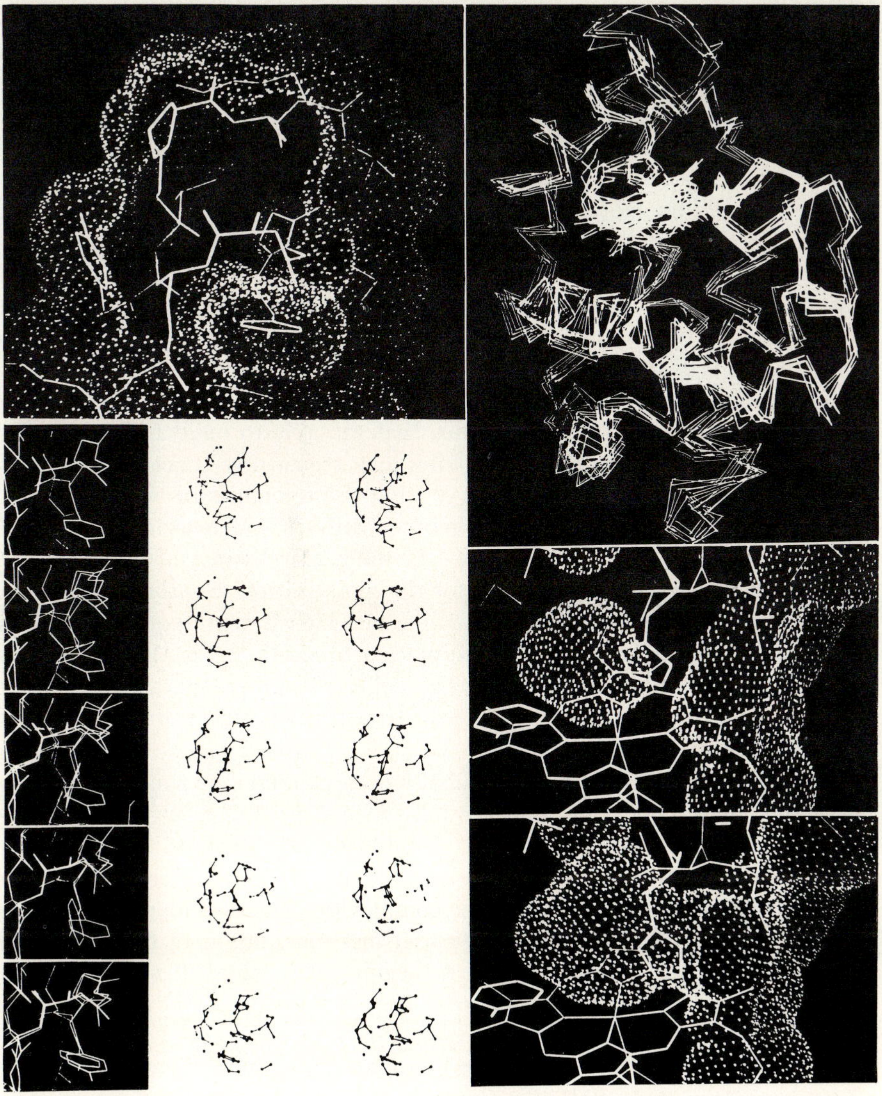

Fig. 10.23. Chiral biomesogenic protein dynamics (*top* to *bottom* and *left* to *right*): rotational isomerizations of Tyr-35 of bovine pancreatic trypsin inhibitor (see also Fig. 10.12), mediated by structural changes in the environing biomesogenic protein matrix – overall view and "film" in skeleton and stereo-presentation [4, 6, 7, 108–110] – myoglobin mediating by chiral biomesogenic dynamics of its protein matrix an oxygen-guest to its store at the heme group: myoglobin motions on a picosecond scale, oxygen blockade in the static structure and gating channel in the dynamic patterns [110] – modified from the cited references

tions of lyotropic and thermotropic aspects [4, 6, 7, 55] advance new preintelligent action modes of dynamic chiral order-disorder patterns – based on favorable free-energy relationships.

Thermotropics resemble lyotropics in that their flexible, more disordered segments serve the purpose of dynamization of the rigid, more ordered parts. Lyotropics, on the other hand, display, within their complex solvent-solute distributions, far more interactive coherences than had originally been implied by classical views. While for lyotropics, suitable solvent partners – especially water – provide a spectrum of entropy-driven self-organizational forces, thermotropics will profit by comparable entropy effects of order minimizations on the complex domain interfaces between rigid cores and flexible terminals. Within this picture, the solvent-like labilizing areas of interacting biopolymer organizations appear as special expressions of more general mobility characteristics of mesogens, that are able to build up within their dynamic chiral order-disorder distributions transiently functional and acting chiral order-disorder patterns. Within the dynamics of water mediated protein, nucleic acid and membrane organizations – submerging the individualities of their respective partners – dynamic, chirally instructed parts exert functions of partial solubilization for the mobilization of rather static chiral solute areas, the preintelligent handling of which is a prerequisite for some new qualities as, for instance, information processing, functional catalysis, semipermeable compartmentalization, collective and cooperative operation modes, and organizational behavior of transiently acting chiral domain systems (Figs. 10.21–25).

Our perceptions, that had originally been attracted by the unifying principles of huge artifical and preferentially achiral molecular ensembles and the overwhelming symmetries of their mesophase relationships, are redirected into the limited areas of cooperatively processed chiral phase-domain systems, that govern with increasing complexity a more and more precisely tuned and refined, highly sophisticated instrumentation of chiral domain-modulated phase transitions. The rather primitive physical views of simple geometric abstractions submerge even on the molecular level into extremely differentiated chiral order-disorder distributions. The chiral mesogen individualities of the grand supra-chirally organized amphiphilic patterns of life display within their molecular imprints the prerequisites for the projection of individual molecular facilities into the structural and functional amplifications of cooperative dynamic chiral mesogen-domain ensembles. The chiral order-disorder designed individual appears as a holographic image of the whole.

The developed chiral amphiphilic patterns – amphiphilic both in terms of their order-disorder dialectics and the spatio-temporal coherences of their

Fig. 10.24. Chiral biomesogen dynamics in information and functional processing (from *top* to *bottom*): complex structure-motion linguistics of a small molecule effector (β-casomorphin-5) in addressing neuroendocrine, cardiovascular/cardiotonic and immunomodulatory subroutines of general regulation programs [111]; cholesterol regulations in functional membranes [6]; valinomycin interplays, mediating K^+-ions across the membrane (see also Fig. 10.16) [4, 6, 7, 75–78]; biomesogenic sig-

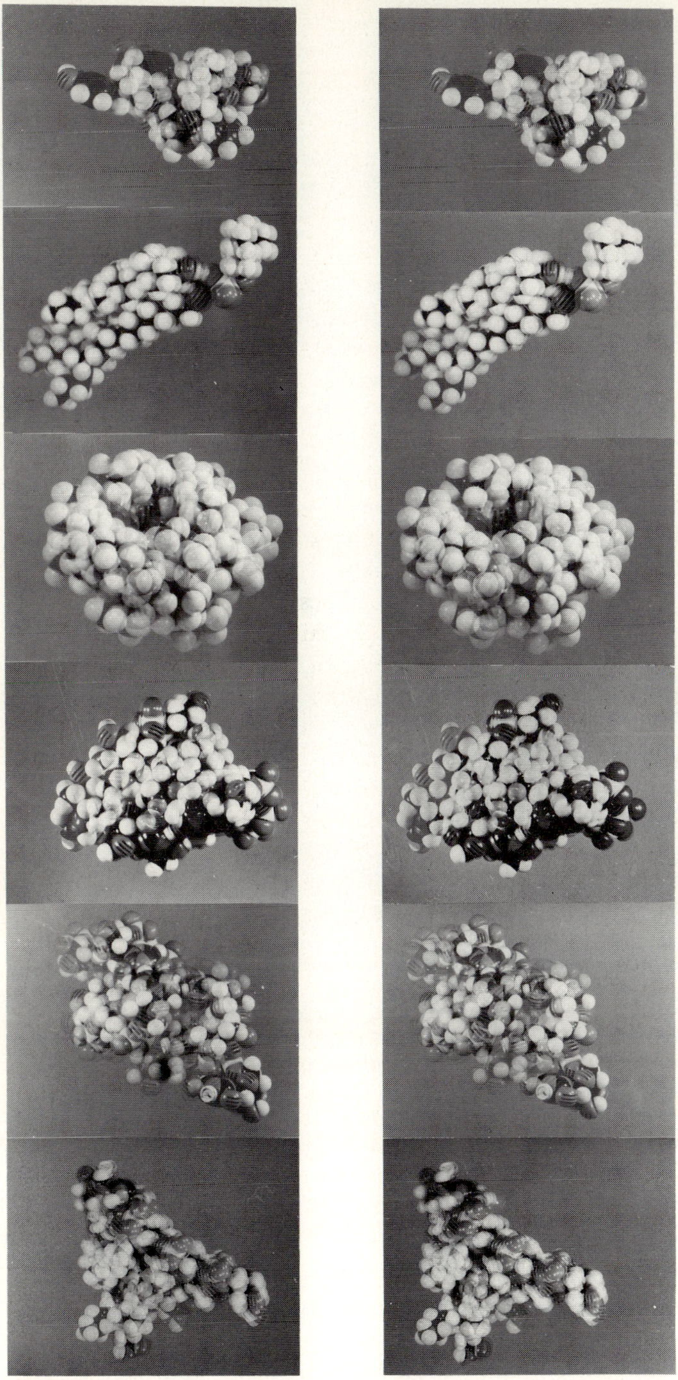

nal amplification in the chromatin: steroid hormone efficiencies in domain regulations (see also Fig. 10.21) [4, 6, 7, 31]; molecular hystereses and memory imprints in polynucleotide triplexes [4, 6, 7, 70–74]; intimacies of nucleic acids and proteins along nucleation trajectories of the nucleoproteinic system in self/nonself-recognition and discrimination [4, 6, 7, 43]

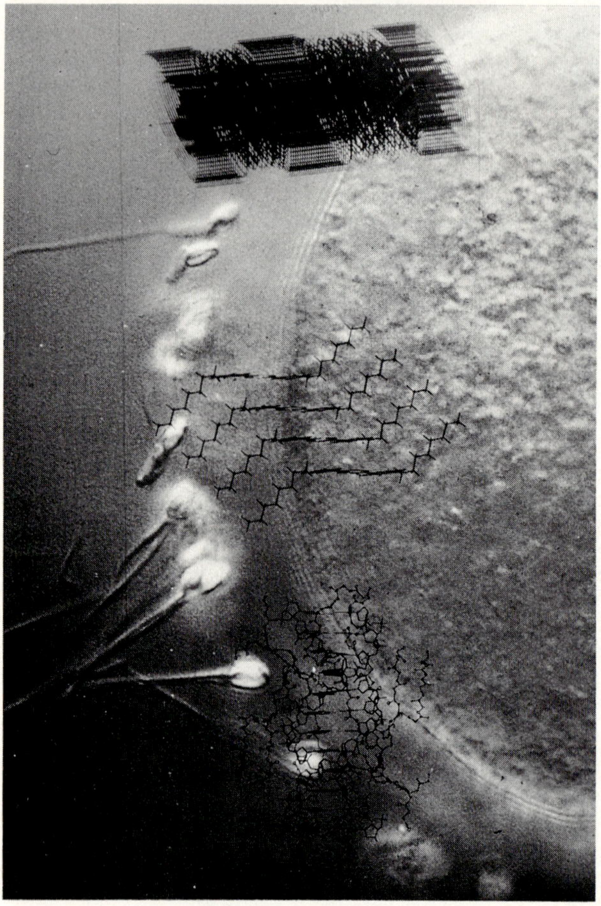

Fig. 10.25. Order-disorder distributions and pattern formations in artificial mesogens and native biomesogens forwarding spatial chiral order-disorder patterns into spatio-temporal coherences (*top* to *bottom*): amphotropic B-DNA/prealbumin complex [112], reminding in order-disorder distributions of the simple achiral thermotropic n-alkoxybenzoic acid dimers [113]; periodic and chaotic oscillations of multiple oscillatory states in glycolysis [114] – modified from the cited references. All this presented before the background of the intimacies of the very event – modelling in all its interfacial relationships early evolutionary origins of the nucleoproteinic system

systems and their sensitive phase- and domain-transition strategies characterize the picture of today's biomesogen organizations. Synergetics within complex solvent-solute subsystem hierarchies [4, 5, 55, 117] and their mutual feed-back found new aspects of chirally determined non-linear dynamics with processings – are acting with mesogen strategies (Figs. 10.5, 15, 21–25), ranging from the localized motions of molecular segments, their coupling to collective processes in the surroundings (Fig. 10.23) up to the interdependences of complex regulations (Figs. 10.21–25) [4, 6, 7, 52, 53, 107–111, 114]. Statics and dynamics of asymmetrically liberated and directed multi-solvent-solute a number of preintelligent operation modes, such as, for instance, reentrant phenomena, molecular hystereses and memory imprints, oscillations and rhythm generation, complex structure-motion linguistics and signal amplifications, self/nonself-recognitions and discriminations, and general optimization strategies (Figs. 10.21–25) [4, 6, 7, 55, 70–74, 108–111, 114–121].

Fig. 10.26. Chiral message – dynamic order patterns

An almost infinite number of examples characterize complex motions of the grand process, that are stabilized just in their dynamics; dynamics that continue asymmetrically the spatial chiral order-disorder prerequisites of biomesogenic dynamic order patterns into the spatio-temporal coherences of ordered dynamics (Figs. 10.22–26). The contradictions between crystalline order and liquid disorder, that rendered Lehmann's "liquid crystals" so extremely suspect to his contemporaries, represent and rule in fact the survival principles of the grand process. Biomesogens, that display within their chiral molecular imprints and designs not only the richness of experienced affinity patterns but also the facilities of their play-educated dynamics, link together entropy and information within a delicate mesogen balance [4, 6, 7, 53, 111].

In this connection, Schrödinger's question [27] for the two so extremely different views of "order" appears dramatically actualized. We seem to be badly in need of new theoretical treatments of old terms, that have somewhat changed their meaning when switching from statics to dynamics. What seems to be helpful, are extensions of classic static descriptions to new self-stabilizing dynamic states. The abstract and mathematically perfect order of an idealized crystal (Schrödinger's "dull wallpaper" in all its symmetries) and the strange order of the grand biomesogenic life-patterns (Schrödinger's "beautiful Raffael-gobelin" in all its asymmetries) are competitors for supreme roles. Covering the classic views of order and disorder as borderline cases, the new qualities of dynamic chiral order and chirally ordered dynamics should be put on the agenda. Within the elementary-particle derived "uncertainty principles" of biological organizations [6, 122], the chiral message of the inherent qualities of our universe might play a decisive part here.

10.5 Outlook

Life patterns are distinguished by the chiral message of our universe. Their evolution elaborated, advanced and optimized an originally strange inherence of the grand process from an early weak signal to an essential of the nucleoproteinic system and the leitmotif of life pattern developments.

Life patterns evolved as early amphiphilic patterns from the interface dialectics between order and disorder within a chaotic scenario on the primordial earth. The chiral amphiphilic molecular designs of their (bio)mesogenic constituents provided decisive prerequisites for the projection of individual molecular facilities into the structural and functional amplifications of cooperative and dynamic (bio)mesogen domain ensembles. By directional nonlinear dynamism, the (bio)mesogenic patterns arrived at a state of recognition and responsiveness. Competing individualization inspired complex information processing and forwarded autocatalytic propagation. Sensitized to the parity-violating influences and discriminations of the electroweak force, the (bio)mesogenic patterns managed the transition from an almost racemic outset to further homochiral developments by the break of chiral symmetry in open, autocatalytic non-equilibrium systems.

The chiral biomesogenic order-disorder patterns of life gained consciousness by adaptation to reality, variation and processing. Experiencing optimization strategies, they developed a singular creativity, and – by feedback with it – to the grand process itself.

Advancing forward so-far unknown horizons, life patterns will succeed, however, only in liability to life.

10.6 References

1 Burns JO (1990) Kosmologie und Teilchenphysik. Spektr Wiss Verlagsges Heidelberg, 10

2 Kelker H (1988) Mol Cryst Liq Cryst *165:* 1; (1986) Naturwiss Rundschau *39:* 239; (1986) In: Sackmann H (ed) Zehn Arbeiten über flüssige Kristalle – 6. Flüssigkristallkonferenz sozialistischer Länder. Wiss Beitr Martin-Luther-Univ 1986/52 (N17), 193; (1973) Mol Cryst Liq Cryst *21:* 1

3 Trefil S (1983) Big Bang Physics. Scribner's New York

4 Hoffmann S (1989) In: Braun D (ed) Polymers and biological function. Angew Makromol Chem *166/167:* 81; (1989) Wiss Z Univ Halle *38/H4:* 3; (1987) Z Chem *27:* 395; (1985) In: Blumstein A (ed) Polymeric liquid crystals. Plenum Publ Co New York, 423

5 Haken H (1981) Naturwissenschaften *68:* 293

6 Hoffmann S (1978) Molekulare Matrizen (I Evolution, II Proteine, III Nucleinsäuren, IV Membranen). Akademie-Verlag Berlin

7 Hoffmann S Witkowski W (1978) In: Blumstein A (ed) Mesomorphic order in polymers and polymerization in liquid crystalline media. Am Chem Soc Symp-Ser *74:* 178

8 Pasteur L (1886) In: Leçons de chimie professées en 1860 par MM. Pasteur, Cahours, Wurtz, Berthelot, Sainte-claire Deville, Barral et Dumas, Hachette, Paris 1861, Sur la dissymétrie moléculaire, Collection Epistème Paris

9 Janoschek R (1986) Naturwiss Rundschau *39:* 327
10 Hegstrom RA Kondepudi DK (1990) Scientific American *262:* 98
11 Freedman DZ Nieuwenhuizen P van (1988) In: Teilchen, Felder und Symmetrien. Spektr Wiss Verlagsges Heidelberg, 170; (1990) In: Kosmologie und Teilchenphysik. Spektr Wiss Verlagsges Heidelberg, 120
12 Bouchiat M-A Pottier L (1988) In: Elementare Materie, Vakuum und Felder. Spektr Wiss Verlagsges Heidelberg, 130
13 Weinberg S (1980) Science *210:* 1212
14 Mason S (1986) Trends Pharmacol Sci 20; (1985) Chem Brit *21:* 538; (1986) Nouv J Chim *10:* 739
15 Tranter GE (1986) Nachr Chem Techn Lab *34:* 866; (1985) Nature *318:* 172; (1985) Mol Phys *56:* 825; (1986) J theor Biol *119:* 469
16 Kondepudi DK Nelson GW (1985) Nature *314:* 438
17 Wainer IW Caldwell J Testa B (ed) (1989) Chirality *1:* 1
18 Mainzer K (1988) Chimia *42:* 161
19 Schadt W Petrzila M Gerber PR Villiger A (1985) Mol Cryst Liq Cryst *122:* 241
20 Langridge R Ferrin ThE Kuntz ID Conolly ML (1981) Science *211:* 661 – courtesy of Laurence H Hurley
21 Nir S Garduno R Rein R Coeckelenbergh Y MacElroy RD Egan JT (1977) Int J Quant Chem Quant Biol Symp *4:* 135
22 Katchalsky A (1973) Naturwissenschaften *60:* 215; (1970) Nature *228:* 636
23 Miescher F (1897) Über die chemische Zusammensetzung der Eiterzellen, Hoppe-Seyler's med Unters 1871; Die histochemischen und physiologischen Arbeiten (FCW Vogel, ed) Leipzig
24 Haeckel E (1917) Kristallseelen. Kröner-Verlag Leipzig
25 Fischer E (1894) Ber Dtsch Chem Ges *27:* 2985, 3189
26 Eschenmoser A (1988) Angew Chem *100:* 5
27 Schrödinger E (1944) What is life. Cambridge University Press New York
28 Chargaff E (1950) Experientia *6:* 201; (1965) On some of the biological consequences of base-pairing in the nucleic acids in development and metabolic control mechanisms and neoplasia. Williams and Wilkins Baltimore; (1970) Experientia *26:* 810; (1974) Building the tower of Babble. Nature *248:* 776
29 Pauling L Corey RB (1951) Proc Nat Acad Sci USA *37:* 205, 735
30 Pauling L (1960) The nature of the chemical bond. Cornell University Press Ithaca-New York
31 Rawn JD (1983) Biochemistry. Harper & Row Publ New York
32 Watson JD Crick FHC (1953) Nature *177:* 964
33 Watson JD (1968) The double helix. Athenaeum
34 Guschlbauer W (1988) In: Encyclopedia of polymer science and engineering. Wiley & Sons New York *12:* 699
35 Dickerson RE Geis I (1971) The structure and action of proteins 1971; Struktur und Funktion der Proteine. Verlag Chemie Weinheim
36 Chadrasekaran R Mitra AK (1983) In: Srinavasan R Sarma RH (eds) Conformations in biology. Adenine Press New York, 91
37 Urry DW Venkatachalan CM Lang MM Prasad KU, ibid, 11
38 Chou KC Pottle M Nemethy G Ueda Y Scheraga HA (1982) J Mol Biol *162:* 89
39 Scheraga HA Chou K-Ch Némethy G (1983) In: Srinavasan R Sarma RH (eds) Conformations in biology. Adenine Press New York, 1
40 Jaenicke R (1987) Progr Biophys Mol Biol *49:* 117; (1988) Naturwissenschaften *75:* 604
41 Mutter M Vuilleumier S (1989) Angew Chem *101:* 551
42 Katchalski-Katzir E (1988) Makromol Chem Macromol Symp *19:* 1

43 Hoffmann S Witkowski W (1984) In: Possin H (ed) Wirkstofforschung '82. Wiss Beitr Martin-Luther-Univ Halle 1984/4(S40), 57, 102
44 Jurka J Smith TF (1987) J Mol Evol *25:* 15
45 Dickerson RE (1989) J Mol Biol *205:* 787
46 Barton JK (1988) Chem Eng News 30
47 Arnott S Chandrasekaran R Banerjee AK He R Walker JK (1983) J Biomol Struct Dyn *1:* 437
48 Olson WK (1977) Proc Acad Sci USA *74:* 1775
49 Sarma MH Gupta G Dhingra MM Sarma RH (1983) J Biomol Struct Dyn *1:* 59
50 Sobell HM (1985) In: Jurnak McPherson (eds) Biological macromolecules and assemblies. Wiley & Sons New York *2:* 172
51 Clementi E (1983) In: Clementi E Sarma RH (eds) Structure and dynamics: nucleic acids and proteins. Adenine Press New York, 321
52 Hoffmann S (1985) In: Beránek J Piskala A (eds) Plenary lectures – Symp Chem Heterocycl Compounds (VIIIth) and of Nucleic Acids Components (VIth). Czechoslovak Acad Sci Inst Macromol Chem Press Prague, 48; (1984) Nucleic Acids Symp Ser *14:* 7
53 Hoffmann S (1983) In: Geissler E Scheler W (eds) Darwin today. Akademie-Verlag Berlin, 192; (1987) In: Scheel F (ed) VI. Int Tagung Grenzflächenaktive Stoffe. Akademie-Verlag Berlin, 545
54 Onsager L (1949) Ann N Y Acad Sci *51:* 62Z
55 Ringsdorf H Schlarb B Venzmer J (1988) Angew Chem *100:* 118; (1988) Angew Chem Int Ed *27:* 113
56 Watanabe J Ono H Uetmatsu I Abe A (1985) Macromolecules *18:* 2141
57 Samulski ET Tobolsky AV (1970) In: Johnson JF Porter RS (eds) Liquid crystals and ordered fluids. Plenum Publ. Co New York, 167
58 Iizuka E (1978) Polymer J *10:* 235; (1988) Adv Biophys *24:* 1
59 Rill RL (1986) Proc Nat Acad Sci USA *83:* 342; together with Strzelecka TE Davidson MW (1988) Nature *331:* 457
60 Livolant F Bouligand Y (1988) J Phys *47:* 1813; (1989) Mol Cryst Liq Cryst *166:* 91; Livolant F (1986) J Phys *45:* 1605; (1984) Eur J Cell Biol *33:* 400
61 Eigen M (1971) Naturwissenschaften *58:* 465; (1987) Cold Spring Harbor Symp Quant Biol *52:* 307; (1986) Chem Scripta *26B:* 13; (1985) Ber Bunsenges Phys Chem *89:* 658
62 Kuhn H (1983) In: Geissler E Scheler W (eds) Darwin today. Akademie-Verlag Berlin, 171
63 Schuster P, ibid 166; (1986) Physica *22D:* 100
64 Sarin PS Gallo RC (eds) (1980) Inhibitors of DNA and RNA polymerases. Pergamon Press Oxford-New York
65 Stryer L (1990) Biochemie. Spektr Wiss Verlagsges Heidelberg
66 Cech ThR (1987) Spektrum Wiss 42; (1988) J Am Med Assoc *260:* 3030; Sullivan FX Cech ThR (1986) J Mol Biol *189:* 143
67 Uhlenbeck OC Haseloff J Gerlach L (1988) Nature *334:* 585
68 Waring RB Towner P Minter SJ Davies RW (1986) Nature *321:* 133
69 Orgel LE (1986) J theor Biol *123:* 127
70 Hoffmann S (1988) In: Seliger H Secrist A (eds) 2nd Swedish-German Workshop Modern Aspects Chem Biochem Nucleic Acids and their Comp. Nucleosides and Nucleotides *7:* 555
71 Hoffmann S (1990) Mitteilungsbl Chem Ges DDR *37:* 45
72 Neumann E Katchalsky A (1970) Ber Bunsenges Phys Chem *74:* 868; (1972) Proc Nat Acad Sci USA *69:* 993
73 Neumann E (1973) Angew Chem *85:* 430; (1973) Angew Chem Int Ed *12:* 356

74 Hoffmann S Witkowski W Rüttinger HH (1974) In: Sedlácek B (ed) Heterogeneities in Polymers – 4th Disc Conf Macromolecules. Czechoslovak Acad Sci Prague, 37; (1976) In: Sackmann H (ed) 1st Liquid Crystal Conf Soc Countries Halle, 36; (1975) Z Chem *15:* 149; Hoffmann S (1979) Z Chem *19:* 241

75 Ovchinnikov YuA Ivanov VT (1975) Tetrahedron *31:* 2177

76 Ovchinnikov YuA (1987) Bioorganicheskaya chimya. Prosvyeschenye Moscow

77 Etchebest C Lavery R Pullman B (1982) Stud Biophys *90:* 7

78 Urry DW (1984) J Protein Chem *3:* 403

79 Carter CW Kraut J (1974) Proc Nat Acad Sci USA *71:* 283

80 Church GM Sussman JL Kim S-H (1977) Proc Nat Acad Sci USA *74:* 1458

81 Hoffmann S (1981) In: Possin H (ed) Wirkstofforschung 1980. Wissensch Publ Martin-Luther-Univ Halle 1981 *2:* 35; (1989) Z Chem *29:* 173, 449

82 Anderson WF Ohlendorf DH Takeda Y Matthews BW (1981) Nature *290:* 754

83 Ohlendorf DH Anderson WF Takeda Y Matthews BW (1983) J Biomol Struct Dyn *1:* 553

84 Gibson TJ Postma JPM Brown RS Argos P (1988) Protein Eng *2:* 209

85 Kim S-H (1983) In: Mizoguchi K Watanabe I Watson JD (eds) Nucleic acids research: future developments. Academic Press New York, 165

86 Reinitzer F (1888) Mh Chem *9:* 421

87 Knoll PM (1981) Fridericiana – Zeitschrift der Universität Karlsruhe, 43

88 Lehmann O (1907) Die scheinbar lebenden Kristalle. Schreiber-Verlag Esslingen; (1921) Flüssige Kristalle und ihr scheinbares Leben – dargestellt in einem Kinofilm. Voss-Verlag Leipzig; (1918) Die Lehre von den flüssigen Kristallen und ihre Beziehungen zu den Problemen der Biologie. Bergmann-Verlag Wiesbaden; (1918) Ergebnisse der Physiologie *16:* 255

89 Vorländer D (1908) Kristallin-flüssige Substanzen. Enke-Verlag Stuttgart; (1924) Chemische Kristallographie der Flüssigkeiten. Akademische Verlagsges Leipzig

90 Hoffmann F, personal communications

91 Sackmann H (1986) In: Sackmann H (ed) Zehn Arbeiten über flüssige Kristalle. Wiss Beitr Martin-Luther-Univ 1986/52(N17), 193

92 Demus D (1988) Mol Cryst Liq Cryst *165:* 45

93 Samulski ET (1985) Faraday Disc Chem Soc *79:* 7

94 Skarp K Handschy MA (1988) Mol Cryst Liq Cryst *165:* 439

95 Leuchtag HR (1987) J theor Biol *127:* 321, 341

96 Collette JW Miller MS (eds) (1989) Advanced Materials. Angew Chem Adv Mater *101:* 654

97 Kelker H Hatz R (1980) Handbook of liquid crystals. Verlag Chemie Weinheim

98 Gray GW (1962) Molecular structure and the properties of liquid crystals. Academic Press New York

99 Chandrasekhar S (1977) Liquid crystals. Cambridge University Press New York

100 Litster JD Birgeneau J Physics Today 1982, 1

101 Schnering HG v. Nesper R (1987) Angew Chem Int Ed *26:* 1059; (1987) Angew Chem *99:* 1097

102 Blum Z Lidin S (1988) Acta Chem Scand *B42:* 417

103 Dörfler H-D Brezesinski G Hoffmann S (1980) Stud Biophys *80:* 59

104 Hoffmann S Jaenecke G Brandt W Kumpf W Weißflog W Brezesinski G (1986) Z Chem *26:* 284

105 Thondorf I Lichtenberger O Hoffmann S (1990) Z Chem *30:* 171

106 Meister W-V Ladhoff A-M Kargov SI Burckhardt G Luck G Hoffmann S (1990) Z Chem *30:* 213

107 Bajer A (1983) In: Alberts B Bray D Lewis J Raff M Roberts K Watson JD, Molecular biology of the cell. Garland Publ New York

108 McCammon JA Lee CY Northrup SH (1983) J Am Chem Soc *105:* 2232

109 Karplus M McCammon JA (1986) Scientific American *254:* 42

110 McCammon JA Harvey SC (1987) Dynamics of proteins and nucleic acids. Cambridge University Press New York
111 Hoffmann S (1990) Z Chem *30:* 94; (1989) Wiss Z Univ Halle *38/H5:* 121
112 Blake CCF (1978) Endeavour *2:* 137
113 Bryan RFP Hartley P Miller W Shen M-S (1980) Mol Cryst Liq Cryst *62:* 281
114 Hess B Markus M (1987) Trends Biochem Sci *12:* 45
115 Petrosian V (1982) Nature *298:* 805
116 Micciancio S Vassallo G (1982) Il Nuovo Cimento *1:* 121
117 Palma MU (1983) In: Clementi E Sarma RH (eds) Structure and dynamics of nucleic acids and proteins. Adenine Press New York, 125
118 Frühbeis H Klein R Wallmeier H (1987) Angew Chem *99:* 413
119 Wolken JJ (1984) In: Matsuno K Dose K Harada K Rohlfing DL (eds) Molecular evolution and protobiology. Plenum Press New York, 137
120 Frauenfelder H (1986) In: Clementi E Chin S (eds) Structure and dynamics of nucleic acids, proteins and membranes. Plenum Publ Co New York, 169
121 Weizsäcker C-F v. (1986) Nova Acta Leopoldina (Neue Folge) *37/2:* 5
122 Cramer F (1979) Interdisciplinary Science Reviews *4:* 132; see also: "Denn nur also beschränkt war je das Vollkommene möglich" – Eine wissenschaftliche Interpretation von Goethes "Metamorphose der Tiere" – Preprint 1989 (kind information by Hartmut Seliger)

Subject Index

Printing and binding: Druckerei Triltsch, Würzburg